6th International Conference of the Indonesian Chemical Society (ICIC 2017)

Stimulating of Advanced Perspective and Current Concepts on Chemistry Field

Journal of Physics: Conference Series
Volume 1095

Palembang, Indonesia
17 – 18 October 2017

ISBN: 978-1-5108-7341-4
ISSN: 1742-6588

Printed from e-media with permission by:

Curran Associates, Inc.
57 Morehouse Lane
Red Hook, NY 12571

Some format issues inherent in the e-media version may also appear in this print version.

Printed by Curran Associates, Inc. (2018)

Institute of Physics
Dirac House, Temple Back
Bristol BS1 6BE UK

Phone: 44 1 17 929 7481
Fax: 44 1 17 920 0979

techtracking@iop.org

Additional copies of this publication are available from:

Curran Associates, Inc.
57 Morehouse Lane
Red Hook, NY 12571 USA
Phone: 845-758-0400
Fax: 845-758-2633
Email: curran@proceedings.com
Web: www.proceedings.com

Table of contents

Volume 1095

The 6th International Conference of the Indonesian Chemical Society

17–18 October 2017, Palembang, South Sumatra, Indonesia

Accepted papers received: 29 August 2018
Published online: 10 October 2018

Preface

Papers

IOP Conf. Series: Journal of Physics: Conf. Series **1095** (2018) 011001 doi:10.1088/1742-6596/1095/1/011001

Preface

The 6th International Conferences of the Indonesian Chemical Society (ICIC) 2017 was organized by the Department of Chemistry, Faculty of Mathematics and Natural Sciences, University of Sriwijaya in collaboration with the Indonesian Chemical Society. The theme for this conference was "*Stimulating of Advanced Perspective and Current Concepts on Chemistry Field*".

The goals of the conference were to provide a vehicle for the state of the art research results and trends in the chemistry field, to offer interaction, discussion and possible collaboration among chemists and the public about chemistry, to increase awareness of policy makers and the public on chemistry's rule in national development.

Therefore, this conference is a great opportunity not only for sharing knowledge and experience in chemical research, but also for starting a long and fruitful cooperation and friendship among Academicians, Researchers and practitioners of Chemistry.

In this conference, there were four keynotes speakers from USA, Japan, the Solomon Island, and Indonesia; and eight invited speakers. The 45 articles published here were picked from 229 participants who delivered them as oral or poster presenters.

Finally, I would like to thank to all participants and their respective institutions that have made this conference possible and I wish you all a pleasant meeting. I also would like to thank and congratulate the organizing committee for their dedication and tremendous efforts in organizing the conference.

Hermansyah,Ph.D.
Chairman of Organizing Committee
The 6th International Conference of Indonesian Chemical Society

Peer review statement

All papers published in this volume of *Journal of Physics: Conference Series* have been peer reviewed through processes administered by the proceedings Editors. Reviews were conducted by expert referees to the professional and scientific standards expected of a proceedings journal published by IOP Publishing.

Characterization of Durian Seed Flour (*Durio zibhetinuss l.*) and Estimation of its Self Life with Accelerated Self Life Testing (ASLT) Moisture Critical Method

A H Mulyati[1], D Widiastuti[1], L M Oktaviani[1]
[1] Chemistry Department, Faculty of Mathematics and Natural Sciences, Pakuan University .Jl. Pakuan P.O Box 452 Bogor, West Java 1614.

E-mail: adeherimulyati@yahoo.com.

Abstract. The content of starch in the seeds of durian is high enough so that they can be processed into flour which has a high economic value and becomes a potential product diversification of wheat flour. Durian seed flour has a yellowish white color, a smooth texture, a slightly acidic odor due to a 2day deposition process that will produce a sour aroma from separate mucus. Durian seed flour has 13.42% moisture, 4.44%, ash, 9.08% protein, 0.55% fat and 72.49% carbohydrate. Its microbiological characteristics include 6.4 x 10^3 colony/gram total plate count, 5.0 x 10 colony/gram mold, <3 APM/gram *Eschericia coli*, and <100 colony/gram *Bacillus cereus*. For shelf life testing, the moisture initial is 0.1213% and moisture critical is 0.1607% with slope isothermal moisture sorption curve of 0.255. From BET (Brunaurer-Enmet-Teller) curve the moisture content which gives maximum shelf life is 2.80%. The shelf life of durian seed flour in LDPE plastic packaging is 134 days, in HDPE plastic packaging 230 days, and PP plastic packaging 536 days.

Keywords: Durian seed flour, ASLT, Shelf life, Isothermal Moisture Sorption

1.Introduction

The import of wheat flour is very high because wheat flour produced by local producers is not enough to meet the needs of wheat flour in Indonesia. Durian seed carbohydrate content is very high so it can be processed into flour with a high economic value and a potential as raw material of flour diversified products [1,2,3,4]. So far, the utilization of durian seeds has only been limited as raw material of chips with a relatively low market share, so it is expected that the waste of durian seeds can be processed into flour and various products and also increase the economic value of the durian seeds.

Prior to consumption, durian seed flour should go through a series of tests to ensure quality and feasibility as a flour diversified product. The parameters tested are physical, chemical and microbiological properties and also shelf life of durian seed flour. The inclusion of the expiry date gives consumers the information about the time limit of the use of the product. Manufacturers and product distributors also benefit from the availability of information about shelf life [5].

One way to find out the level of product durability during the storage period is through the determination of shelf life so it can be known how long a product can be stored before going down to its critical point. The shelf life of food products can be predicted and then the expiry date can be set by using two concepts of food storage product study, namely the Extended Storage Studies (ESS) and the Accelerated Storage Studies (ASS). Assessment of shelf life can be done in Accelerated Shelf Life Testing which can then predict the actual shelf life [6]. The determination of shelf life with the accelerated conditions most commonly used in the food industry is the Arrhenius method and the critical water content method. The determination of shelf life of critical moisture content method can be done by the assessment of isotherm of water sorption (ISA) that is the curve of relation between air storage RH and equilibrium moisture content at a product that is allowed to reach equilibrium during storage. Air storage RH can be arranged by balancing air using a suitable saturated salt solution [7]. The critical moisture method is generally used for food products with tipping point, congestion, and food damage based on their water activity. One of the most important determinants of product shelf life is product packaging. This research, therefore, performs physical n, chemical and microbiological characterization of durian seed flour as well as the determination of its shelf life with some types of plastic packaging such as high density polyethylene (HDPE), low density polyethylene (LDPE), and Polypropylene (PP) by ASLT method.

2. Materials and Methods

2.1. Durian Seeds

Durian seeds are obtained from local cultivars in sub district Jasinga, Bogor, West Java and Sibolga area, North Sumatra.

2.2. The Making of Durian Seed Flour and its Characteristics

The durian samples are determined and then the durian seeds are collected for sorting. The process of durian seed flour making is with a wet method started by the separation of durian seed starch. It is then analyzed physically (color, taste and odor) and chemically, including water content (SNI 3751-2009), ash content (SNI 3751-2009), protein content (SNI 3751-2009), fat content (AOAC, 2005), carbohydrate content (AOAC, 2005), Total Plate Count (SNI 3751-2009), Coliform and *Escherichia coli* contamination (SNI 3751-2009), molds (SNI 3751-2009) and *Bacillus cereus* (SNI 3751-2009).

2.3. Shelf Life Estimation of Durian Seed Flour With ASLT Method

The determination of shelf life of durian seed flour by ASLT method uses critical moisture content (Labuza) method with different plastic packs: high density polyethylene (HDPE), low density polyethylene (LDPE), and polypropylene (PP) through various stages including the measurement of moisture initial (Mi), moisture critical (Mc), moisture equilibrium (Me) and the determination of isothermal sorption model, permeability of packaging to moisture and BET moisture content.

The moisture equilibrium (Me) and Aw values are included in the isothermal sorption equation models of Chen Clayton, Henderson, Hasley, Caurie, and Oswin. The five models are then evaluated for the Mean Relative Deviation (MRD) values. If the MRD value is <5 then the isothermal sorption model can describe the actual or very precise state. If it is 5 <MRD <10 then the model is rather precise to describe the actual state and if MRD is > 10 then the model does not exactly describe the actual condition.

Silica gel is inserted in a WVP cup and then it is sealed with a package of which its moisture permeability will be determined. The silica gel and the cup which has been covered by a packaging are weighed and put in a closed jar containing the NaCl solution. The determination of the permeability of the packaging is carried out at a temperature of 28℃and 75.62% RH. To adjust the room RH in the jar to reach 75.62% NaCl solution is then used. The permeability of the packaging is determined by the formula [8]: data obtained from the determination of the ISL curve is aw and moisture equilibrium. While the Mo BET requires aw / (1-aw) data M. Next, the linear regression curve is made with aw as the X-axis and aw / (1-aw) M as the Y-axis, so that a linear equation is obtained.

3. Results and Discussion

3.1. Characteristics of Durian Seed Flour

Durian seed flour has a yellowish white color, a smooth texture, a slightly acidic odor due to a 2day deposition process that will produce a sour aroma from separate mucus. Durian seed flour has 13.42% moisture, 4.44%, ash, 9.08% protein, 0.55% fat and 72.49% carbohydrate. Compared to other types of flour, namely cassava flour, protein content in durian seed flour is higher. Proteins contribute significantly to the physical properties of the food through its ability to make and stabilize gels, foam, emulsions, and fibril structures [9]. Fat content of more than 0.5% can be caused by residual malvalic acid that is carcinogenic cyclopropene fatty acid which exists between the outer skin and the epidermis of durian seeds. Therefore, the process of washing, soaking, and drying is essential to make durian seed flour a commodity for consumption and a substitution of wheat flour [10]. Moreover, microbiological characterization is performed to determine the level of hygiene in food product manufacturing as an indicator of food sanitation, and to estimate its shelf life according to SNI 01-

3751-2009. Its microbiological characteristics include 6.4 x 10^3 colony/gram total plate count, 5.0 x 10 colony/gram mold, <3 APM/gram *Eschericia coli*, and <100 colony/gram *Bacillus cereus*.

Table 1. Characteristics of Durian Seed Flour

Parameter	Unit	Durian Seed Flour	01-3751-2009 SNI Wheat
Rendement	%	26,23	-
Color		yellowish white	White
Texture		smooth texture	Powder
Odor		slightly acidic	Normal
Water	%	13.42	Maximum 14.5
Ash	%	4.44	Maximum 0.70
Protein	%	9.08	Minimum 7.0
Fat	%	0.55	-
Carbohydrate	%	72.49	-
Total Plate Count	colony/g	6.4 x 10^3	Maximum 1 x 10^6
Molds	colony/g	5.0 x 10	Maximum 1 x 10^4
Eschericia coli	APM/ g	3	Maximum 10
Bacillus cereus	colony/g	<100	

3.2. Estimation of Durian Seed Four Shelf Life with ASLT Method

The estimation of shelf life of durian seed flour products is calculated through the Labuza equation (1985) that is the shelf life at 78% RH storage [8]. Moisture content in food products determines acceptability, freshness, texture, resistance to microorganisms, and shelf life of the products. In dry food products such as flour, moisture content is a critical characteristic that affects consumer acceptance because it determines textures (clumping and wet). Durian seed flour is slightly hygroscopic (easy to clot) so it needs to determine the point of damage based on critical moisture content (Mc).

Table 2. *Moisture Initial (Mi) and Moisture Critical (Mc)*

Parameter	Unit	Results
Moisture initial (Mi)	% (dry based)	0.1213
Moisture critical (Mc)	% (dry based)	0.1607

Moisture equilibrium is required to obtain the isothermal sorption curve of the product, by conditioning the durian seed flour in some type of saturated salt solution with different relative humidity. The achievement of equilibrium conditions between samples and the environment is characterized by constant sample weights. A constant weight is shown by the difference of consecutive weighing of not more than 2 mg / g of samples stored at RH below 90% and not more than 10 mg / g of samples stored at RH above 90% [11].

Table 3. Equilibrium Moisture Content in Different Conditions of Aw

Saturated Salt	RH	Aw	Moisture Content (% wet based)	Moisture Content (% dry based)
NaOH	6.9	0.069	4.4382	0.0464
MgCl2	32.4	0.324	7.4843	0.0809
KI	69	0.690	11.3943	0.1286
NaCl	75.5	0.755	13.8462	0.1607
KCl	84	0.840	18.5341	0.2275
BaCl2	90.3	0.903	23.4764	0.3068

The equilibrium weight starts to be gained on the 5th storage day for samples stored under low RH conditions, as in RH 7-69 storage. The higher the relative humidity value of storage, the longer it takes to reach a state of equilibrium with the environment. The equilibrium weight of the durian seed flour samples shows the phenomenon of weight gain except in low storage RH that is at a condition of 6.9% RH. In the flour stored at a relative humidity condition of 6.9%, the process that occurs is the release of water vapor from samples to the environment (desorption phenomenon), this is because the samples under low RH conditions have a higher aw than the relative humidity, reaching the equilibrium of durian seed flour will release moisture. While the durian seed flour stored in the relative humidity of 32%, 69%, 76%, 84% and 90% increase weights (adsorption phenomenon) because of the lower aw than the relative humidity.

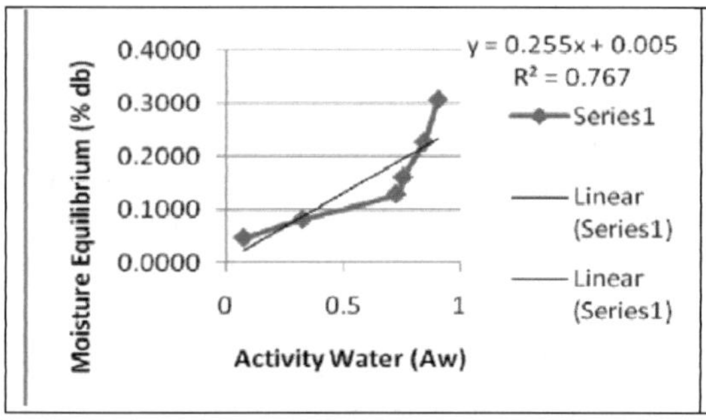

Figure1. Curve of Isothermal Moisture Sorption (Hasley)

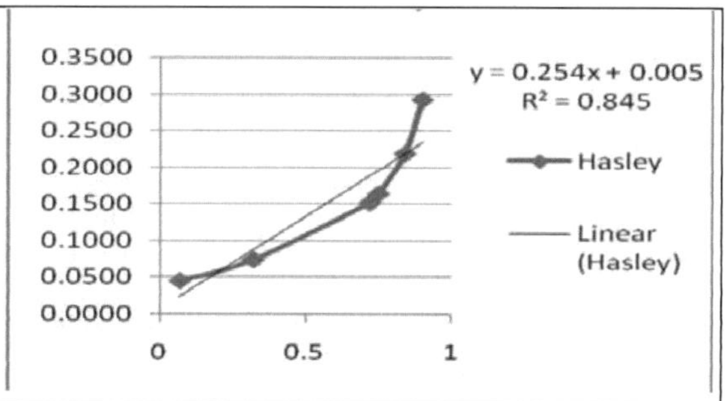

Figure 2. Curve of Isothermal Moisture Sorption

The Hasley equation model is chosen as the model has the most squeezed curves, the isothermal sorption curves describe the entire isothermal sorption curve of crackers rather precisely (5 <MRD <10). While the model equations of Chen Clayton, Oswin, Henderson and Caurie describe the overall isothermal sorption curves that are not appropriate because MRD is > 10. The slope value on the isothermal sorption curve is 0.254 based on the Hasley equation model. The relation dots between water activity and moisture equilibrium make a linear equation: y = a + bx. A value of 0.254 is determined as a linear slope that passes through the moisture initial and moisture equilibrium at each storage RH. This is in accordance with Labuza's statement that the linear area for determining the isothermal sorption slope curve is taken in the area that passes through the Mi (Moisture Initial).

Moisture content can be determined by using the Brunauer-Ermet-Teller (BET) equation, with a linear regression curve connecting aw of [aw / (1- aw) M]. If the slope and intercept data are included in the formula, moisture content durian seed flour of 2.80% (dry based) powder is obtained.

The determination of permeability is performed at temperature 280 ℃ by using NaCl solution having RH equal to 75.5%. The outer air pressure (P out) is determined by multiplying the vapor pressure at the storage temperature with RH. According to the water vapor table of Labuza (1985), the vapor pressure at 28 ℃ is 28.349 mmHg. So P out in this research is 21.4035 mmHg.

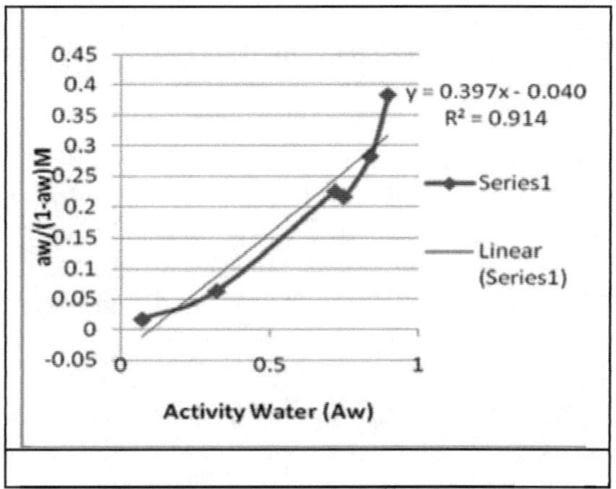

Figure 3. Brunauer-Ermet-Teller (BET) Curve

In general, PP plastic has the lowest permeability. Compared to polyethylene, the polypropylene properties which are the reasons why many are used as packaging materials are the low price, flexibility, appearance, smoothness, inert to chemicals, and low moisture permeability. Polypropylene has better tensile strength and clarity as well as lower water vapor and gas permeability [12]. Other properties of polypropylene are not reacting with materials, able to reduce the contact between materials with oxygen, not toxic and able to protect the material from contaminants.

The isothermal sorption model chosen for this durian seed flour product is the Hasley equation model (5 <MRD <10) [13]. The slope value is determined from the linear area of the isothermal sorption curve predetermined from the isothermal sorption curve of the Hasley model of 0. 2542. The shelf life of the product is calculated under storage conditions at 70% RH.

Table 4. Shelf Life of Durian Seed Flour in Different Plastic Packaging

Type of Packaging	k/x	Shelf Life (day)
LDPE	1.098681334	134
HDPE	0.640897445	230
PP	0.274670334	536

Table 4 showed that shelf life of durian seed flour packaged in LDPE packaging is 134 days, in HDPE packaging 230 days and in PP packaging 536 days. Durian seed flour packaged with PP plastic packaging with a thickness of 0.05 mm has a longer shelf life, 536 days. This is because the PP packaging has a lower permeability so that durian seed flour takes longer to reach its critical moisture content than it does in other packs with higher permeability. As the lower permeability of packing, the higher ability of protection against the absorption of water vapor, the shelf life of food products in the packaging is longer. The storage of durian seed flour with PP plastic packaging with a thickness of 0.05 mm at 28℃ and 70% RH is advisable to be at 2.80% moisture content.

4. Conclusion

Durian seed flour produced from local durian seeds contains 13.42% water, 4.46% ash, 0.55% fat, 9.08% protein and 72.49% carbohydrate (by difference). Besides, the microbiological analysis results in 6.4 x 103 colony/gram total plate number, 5.0 x 10 colony/gram molds, 3 APM / gram *Escherichia coli*, and <100 colony/gram *Bacillus cereus*. The shelf life of durian seed flour packaged in LDPE packaging is 134 days, in HDPE packaging 230 days and in PP packaging 536 days.

Acknowledgments

This Research is supported by the Ministry of Research and Technology Directorate General of High Education with project reference number 1598/K4/KM/2017 and contract number 02/LPUP/VII/2017.

References

[1] Hutapea and Paulina 2010 Effect of Water Immersion flour Limestone On Seeds Quality Durian. Faculty of Public Health, (Medan-USU).
[2] Magdalena and Kristiana 2010 Effect of Balance Flour Durian Pork Beans Against Physical and acceptability Naget Andaliman Seasoning. Faculty of Animal Husbandry (Bandung – UNPAD).
[3] Mat Amin and Amiza 2009 Proximate composition and pasting properties of durian (Durio zibethinus) seed flour. School of Food Science and Technology (Malaysia – UTM).
[4] Prasetyaningrum, A and Moh Djaeni 2010 Aspects of Nutrition and Technoeconomy. Riptek, 4 (11) 37.
[5] Larasati and Sita Annisa 2013 *Estimation of Shelf Life Flour Aloe Method of Critical Water Content. Faculty of Agricultural Technology* (Bogor – IPB).
[6] Floros J D and V Gnanasekharan 1993 *Shelf life prediction of packaged foods: chemichal, biological, physical, and nutritional aspects. G. Chlaralambous* (London – Elsevier).
[7] Hariyadi P 2006 *Handouts and Module Estimation and Determination of Shelf Life of Food Products. SEAFAST Center* (Bogor- IPB).
[8] Labuza, T P and MK Schmidl. 1985 *Accelerated shelf life testing of foods. Food Technology.*
[9] Belitz, H D, Grosch, W, and Schieberle,P. 2008. *Food Chemistry 4th revised and extended ed.* (Munchen – Springer).
[10] Brown W E 1992 *Plastic in Food Packaging.* Marcel Dekker, Inc, New York.
[11] Adawiyah. 2006. *Water Sorption Relation, Glass Transition Temperature, and Water Mobility As Well As Its Influence On Product Stability In Food Models [dissertation]* (Bogor - IPB)
[12] Suyitno 1990 *Packaging Materials. PAU Food and Nutrition* (Yogyakarta.- UGM)
[13] Hasnaini 2012. *Estimation of Shelf Life Crackers Rame 'Seaweed (Euchema cottoni L) Method Using Accelerated Shelf Life Testing. Faculty of Food Science* (Makassar- UNHAS).

Synthesis and characterization of Schiff base 4,4-diaminodiphenyl ether-vanillin possessed of free primary amine

A Fatoni[1], P L Hariani[2], Hermansyah[2] and A Lesbani[2],

[1]Bhakti Pertiwi School of Pharmacy Science
Jl. Ariodillah 3 No. 22 Palembang 30128 South Sumatera, Indonesia
[2]Department of Chemistry, Faculty of Mathematic and Natural Sciences, Sriwijaya University,
Jl. Palembang-Prabumulih Km. 32 Indralaya Ogan Ilir South Sumatera, Indonesia

E-mail : ahmad_fatoni@stifibp.ac.id

Abstract. The synthesis and characterization of Schiff base 4,4 diaminodiphenyl ether-vanillin possesed of free primary amine has been done.The aims of this research weresynthesis Schiff base 4,4 diaminodiphenyl ether-vanillin possesed of free primary amine from reaction between 1 mol of 4,4 diaminodiphenyl ether and 1 mol of vanillin which affected by temperature of synthesis (30-90^{0}C)and its characterization using FTIR spectroscopy and X Ray diffraction analysis. The result showed that the functional group of all product of Schiff base 4,4 diaminodiphenyl ether-vanillin possesed of free primary amine was not affected by temperature of synthesis. The functional group of –C=N in this Schiff base compound appeared at wave number 1597 cm^{-1} and its consequences, the aldehyde group (C=O) was not detected at 1666 (stretching) and 632 cm^{-1} (bending vibration). The wave number between 3387-3441 cm^{-1} was stretching vibration of primary amine (N-H) and stretching vibration of –C-N from aromatic amine (NH_2-aromatic) can be found at wave number between 1280-1288 cm^{-1}. Diffractogram of 4,4 diaminodiphenyl ether, vanillin and Schiff base 4,4 diaminodiphenyl ether-vanillin were 11.91^0 - 50.39^0,13.13^0 and 39.99^0 and 10.13^0 - 50.6^0 respectively.

Keywords : Schiff base, 4,4 diaminodiphenyl ether-vanillin, free primary amine

1. Introduction

4,4 diaminodiphenyl ether or 4,4 oxydianilline is diamine monomer because has two primary amine (-NH_2). Chemical structure of 4,4 diaminodiphenyl ether is $C_{12}H_{12}N_2O$ and has molecular weight 200,24 gram/mol [1]. 4,4 diaminodiphenyl ether is heterocyclic compound because has one or more oxazine ring [2]. This primary amine at 4,4 diaminodiphenyl ether compound can be modified to be derived4,4 diaminodiphenyl ether and used as biological and medical area [3].

Vanillin (3-methoxy-4-hydroxybenzaldehyde) obtained from vanilla pods is flavouring agent and used such as food, drink and cosmetic [4]. Vanillin has chemical structure $C_8H_8O_3$, molecular weight 152.15 gram/mol [5] and trisubstituted with –CH=O, -OCH_3 and –OH groups [6].

Schiff bases, named after the German chemist, Hugo Schiff found the Schiff base reaction in 1864 and this reaction can be done at room temperature without activation and water (as side) product [7]. Schiff bases (–RC=N–) compound can be formed by reaction between aldehydes and ketones group[8] as seen in Fig. 1.

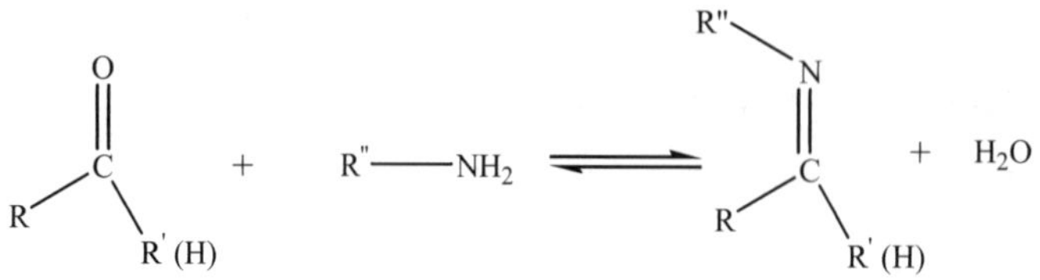

Figure. 1. Schematic illustration of Schiff base reaction

Aromatic Schiff-Bases are a large class of organic compounds with imino / azomethine (–HC= N–) group and formed by reaction between aromatic aldehyde and amine [9]. Aromatic aldehyde group (-CH=O) in vanillin can be reacted by primary amine to form Schiff base compound [10-13]. Primary amine from 4.4 diaminodiphenyl ether compound can be reacted by aromatic aldehyde group (-CH=O) in vanillin to form Schiff base (-N=CH-) group with mol ratio aldehyde and primary amine were 2 : 1 [14].

In this paper, we made Schiff base (-N=CH-) compound from aromatic aldehyde group of vanillin and one primary amine of 4,4 diaminodiphenyl etherwith mol ratio aldehyde and primary amine were 1 : 1 to form schiff base 4,4 diaminodiphenyl eter-vanilin which affected with temperature variation and this Schiff base compound was still possessed of free primary amine. The free primary amine can be used as active site in adsorption process [15] and linkages if reacted with formaldehyde [16-18]. The final product named Schiff base 4,4 diaminodiphenyl ether–vanillin possessed of free primary amine. Characterization of this product has been analyzed by FTIR spectroscopy and X Ray diffraction analysis.

2. **Experimental section**

2.1. *Materials*

Methanol (Merck), vanillin (Merck) and 4,4 diaminodiphenyl ether (Aldrich). All in analytical grade.

2.2. Instrumentation

The instruments used for characterization included Fourier Transform Infrared (FTIR) spectrophotometer (Shimadzu Prestige-21) for identifying the presence of functional groupsand XRD Spectrophotometer (Shimadzu XRD-6000) for evaluating the crystalline level.

2.3. Procedure

2.3.1. Synthesis Schiff base 4,4 diaminodiphenyl ether –vanillin [14]).

4,4 diaminodiphenyl ether (4 g, 0,02 mol) was dissolved in methanol (20 mL)at beaker glass 250 mL. In this beaker glass, added vanillin solution (3.04 g, 0.02 mol in10 mL methanol). The mixture was stirred continously at temperature 90^0, 75^0, 60^0, 45^0 and 30^0 C respectively for 1 hour and cooled at room temperature until yellow powder formed. The yellow powder washed with 10 mL methanol and dried in oven at 55^0C until constant weight.

2.3.2.The functional group and physical structure analysis

The chemical structure of this Schiff base is confirmed by FT.IR and XRD spectra.

3. **Result and discussion**

Chemical structure of Schiff base 4,4 diaminodiphenyl ether –vanillin possessed of free primary amine proposed as seen in Fig. 2.

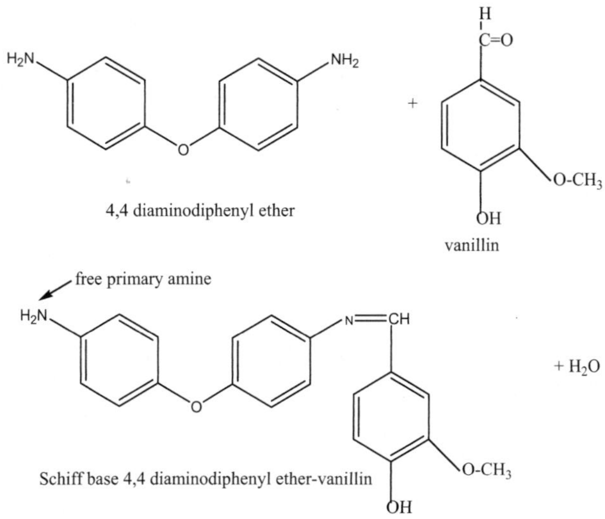

Figure. 2. Chemical structure of Schiff base 4,4 diaminodiphenyl ether –vanillin

3.1. *Presence of functional group with FT.IR*

3.1.1.Synthesis Schiff base 4.4 diaminodiphenyl ether –vanillin

Reaction between aldehyde group (vanillin) and primary amine (4,4 diaminodiphenyl ether) formed Schiff base compound with mol ratio (aldehyde and primary amine group) 1:1. The spectra FT. IR of vanillin, 4,4 diaminodiphenyl ether and Schiff base 4,4 diaminodiphenyl ether –vanillin are shown in Fig. 3 and 4.

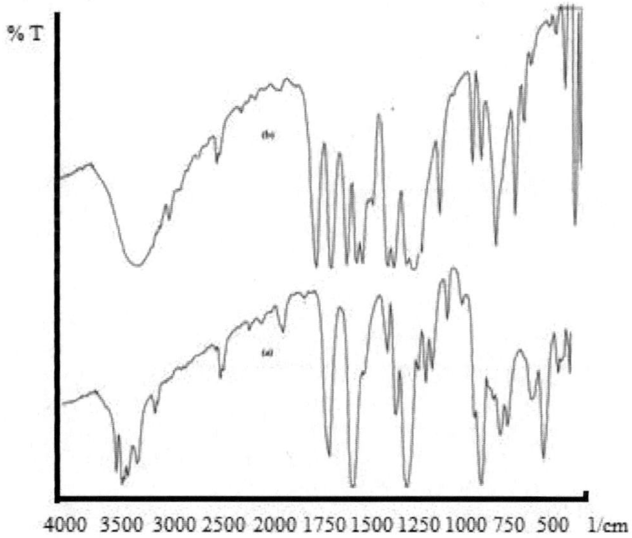

Figure. 3. FTIR spectra of (a) 4,4 diaminodiphenyl ether dan (b) vanillin

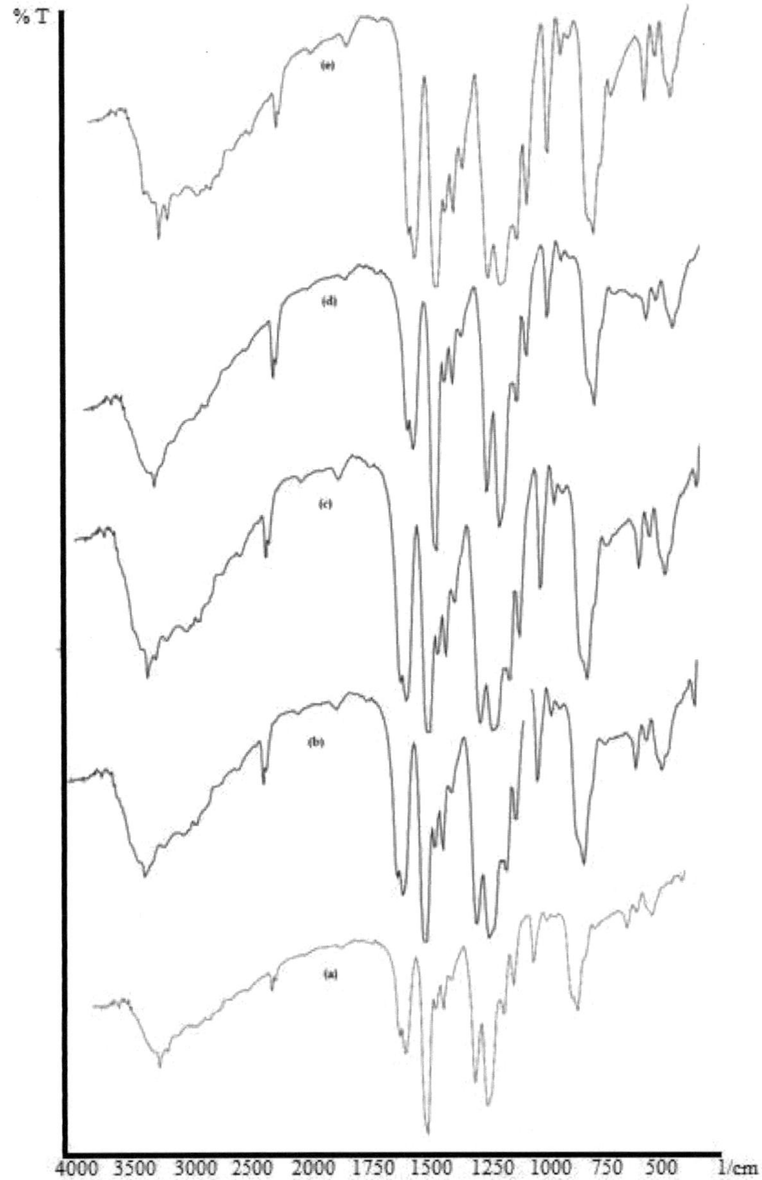

Figure. 4. FTIR spectra of Schiff base 4,4 diaminodiphenyl ether-vanillin compound synthesized at temperature (a) 30^{0}C , (b) 45^{0}C, (c) 60^0 C, (d) 75^{0}C dan (e) 90^{0}C

The main bands in 4,4 diaminodiphenyl ether (Fig. 3.a) shows that asymmetric and symmetric stretching vibration of primary amine (-NH_2) at (3441 and 3387) cm^{-1} respectively [19,20]. Stretching vibration of C-O (C-O-C) at (1219 , 1087 and 1010) cm^{-1}, stretching vibration aromatic ring (C=C-C) at (1442 and 1504) cm^{-1} , stretching and bending vibration C-H aromatic ring at 3016- 3039 and 732-825 cm^{-1} respectively [21,22]. Bending vibration N-H from primary amine at 1620 cm^{-1}and stretching vibration C-N from aromatic amine (NH_2-Aromatic) at1273 cm^{-1}[21].

FT.IR spectra of vanillin (Fig. 3.b) are follows : stretching and bending vibration -OH group at 3178 and between 1265 -1296 cm^{-1} respectively [23]. Stretching vibration –C-H at 2847-3024 cm^{-1} and stretching and bending vibration of aldehyde group (C=O) at 1666 and 632 cm^{-1} respectively

[23,6]. Stretching vibration $O\text{-}CH_3$ at 1026 cm^{-1}[6] and stretching vibration C=C-C group at region 1427-1589 cm^{-1}[21,22].

Fig. 4.a, b, c , d and e are the FT.IR spectra of Schiff base 4,4 diaminodiphenyl ether-vanillin which synthesized at 30^{0}C , 45^{0}C, 60^{0}C, 75^{0}C and 90^{0}C respectively.The functional groups of Schiff base 4,4 diaminodiphenyl ether–vanillin possessed of free primary amine (a-e) had similarity although synthesized at differently temperature. All spectra shows that stretching vibration asymmetric of primary amine (N-H) at 3387-3441 cm^{-1} and overlap with stretching vibration –OH, but the band of –OH group also at 3224 cm^{-1}[19,20,21]. Bending vibration of OH group at 1280 cm^{-1}[23]. Stretching and bending vibration of C-H group at 3008 and 833 cm^{-1} respectively and stretching vibration C-O (C-O-C) at 1226-1234 cm^{-1}[6,22].Stretching vibration C=C-C group at 1504 $cm^{-1}$and stretching vibration C-N from aromatic amine (NH_2-Aromatic) at 1280 cm^{-1}[21]. Stretching vibration of $O\text{-}CH_3$ at 1033 cm^{-1}[6]and a new peak appeared at 1597 cm^{-1} is a azomethine group (-N=CH-, schiff base) [21,24,25].The functional group of aldehyde group (C=O) in Fig.4a-e was not detected again after the functional group of azomethine group appeared.

3.2. Analysis of physical structure with XRD

The physical structure of 4,4 diaminodiphenyl ether, vanillin and Schiff base 4,4 diaminodiphenyl ether-vanillin (at 75^{0}C) are shown in Fig. 5. Fig. 5a is diffractogram of vanillin. Vanillin powder is a crystalline material and had a sharp peak at 2θ = 13.13^0[26]. Diffractogram of 4,4 diaminodiphenyl ether as seen at Fig. 5b and had 2θ = 11.91^0 - 50.39^0. If compared with 4,4 diaminodiphenyl ether and vanillin, diffractogram of Schiff base 4,4 diaminodiphenyl ether-vanillin had 2θ = 10.13^0 - 50.6^0.

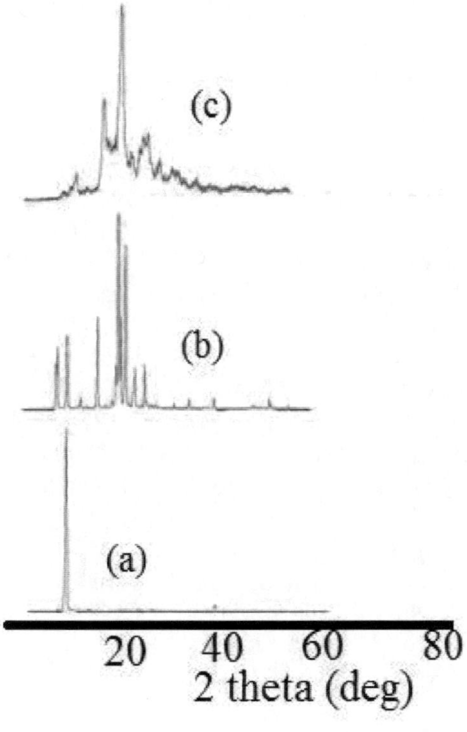

Figure. 5. XRD patterns of (a). vanillin, (b). 4,4 diaminodiphenyl ether and (c). Schiff base 4,4 diaminodiphenyl ether – vanillin synthesized at 75^{0}C.

4. Conclusion

Schiff base 4,4 diaminodiphenyl ether – vanillin possessed of free primary amine was resulted by reaction between 1 mol 4,4 diaminodiphenyl ether and 1 mol vanillin. The functional group of Schiff base 4,4 diaminodiphenyl ether – vanillin was not affected by temperature of synthesis. Presence of functional group –N=CH- (azomethine, Schiff base) at1597 cm^{-1} and primary amine (-NH_2) at 3387 cm^{-1}. The physical structure of Schiff base 4,4 diaminodiphenyl ether – vanillin possessed of free primary amine had $2\theta = 10.13^0 - 50.6^0$.

Refferences

[1] Gangolli S D 1999*The Dictionary of Substances and their Effects (DOSE) : O-S* Vol 6 second edition (Cambridge : The Royal Society of Chemistry) p75.
[2] Liu Y, Zhang J, Li Z, Luo X and Run M 2014 *Polymer* **55** 1688
[3] Singh V K, Kadu R and Roy H 2014 *Eur. J. Med. Chem.***74** 552
[4] Wang G, Li P, Peng Z, Huang M. and Kong L. 2011 *Adv. Mat. Res.***335-336** 474
[5] Shakeel F, Haq N and Siddiqui N A, 2015 *Food Chem.***180** 244.
[6] Balachandran V and Parimala K. 2012 *Spectrochim. Acta Part A : Mol. Biomol. Spectrosc.***95** 360.
[7] Jia Y and Li J 2015 *Chem. Rev.***115** 1597
[8] Jin X, Wang J and Bai J 2009 *Carbohydr. Res* **344** 825.
[9] Wang Z, Gao J, Wang J, Jin X, Zou M, Li K. and Kang P 2011 *Spectrochim. Acta Part A: Mol. Biomol. Spectrosc.***83** 511
[10] Sun H , Kanehashi S, Tsuchiya K and Ogino K. 2016 *Chem. Lett.***45** 439
[11] G¨ulcan M and S¨onmez M 2011 *Phosphorus, Sulfur, and Silicon* **186** 1962
[12] Kaya I., Bilici A and Gul M 2008*Polym. Adv. Technol.***19** 1154.
[13] Wu W, Xua Y, Yin J L and Shujie L 2014 *Synth.Reac. Inor. Met.Org. Nano-Met. Chem.* **44** 877
[14] Cucos P, Tuna F, Sorace L, Matei I, Maxim C, Shova S, Gheorghe R, Caneschi A, Hillebrand M and Andruh M 2014 *Inor. Chem.***53** 7747.
[15] Wang X, Deng W, Xie Y and Wang C 2013 *Chem. Eng. J.***228**238.
[16] Fajardo, A R, Lopes L C, Rubira A F and Muniz E C 2012 *Chem. Eng. J.* **83** 258.
[17] Li M, Xu J, Li R., Wang D, LiT, Yuan M and Wang J 2014 *J. Colloid Interface Sci.* **417** 133.
[18] Monier M 2012 *Int. J. Biol. Macromol.***50** 779
[19] Chi N T Q, Luu D X and Kim D 2011*Solid State Ionics* **187** 80
[20] Uysal S. and Uca I 2010 *J. Inclution Phenom. Macrocyclic Chem.***68**169.
[21] Coates J 2006 Interpretation of Infrared Spectra, A Practical Approach in *The Encyclopedia of Analytical Chemistry* (New York : John Wiley and Sons) p20.
[22] Liu Y, Li Z, Zhang J, Zhang H, Fan H and Run M. 2013 *J.Therm. Anal. Calorim.* **111** 1527.
[23] Peng H, Xiong H, Li J, Xie M, Liu Y, Bai C and Chen L 2010 *Food Chem.***121** 26.
[24] Zhou G, Ruhan A, Ge H, Wang L, Liu M, Wang B, Su H, Yan M, Xi Y and Fan Y 2014 *Burns* **40** 1672.
[25] Li N and Bai R 2005 *Ind. Eng. Chem. Res.***44** 6692
[26] Stroescu M., Stoica-Guzun A, Isopencu G, Jinga S I, Parvulesc O, Dobre T and Vasilescu M. 2015 *Food Hydrocoll.* **48** 69.

IOP Conf. Series: Journal of Physics: Conf. Series **1095** (2018) 012003 doi:10.1088/1742-6596/1095/1/012003

Oil Recovery Tests by using Bio surfactant of Indigenous *Pseudomonas peli* and *Burkholderia glumae* Bacteria from South Sumatera at Various Temperature Conditions

Bambang Yudono[1], Widia Purwaningrum[1], Sri Pertiwi Estuningsih[2], Venny Anggraini[1], Muhammad Said[3]

[1]Department of Chemistry, University of Sriwijaya
[2]Department of Biology, University of Sriwijaya
[3]Department of Chemical Engineering, University of Sriwijaya

Email: yudonob@hotmail.com Phone:+628153821994

Abstract. The isolates of indigenous *Pseudomonas peli* and *Burkholderia glumae* bacteria were obtained from Babat Toman Village, Musi Banyuasin, South Sumatera which have a potential as bio surfactant. The bacteria are thermophile bacteria, and then they were used to produce bio surfactant by using molasses as carbon source. The bio surfactant were tested their ability to the oil recovery of soil contained oil 18.64%. The variables of treatments were temperatures from 40^{0}C to 80^{0}C. The results showed that the optimal oil recovery by using *Pseudomonas peli* and *Burkholderia glumae* were 2.20% and 18.19% at temperatures 80°C and 40°C respectively. The results of optimal oil recovery of each bacterium were analysed using gas chromatography to determine the constituent components of petroleum that can be extracted by using the bio surfactant. The bio surfactant of *Burkholderia glumae* dissolved hydrocarbons at a fraction of temperatures 133; 139; 155; 156.8; 165; 173; 190; 197.5 and 206.3 ^{0}C with the total abundance of soluble oil 25.22%. Furthermore, the bio surfactant of *Pseudomonas peli* dissolved hydrocarbons at a fraction of temperatures 133.3; 134; 139; 140; 145 and 150 ^{0}C with the total abundance of soluble oil 16.94%.

Keywords: bio surfactant, temperature, oil recovery, *Pseudomonas peli, Burkholderia glumae*

1. Introduction

Microbial Enhance Oil Recovery (MEOR) is a technology that utilizes microbial activity by injecting microbes / bacteria into the reservoir to increase oil recovery. This technology is applied in tertiary recovery technology utilizing microorganisms and/or their metabolic products for recovery of residual oil [1]. The metabolic processes of bacteria can produce bio product such as bio surfactant, alcohol, biopolymers, gas, and acidity and can be proliferated if given the biocatalyst and appropriate nutrition. Bio product produced is able to extract the optimal oil recovery by improving displacement efficiency and reduce the viscosity of the oil trapped in the reservoir [2]. However, the use of chemical surfactant or physical processes such as pressurization, water flooding of steaming is generally inappropriate to most oil reservoirs [3]. The use of chemical surfactants for cleaning up oil reservoirs is not favorable in practice, because it is hazardous and costly and will leave undesirable residues which are difficult to dispose without adversely affecting the environment[4].

Low oil production in the tertiary phase because the remaining oil dominated by long chain hydrocarbon compounds, high viscosity and low mobility. High interfacial tensions between the water and oil may also produce high capillary forces retaining the oil in the reservoir rock [5]. Some bacteria may produce bio surfactants having two hydrophilic and hydrophobic active groups, for example *Pseudomonas sp, Bacillus sp,* and *Burkholderia sp* bacteria[6][1].
Bio surfactants have hydrophilic and hydrophobic groups so they can be partitioned between phases in water/oil and decrease the interfacial of surface tension [7]. Therefore, bio surfactant potentially can be used as Microbial Enhanced Oil Recovery (MEOR)[8][9]. Bio surfactant has several advantages over the chemical surfactant such as lower toxicity, higher degradability and better environmental compatibility, higher foaming, higher selectivity and specific activity at extreme pH, temperature and salinity[10].

The use of indigenous bacteria has the advantage of being adapted to the environmental conditions such as temperature, salinity, and pH. The *Pseudomonas peli* and *Bhurkholderia glumae* are thermo tolerant bacteria[11]. It is expected that the bio surfactant produced by the bacteria will also be suitable for the conditions.

This studies, isolation and identification of indigenous bacteria from South Sumatera which producing bio surfactant used to extract crude oil from soil contained oil at various temperature conditions (40-80°C). The bio-products excreted by these strains cause a series of very desirable changes in the properties of the crude oil [12].

2. Materials and Methods

Sterilization Equipment: All heat-resistant equipment sterilized by autoclave at a temperature of 121°C. Other tools sterilized with 70% alcohol.

2.1. Sampling
Samples were collected from 8 different areas of oil contaminated soils, extracted oil reservoirs, oil pipeline leakages and oil sewage and sludge dumps in Babat Toman Village, Banyuasin District of South Sumatera.

2.2. Cultivation media[13]
NA medium 150 mL inserted each 5 mL in a test tube. Tube containing NA tilted medium and allowed to stand for 24 hours. 1 Ose *Pseudomonas peli* inoculated on NA medium that has been hard and incubated at 37°C for 48 hours. The same treatment was done for *Bhurkholderia glumae* bacteria.

2.3. *Regeneration Time of Bacteria* [14]
Medium ZoBell as much as 100 mL plus 10 mL molasses autoclaved for ± 1 hour and cooled at room temperature ± 30°C . The medium was added 2 oses isolates of *Pseudomonas peli.* Bacterial inoculum solution was shook for 18 hours and every 3 hours a solution of 2 mL taken of and put into a test tube. Bacterial inoculum solution in a test tube which been separated from hour to 3, 6, 9, 12, 15, and 18 is inserted into the petri dish. Bacterial growth calculated using a counting chamber was placed on the microscope so that the resulting bacterial growth curve and found the shortest regeneration time of the bacteria. The same treatment as above was performed for both bacterial isolates that *Bhurkholderia glumae* so that the resulting curve obtained bacterial growth and shortest regeneration time. A generation time of bacteria could be calculated by the formula:

$$g = \frac{2\log t}{(\log X_t - \log X_o)}$$

Where:
X_t = final number of cells (cell)
X_0 = initial cell number (cell)
g = regeneration time (cells / hour)

2.4. *Preparation of sample for oil recovery test*
Soil was heated for 24 hours at a temperature of 110°C. Dry soil crushed using a mortar until smooth and screened so that the soil particles size 200 meshes. Sample soil prepared by 30 g of sludge oil and 70 g finely soil put into a glass beaker, and stirred until homogeneous.

2.5. *Measurement of initial TPH of the sample* [15]
Boiling flask of 250 mL size with sokhlet extraction tool which going to be used, dried in an oven, and cooled in a desiccator. A total of 10 g soil weighed in a filter paper. Filter paper containing the soil samples were placed in sokhlet extraction tool, further on it mounted condenser apparatus and below the boiling flask using n- hexane solvent by volume wetting the entire filter paper containing the soil samples and oil was then extracted using reflux. Reflux is performed until the solvent went back down to the boiling flask colorless. Solvent present in the boiling flask is distilled and collected, and then the boiling flask containing the extraction is heated in an oven at 100°C for one hour and then stored in a desiccator. After that, the flask along with the oil is weighed to constant weight. Percentage of Total Petroleum Hydrocarbons (TPH) petroleum is calculated using the formula:

$$\%TPH = \frac{W1-W2}{W3} x\ 100\%$$

Where:

W_1 = weight boiling flask and extract petroleum (g)
W_2 = weight of the empty boiling flask (g)
W_3 = weight of the soil sample (g)

2.6. *Zobell liquid medium preparation* [16]

Each ingredient is weighed according to the composition. All materials incorporated into the Erlenmeyer and dissolved by adding 1 L of distilled water and then inserted magnetic stirrer that serves to homogenize the solution and then covered with cotton and boiled using a hot plate. Erlenmeyer mouth covered with aluminum foil, and then sterilized in an autoclave for 20 minutes at a temperature of 121°C .

2.7. *Starter Bacteria preparation* [17]

Bacterial culture as much as 5 ose inoculated into each 100 mL of liquid medium and shaken ZoBell using a shaker for 24 hours at 120 rpm. Each subculture bacteria inoculated into Erlenmeyer containing 100 ml of liquid medium ZoBell based on the shortest regeneration time while still shaken using a shaker. *Pseudomonas peli* and *Bhurkholderia glumae* regeneration times are 12 hours.

2.8. *Crude bio surfactants production* [18]

Medium ZoBell as much as 100 mL mixed with 5 % (v / v) molasses into the flask. Media moved into Erlenmeyer, and then into the medium is inoculated 10 % starter bacteria. Culture aerated for 24 hours, and then incubated with a temperature of 40°C for 24 hours

2.9. *Effect of Temperature on Oil Recovery* [19]

Erlenmeyer already containing 10 g of soil contaminants coupled with 200 mL of crude bio surfactant of *Pseudomonas peli* then aerated and heated to a temperature variation of 40, 50, 60, 70, 80°C for 7 days and then filtered using a filter paper. The resulting precipitate washed with distilled water and dried at a temperature of 100°C . The precipitate was dried and calculated so the percentage TPH. The same treatment is also performed using bacteria *Bhurkholderia glumae.*

2.10. *Initial components in sample analysis by using Gas Chromatography*[15]

Extract oil from sample diluted with n- hexane at a ratio of 1: 9. The results of such dilution are injected into the GC apparatus with injector temperature of 290°C. Sample 2,5 μL is injected at 40°C on hold for 5 minutes, then the temperature is increased every minute to the temperature rise 5°C to 50°C, then the temperature was raised 10°C per minute up to 280°C and hold for 5 minutes. The mobile phase used is Helium. This procedure is also used to analyze of sample components after dissolved into crude bio surfactants.

3. Results and discussion

3.1. *Regeneration time Pseudomonas peli and Burkholderia glumae bacteria*

The growth curve was conducted to determine the optimum growth rate and optimum formation of metabolites produced by a microbe. The growth curve of bacteria was made with the intention to see the growth phases of the respective bacteria, namely the phase lag, exponential, stationary and death.

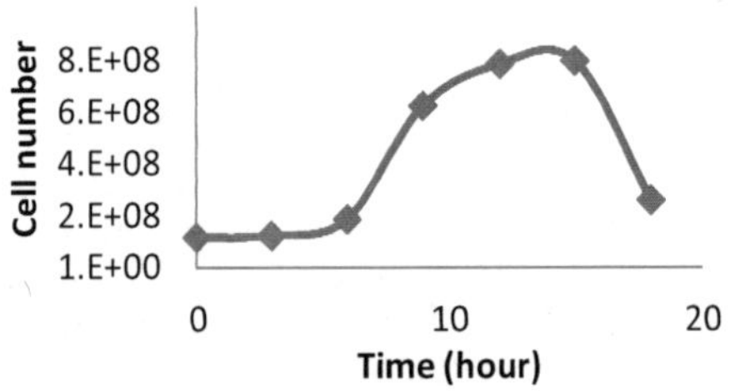

Figure1. Growth curve of bacteria *Pseudomonas peli*

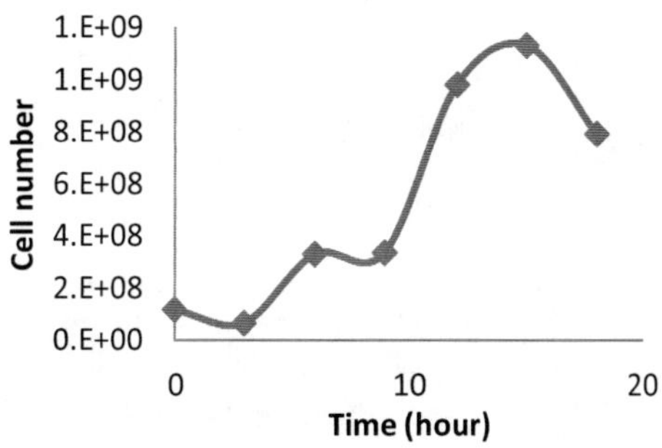

Figure 2. Growth curve of bacteria *Burkholderia glumae*

The shortest regeneration time of *P. peli* and *B. glumae* are 2.5 cells/hour and 2.35 cells/hour during 12 hours respectively. The shortest regeneration time is achieved when the growth rate rise sharply. This growth rate constant shows the number of bacteria that grow per unit of time in culture that is growing exponentially. Judging from both bacteria constant growth rate is highest in *B. glumae* i.e. 0.295 $hour^{-1}$, whereas *P. peli* growth rate constant was 0.277 h^{-1}. *Peli P.* and *B. glumae* used in oil recovery treatment should be based on a growth curve. Both of these bacteria have the same exponential phase so in this study the bacteria are combined with each other. Thus, each of the bacteria will not have competition carbon source during its growth[20] [7].

3.2. Measurement of Total Petroleum Hydrocarbons in initial sample

Measurement of Total Petroleum Hydrocarbon (TPH) is performed three repetitions on the soil sample is mixed with petroleum sludge generate initial TPH value by an average of 18.64 % from the three repetitions. Measurement of initial TPH samples land based on solubility in n- hexane extraction. After extraction the solvent is evaporated, subsequently determined by gravimetric TPH, it means physically separated analyses from the sample or from the solvent. Stages measurement initial TPH content of soil samples carried by solvent extraction using soxhletation. TPH is measured to determine the percentage of hydrocarbons in sample.

3.3. Effect of Temperature on the Recovery of Oil on Each Bacteria

Crude bio surfactant of *Pseudomonas peli* and *Burkholderia glumae* applied to oil recovery on soil contained petroleum with temperature variation of 40°C to 80°C with 5% molasses concentration and incubated for 7 days. Both of these bacteria are able to live in temperatures of 40°C to 90°C or in the grouping of thermophilic bacteria[18][12][5].

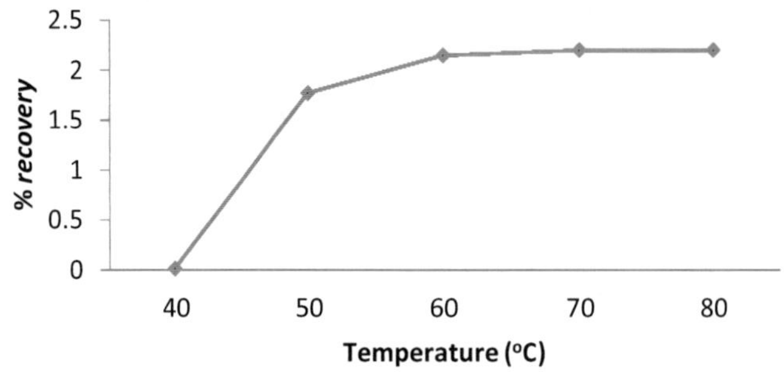

Figure 3. Percentage of oil recovery *Pseudomonas peli*

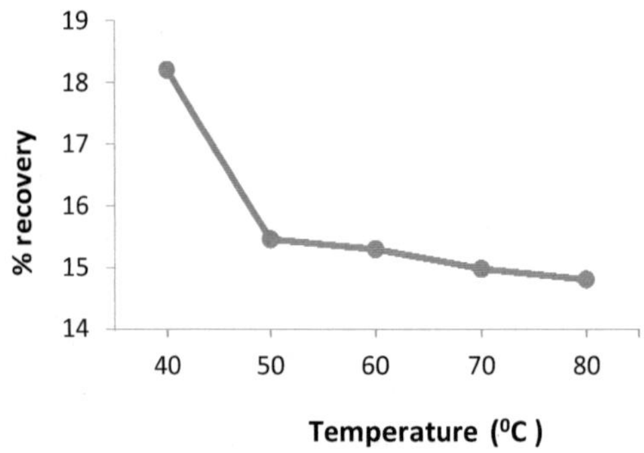

Figure 4. Percentage of recovery of oil bacteria *Burkholderia glumae*

Based on Figure 3. *P. peli* shows that the higher temperature of crude bio surfactant, the percentage recovery of petroleum also increased. At temperatures of 40; 50; 60; 70; and 80 °C percentage recovery of petroleum from *P. peli* 0.16; 1.77; 2.15; 2.20; and 2.20 % respectively. However, bio surfactant produced by these bacteria is relatively small. Contrary, in Figure 4 *B. glumae* shows that the higher the temperature of crude bio surfactant, the percentage recovery of petroleum decreased. At a temperature of 40 ; 50 ; 60 ; 70 ; and 80 °C percentage recovery of oil from *B. glumae* i.e. 18.19 ; 15.45 ; 15.29 ; 14.97 ; and 14.81 % respectively. *B. glumae* produce bio surfactant relatively higher than the *P. peli.*

The results obtained in this study can be prove that the bio surfactant produced by *B. glumae* amounted to 81.25% and *P. peli* bio surfactant produced relatively little that is equal to 33.75% [3]. At different temperatures by both bacteria in producing petroleum recovery percentages because, each bacterium has a specific temperature to produce bio surfactant. The highest population of *P. peli* range of $2x10^6$ cells at 90°C, while *B. glume* highest population range 6 x 10^6 at 40°C [3]. The ability of the bacteria can be seen from the clear zone on an agar medium[5][21] [22]. The more extensive the resulting clear zone, the greater the potential for bacteria to produce bio surfactant. The results of the study, said that the clear zone produced by *P. peli* and *B. glumae* by 25.79 and 21.33 mm respectively. However, in this study *B. glumae* produce bio surfactant more than *P. peli.* This may be caused by the influence of temperature in the bio surfactant produced *P. peli* less than optimal. The ability of bacteria to produce bio surfactant relates to the ability of bacteria using hydrocarbons as substrates. Microorganisms with great bio surfactant generally have a large capacity also in solving hydrocarbons [21] [23].

3.4. Analysis components of crude oil samples before and after treated crude bio surfactants at variation of temperatures

GC analysis provides information on the number of components making up the petroleum before and after treatment by using crude bio surfactant. GC analysis results are shown in Figure 5.

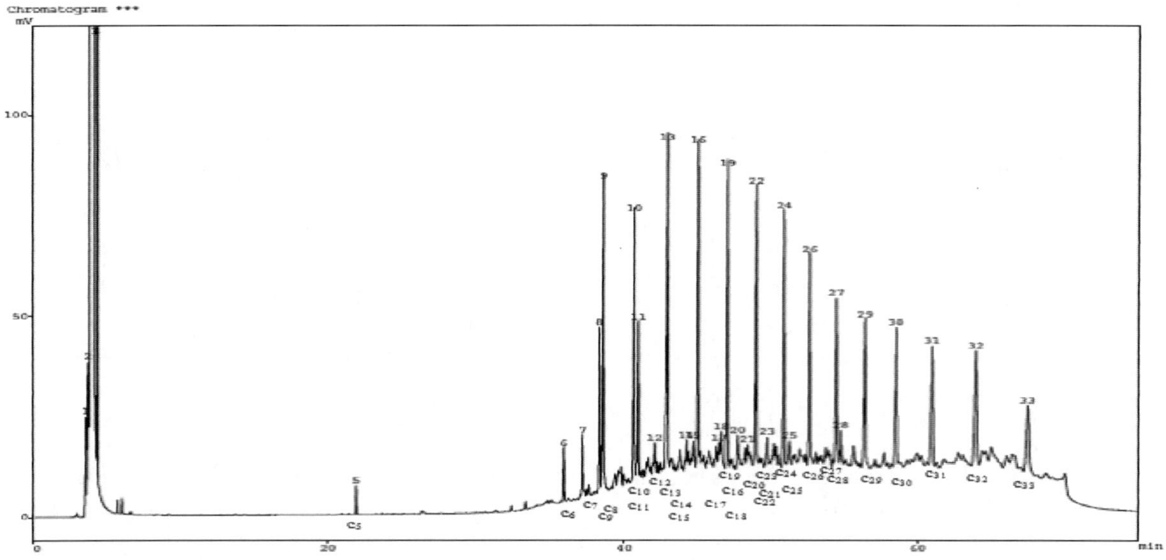

(a)

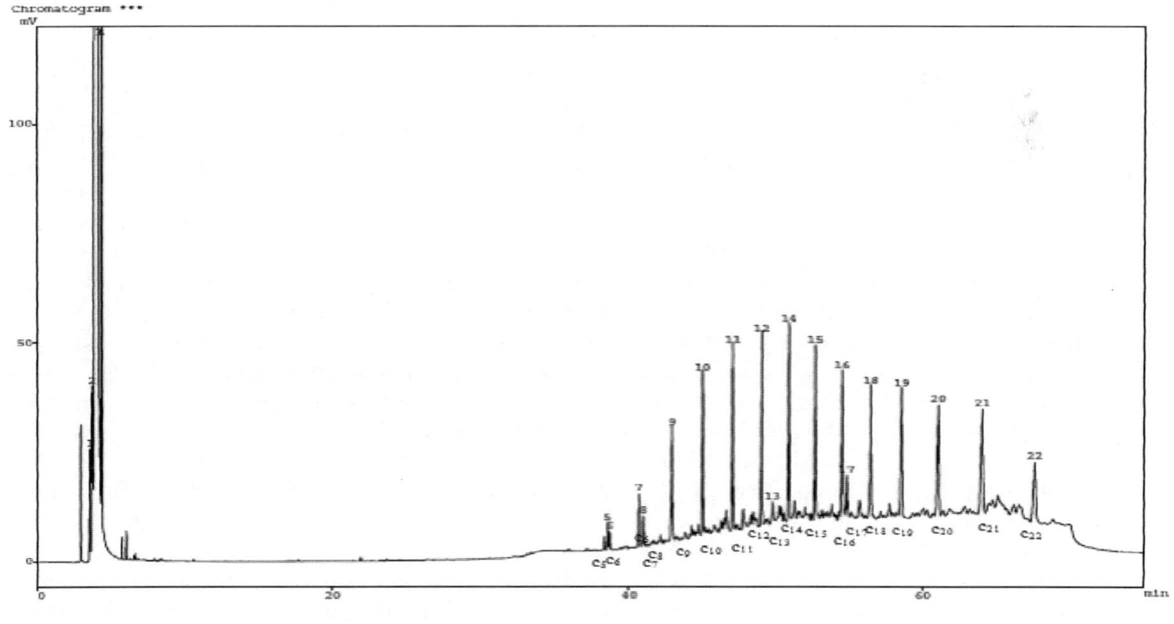

(b)

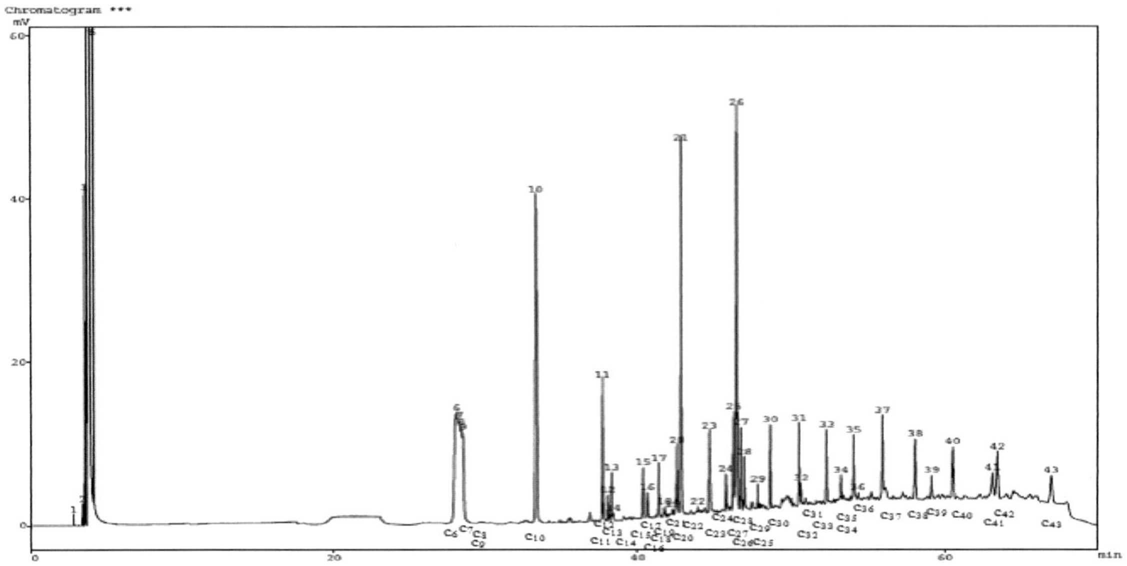

(c)

Figure 5. (a) The chromatogram of petroleum oil at initial condition, (b) Chromatogram of sample after treatment by using *Pseudomonas peli*, and (c) Chromatogram of sample after treated by using Bacteria *Burkholde glumae*.

Figure 5 shows 33 peaks of chromatogram of oil prior to treatment. Based on the total retention time for 67 minutes, the highest peak in the chromatogram retention time of 38.561; 40.652; 42.876; 44.99; 47.011; 48.94; 50.799; 52.58 while the other had a retention time of chromatogram with low peak. Samples of soil and oil after the addition of bio surfactant of *Pseudomonas peli* with the best test temperature is 80°C has 22 peaks of chromatogram with retention time for 67 minutes. The highest peak of the chromatogram at the retention time of 44.988; 47.019; 48.96; 50.826; 52.616; 54.415; 56.347; 58.528; 61.057; 64.061 while the other had a retention time of chromatogram with low peak. For the next sample after the addition of bio surfactant of *Burkholderia glumae,* the best test temperature is 40°C has 43 peaks of chromatogram with retention time for 67 minutes. The peaks of chromatogram at the retention times of 33.305; 42.835; 46,434 have high peak while the others had a retention time of chromatogram with low peak. The third chromatogram shows that oil is a multi-component compound [9]. The performance of these two bacteria can be seen on the chromatograms of initial sample and after each treatment by comparing the percentage abundance of compounds between samples. Changes chromatogram peaks can occur because the process of degradation by bacteria. The results may indicate a degradation peaks are missing, emerging, growing peak or reduced peak. Solving hydrocarbon compounds into simpler compounds is largely determined by the structure of the hydrocarbon molecules and types of bacteria[4][14][23].

Degraded hydrocarbon compounds can be demonstrated by comparing the data histogram with peaks identical carbon fraction of percent abundance of each component. Of the three, the chromatogram peaks there are some areas that have a fraction of the carbon chromatogram peaks were identical. The carbon chains are identical peak on the chromatogram TPH before the addition of crude bio surfactant and chromatogram TPH after the addition of crude bio surfactant. The determination of peak abundance percent identical and can be describes the fraction of carbon degraded or dissolved [24].

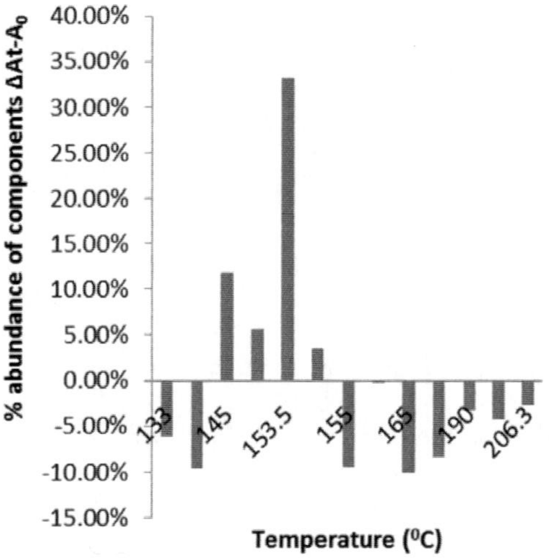

Figure 6. Changes in % Abundance After treatment with the *B. glumae* bacteria

The difference in percent abundance of component is calculated by subtracting the percent abundance peak after the addition of crude bio surfactant (A_t) from the bacterium *Burkholderia glumae* reduced percent of peak abundance prior to the addition of crude bio surfactant (A_0) as shown in Figure 6. From the results of these reductions was found that hydrocarbon compounds are dissolved in bio surfactant of bacteria *Burkholderia glumae* obtained if the result of subtraction is negative, which means that the hydrocarbon compounds soluble bio surfactant[25]. Histogram declining indicates that the short-chain hydrocarbon compounds soluble bio surfactant, so that the chromatogram peaks are short carbon chain is missing. Such changes occur at a temperature of 133; 139; 155; 156.8; 165; 173; 190; 197.5 and 206.3 °C, where the short-chain hydrocarbons dissolved in abundance percentage of 25.22%. Long-chain hydrocarbons increase in temperatures program 145; 149; 153.5 and 154 °C.

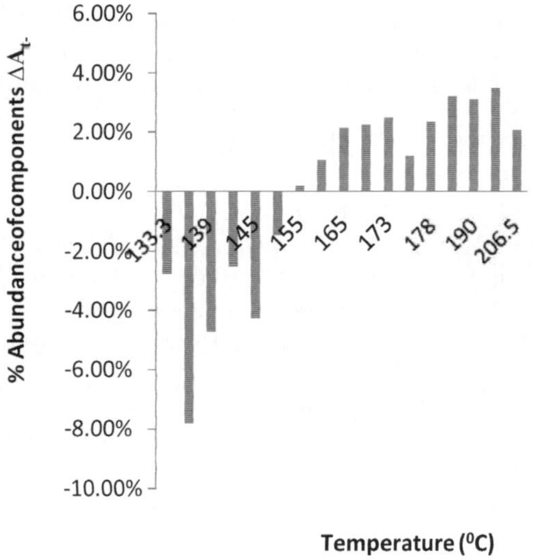

Figure 7. Changes % Abundance After treatment with the *P. peli* bacteria

In Figure 7. The components with temperatures of 133.3; 134; 139; 140; 145 and 150 °C are decreased in percent abundance of components. This indicates that the bio surfactant of *Pseudomonas peli* bacteria can dissolve petroleum hydrocarbons with an abundance percentage of 16.94 %. At temperatures of 160 ; 165 ; 169; 173 ; 174 ; 178 ; 184 ; 190 ; 197 and 206.5 °C percent abundance increased . These circumstances show that the long-chain hydrocarbon compounds remaining in the residue and insoluble in bio surfactant. Meanwhile, if there is a decrease the microbial hydrocarbon is used as an energy source, or dissolved this has led to wide differences in each area at the same retention time[26].

4. Conclusion

Based on this study, we can conclude:

1) The temperature that produces% recovery of oil in *Burkholderia glumae* maximum occurs at a temperature of 40°C by 18.19% compared to the relatively higher than *Pseudomonas peli* occurs at a temperature of 80°C equal to 2,20%.
2) The bacterium *Burkholderia glumae* bio surfactants can dissolve hydrocarbons of petroleum fractions temperatures 133; 139; 155; 156.8; 165; 173; 190; 197.5 and 206.3 °C with total abundance of 25.22%, while the bio surfactant produced by Pseudomonas peli can dissolve hydrocarbons of petroleum fractions temperature 133.3; 134; 139; 140; 145 and 150 °C with a total abundance of 16.94%.

References

[1] Haghighat S, Sepahy A A, Mazaheri Assadi M and Pasdar H 2008 *Int. J. Environ. Sci. Technol.* **5** 385
[2] Feng K, Li R, Chen Y, Zhao B, Yin T and Antonescu C 2015 *PLoS One* **10** 141
[3] Singh A, Hamme J D Van and Ward O P 2007 Biotechnology Advances **25** 99
[4] Rao K J, Chaudhuri R G and Paria S 2014 *Biochem. Pharmacol.* **2** 826
[5] Shibulal B, Al-bahry S N, Al-wahaibi Y M, Elshafie A E, Al-bemani A S and Joshi S J 2014 Scientific World Journal **2014** 309159
[6] Yudono B and Estuningsih S P 2013 *Pros. Semirata FMIPA Univ. Lampung*
[7] Saimmai A and Rukadee O 2012 World J Microbiol Biotechnol **28** 2973
[8] Omoniyi O A 2015 International Journal of Scientific & Engineering Research **6**(1) 1378
[9] Chandankere R, Yao J, Choi M M F, Masakorala K and Chan Y 2013 *Biochem. Eng. J.* **74** 46
[10] Zinjarde S, Apte M, Mohite P and Kumar A R 2014 *Biotechnol. Adv.*
[11] Fanani Z, Yudono B and Situmorang V R 2014 *Pros. Semin. Nas. Lahan Suboptimal* **3** 979
[12] Adhitya I and Yuichi P 2014 J Petrol Explor Prod Technol **4** 189
[13] Yudono B, Said M, Sabaruddin , Napoleon A and Fanani Z 2011 *J. TANAH Trop. (Journal Trop. Soils)* **16** 33
[14] Yudono B, Said M, Sabaruddin ., Napoleon A and Utami M B 2010 *HAYATI J. Biosci.* **17** 155
[15] Yudono B 2009 J. Sustain. Dev. **2** 64
[16] Jain R M, Mody K, Mishra A and Jha B 2012 *Carbohydr. Polym.* **87** 2320
[17] Yu H and Saskatchewan R 2010 *Dissertation* (University of Regina)
[18] Bharali P, Das S, Konwar B K and Thakur A J 2011 *Int. Biodeterior. Biodegradation* **65** 682
[19] Marchant R and Banat I M 2012 *Trends Biotechnol.* **30** 558
[20] Shavandi M, Mohebali G, Haddadi A, Shakarami H and Nuhi A 2011 *Colloids Surfaces B Biointerfaces* **82** 477
[21] Makkar R S and Cameotra S 2011 Colloids and Surfaces B: Biointerfaces **85** 174
[22] Dubey K V, Charde P N, Meshram S U, Shendre L P, Dubey V S and Juwarkar A A 2012 *Bioresour. Technol.* **126** 368
[23] Jain R M, Mody K, Joshi N, Mishra A and Jha B 2013 *Int. J. Biol. Macromol.* **62** 52
[24] Souza E C, Vessoni-penna T C, Pinheiro R and Oliveira D S 2014 *Int. Biodeterior. Biodegradation* **89** 88
[25] Yudono B, Said M, Sabaruddin, Napoleon A and Utami M B 2010 *HAYATI J. Biosci.* **17** 155
[26] Gomaa E Z 2013 J Polym Environ **21** 495

Lipid Extraction Method From Microalgae *Botryococcus Braunii* As Raw Material To Make Biodiesel With *Soxhlet* Extraction

J Boni[1], S Aida[2], K Leila[3]

[1] Applied Master of Renewable Energy Engineering, Politeknik Negeri Sriwijaya, Jl. Srijaya Negara, 30139, Palembang, Indonesia
[2] Energy Engineering Department, Politeknik Negeri Sriwijaya, Jl. Srijaya Negara, 30139, Palembang, Indonesia
[3] Renewable Energy Engineering Study Program, Politeknik Negeri Sriwijaya, Jl. Srijaya Negara, 30139, Palembang, Indonesia

E-mail's : bonijunita21@gmail.com

Abstract. This research was a preliminary research that aimed to determine the number of lipid from microalgae *Botryococcus braunii* through the extraction process and to find out the potentiality of microalgae to be converted into biodiesel. In this research used n-hexane as solvent, then it was conducted a literature study for lipid extraction from *Botryococcus braunii* with *soxhlet* method as companion data result. In this research, lipid extraction was performed on a laboratory scale using 175 ml n-hexane. Microalgae *Botryococcus braunii* were cultivated for 10 days preceding the process of drying and it was followed by the process of lipid extraction with *soxhlet*. The results of this study showed that the extraction with *soxhlet* method had the percentage of lipid yield that was 24% with 45 cycles. The number of lipid yields obtained was affected by the number of solvent cycles during the extraction process. The extraction process run optimally since the number of solvent cycles increased more and more until there was no more lipids extract that could be dissolved by the solvent. Therefore, the best method of extraction was the *soxhlet* method because it had a dilution principle that occurred repeatedly.

Keywords : microalgae *Botryococcus braunii,* biodiesel, lipid

1. Introduction

The use of alternative energy from renewable energy sources can help in reducing the dependence on the use of fossil fuels whose reserves are depleting and their prices are increasing, as well as to safeguard the environment from global climate change and natural degradation resulting from the use of fossil fuels as an energy source [1]. Microalgae can be converted into biodiesel which is one of the renewable alternative energy and is considered as a good alternative energy source when compared with other alternative energy sources such as corn, soybeans, sunflower, palm, coconut and others [2]. Microalgae can absorb CO_2 which comes from burning fossil fuels and other sources from the nature, so that it can reduce emissions [3]. Based on planting area base, the microalgae have the ability to produce oil as biodiesel feedstock 15 - 300 times more than other plants. Microalgae also have a fairly short harvest cycle of 1 day to 10 days depending on the process [4]. The comparison of potential utilization of microalgae into biodiesel with other raw materials can be seen in Table 1.

Microalgae consist of proteins, carbohydrates, and lipids [3]. Lipid in the microalgae can be used to produce biodiesel, while the carbohydrates and proteins can be used to produce bioethanol [6]. There are various types of microalgae that can be converted into biodiesel. Microalgae with high lipid content are *Botryococcus braunii* with lipid content of 25 - 75% (% dry weight biomass) [4]. The lipid content of some microalgae can be seen in Table 2.

Table 1. Comparison of Biodiesel Raw Material Potentiality [5]

Raw Material	Produced Lipid Yield (L/Ha)	Required Area Width (M ha)
Corn	172	1,540
Soybean	446	594
Canola	1,190	223
Jatropha	1,892	140
Coconut	2,689	99
Palm	5,950	45
Microalgae[a]	58,700	4.5
Microalgae[b]	136,900	2

Table 2. Lipid Content of Some Microalgae [4]

Mikroalgae	Lipid Content (% dry weight biomass)
Botryococcus braunii	25 – 75
Chlorella sp.	28 – 32
Crypthecodinium cohnii	20
Cylindrotheca sp.	16 – 37
Dunaliella primolecta	23
Isochrysis sp.	25 – 33
Monallanthus salina	>20
Nannochloris sp.	20 – 35
Nannochloropsis sp.	31 – 68
Neochloris oleoabundans	35 – 54
Nitzschia sp.	45 – 47
Phaeodactylum tricornutum	20 – 30
Schizochytrium sp.	50 – 77
Tetraselmis sueica	15 – 23

Microalgae *Botryococcus braunii* (Figure 1) is a green single-celled microalgae of the *Chlorophyta* microalgae division, these microalgae can be found in lake waters, brackish waters or ponds, and sea. Microalgae *Botryococcus braunii* can be a source of long-chain liquid hydrocarbons (C21 - C33) which can be further processed into alternative feedstock of renewable energy sources replacing fossil energy [7,8].

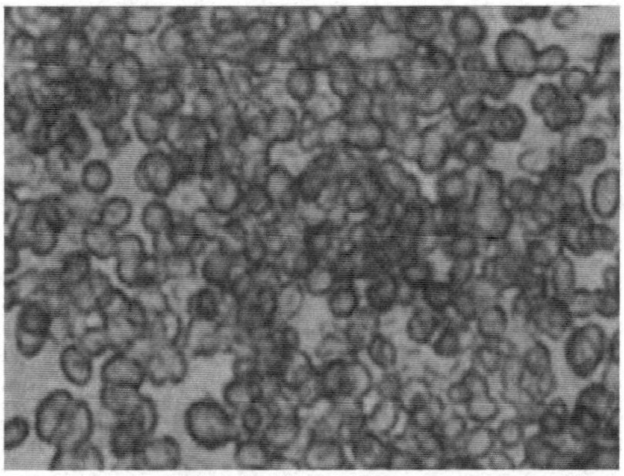

Figure 1. Microscopic Picture of The Mikroalgae *Botryococcus braunii* [9]

Biodiesel from microalgae production process includes cultivation, harvesting, biomass processing, lipid extraction, and transesterification processes. Amongst all the stages, the stages of lipid extraction are important and costly process. The production of microalgae biodiesel mainly depends on the type of lipid extraction process used [10].

This article was a preliminary research that described the effect of lipid extraction with *soxhlet* extraction method on lipid number of microalgae *Botryococcus braunii* from cultivation results within 10 days, then it was conducted a study of previous literature study related to microbial lipid extraction of *Botryococcus braunii* with *soxhlet* extraction method as supporting data for the results of conducted research and the potentiality of microalgae as a source of biodiesel raw materials.

2. Material and Methods

2.1. Microalgae Cultivation Stage

Stages of cultivation, determination of cell number, lipid extraction, and determination of lipid number were done for ± 2 months in Chemical Engineering Processing Unit Laboratory of Politeknik Negeri Sriwijaya. The research was conducted on the laboratory scale. Microalgae *Botryococcus braunii* seeds were obtained from Balai Besar Pengembangan Budidaya Air Payau (BBPBAP) of Jepara, Central Java.

The cultivation stage of microalgae *Botryococcus braunii* was carried out for 10 days in aerated culture bottles. Microalgae *Botryococcus braunii* of 100 ml were cultivated on cultivation medium in the form of 500 ml sea water with 25 - 35 ppt salinity, and nutrition medium of walne was used as much as 1 ml. The lighting process was done by using TL lamp 20 watts for 12 hours daylight and 12 hours dark. During the cultivation stage, the determination of the cell number of microalgae *Botryococcus braunii* with the aid of hemocytometer and microscope was done every day. It was calculated using equation below (1) [11].

$$\text{Number of cells (cell/ml)} = \left(\frac{\text{cell number}}{\text{box number}}\right) \times 25 \times 10^4 \tag{1}$$

2.2. Microalgae Extraction Stage

After cultivation stage, microalgae *Botryococcus braunii* was harvested and dried. The dried *Botryococcus braunii* were extracted by the *soxhlet* extraction method. The solvent used in this extraction was a dry 10% n-hexane. Dry microalgae *Botryococcus braunii* as much as 10 gram was extracted with 175 ml of n-hexane for 4 hours of heating. After the extraction process finished, the resulted lipids were distilled to separate the lipids and n-hexane solvent for 3 hours at 80°C. After that, the lipid was weighed and calculated by using equation (2) [11]:

$$\%\ \text{Yield} = \frac{\text{Extracted lipid weight}}{\text{Initial sample weight}} \times 100\% \tag{2}$$

3. Results and Discussion

This research was begun with the cultivation stage. A total of 100 ml of microalgae *Botryococcus braunii* was cultivated for 10 days on a 500 ml sea water medium with 25 - 35 ppt salinity. Cultivation of microalgae *Botryococcus braunii* was carried out in a culture bottle equipped with an aerator. The cultivation process inside the culture bottle was done indoors with the help of 20 watt TL lamp lights or about 4000 lux with long lighting time for 12 hours of daylight and 12 hours of darkness. During the cultivation stage, the calculation of the number of cells was done every day during the cultivation period by calculating cell density by using chamber hemocytometer and microscope as a tool to see microalgae *Botryococcus braunii*. The result of cell calculation for 10 days cultivation period can be seen in Table 3.

Table 3. Number of Cell Density of Microalgae *Botryococcus braunii*

Days	Number of Cell Density (cell/ml)
0	0.694×10^4
1	2.257×10^4
2	2.430×10^4
3	3.125×10^4
4	3.993×10^4
5	4.340×10^4
6	4.861×10^4
7	5.208×10^4
8	3.645×10^4
9	3.472×10^4
10	2.604×10^4

The research results showed that the greatest number of cell density of microalgae *Botryococcus braunii* was obtained on the 7 day of cultivation which was 5.208×10^4 cells/ml. The calculation data of the number of cell density of microalgae *Botryococcus braunii* are depicted in a graph of Figure 2.

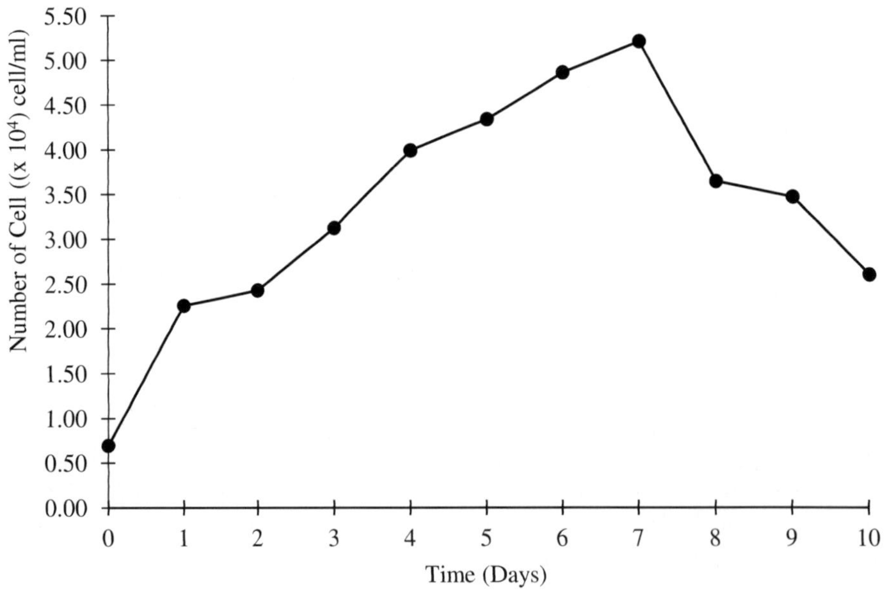

Figure 2. Graph of The Number of Cell Density of Microalgae *Botryococcus braunii*

The 10 day cultivated microalgae *Botryococcus braunii* was taken as 10 grams to be extracted to 175 ml of n-hexane. The extracted result was calculated in % yield which was the comparison between the extracted-lipid weight and the initial sample weight based on formula. After being extracted and distilled, the drying process within the oven was done and the obtained dried yield was as 2.4 grams and the calculation of % yield was done. The result was 24% lipid yield. During the 4 hour extraction, 45 solvent cycles were occurred.

The comparing data were obtained from the literature study of several previous studies. On the literature study, data was selecting by the highest result of % yield in each study. The results of lipid extraction from previous studies and direct study can be seen in Table 4 and Figure 3.

Table 4. The Results of Microalgae *Botryococcus braunii* Lipid Extraction

Ref.	Lead Author	% Yield	Description
-	Research	24	10-day culture, 25-35 ppt salinity, 1 ml walne, 100 ml of microalgae seeds in 500 ml of culture media, 20 watt of TL exposure for 12 hours, 175 ml of n-hexane solvent.
[12]	Febriana (a)	58.23	7-day culture, 45 ppt salinity, 5,000 lux of exposure for 12 hours, 250 ml of n-hexane solvent, 200 ml of microalgae seeds in 800 ml culture media.
[13]	Sari (b)	67	7-day culture, 30 ppt salinity, 10,000 lux of exposure for 24 hours, 250 ml of n-hexane solvent, 200 ml of microalgae seeds in 800 ml of culture media, 5 ml of walne.
[14]	Nagaraja (c)	78	35-day culture, 0.15 molar of NaCl salinity, 12 hours of exposure, 250 ml of microalgae seeds, 250 ml of n-hexane solvent.

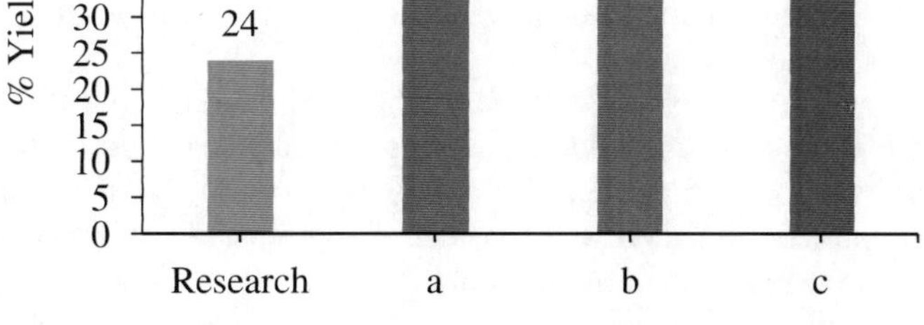

Figure 3. Graph of % Yield of Lipid Result of Direct Study and Literature Study

In the stage of cultivation of microalgae *Botryococcus braunii* in the direct study (Figure 2), the greatest number of cell density was occurred on the 7th day of cultivation period which was 5.028 x 10^4 cell/ml. It meant that on the 7th day, the exponential phase of *Botryococcus braunii* growth phase was occurred. The initial growth phase or lag phase was occurred on the 1st day of cultivation period with the additional nutrition of walne and the culture media which used to be increased frequently until the 6th day of cultivation period. The lowering phase of growth rate started to happen on the 8th day of cultivation period, then on the 9th day of cultivation period, the stationary phase was occurred, in which the number of microalgae *Botryococcus braunii* cells were constant as the number of new cell formation was offset by the number of cell deaths. On the 10th day, there was a decrease in the

IOP Conf. Series: Journal of Physics: Conf. Series **1095** (2018) 012004 doi:10.1088/1742-6596/1095/1/012004

number of cell density of microalgae *Botryococcus braunii* to 2.604 x 10^4 cell/ml which meant that there had been a death phase due to the decreasing number of microalgae *Botryococcus braunii* since nutrients in the culture medium was running out.

After the 10 day cultivation process, microalgae *Botryococcus braunii* was harvested and dried, then extracted using *soxhlet* extraction method. The *soxhlet* method was selected because this method is the best method used to get great numbers of yields with little amount of solvents, faster extraction period, and perfectly extracted samples as it was done in repetition [15]. N-hexane solvent was chosen since lipids are soluble in non-polar solvents only, n-hexane has also been extensively used throughout the world as a solvent for extracting vegetable oils such as lipid microalgae [17].

The results of the direct study extraction yielded 24% yield of lipid with 45 cycles of solvent by using 175 ml of n-hexane solvent. If compared to the literature study results of previous studies (Figure 3), the % yield lipid obtained was the lowest % yield, but this was influenced by several factors (Table 4). The results of the highest % yield of lipid yields were 78% done by (c) with 250 ml of cultivated microalgae *Botryococcus braunii* while the researchers only used 100 ml of it. Researcher (a) and (b) also cultivated with 200 ml of microalgae *Botryococcus braunii* seeds that resulted a lower % yield than researcher (c) did. The direct study used 175 ml of n-hexane solvent for extraction while the researchers (a), (b), and (c) used 250 ml of it.

Researcher (c) conducted the cultivation process for 32 days while researcher (a) and (b) conducted the cultivation for 7 days and the direct study conducted 10 days of cultivation. Researcher (b) had a higher % yield compared to researcher (a) and the direct study since the number of walne as the nutrient during the cultivation process was greater, which was 5 ml. This could affect the yield as nutrients are needed during the growth period, if the intake of nutrients and medium is not sufficient, then the rate of growth will be obstructed.

Exposure in direct research was done with 4,000 lux of exposure, while for researcher (a) with 5,000 lux and researcher (b) with 10,000 lux which meant that the greater the exposure or intensity of light given, the greater the lipid content in microalgae *Botryococcus braunii* would be.

The salinity of culture media during the cultivation period may also affect the number of lipid of microalgae *Botryococcus braunii*. As in the direct study, salinity used during the cultivation period was 25 - 35 ppt which had lower % yield of lipid compared to researcher (a) who cultivated culture media with the salinity of 45 ppt. This suggested that the higher the salinity of culture media from microalgae *Botryococcus braunii* is, the higher the productivity of lipids of *Botryococcus braunii* microalgae during the cultivation would be.

Different conditions (Table 4) will determine the number of lipid to be obtained since the lipid productivity of microalgae *Botryococcus braunii* was influenced by the following factors: nutrition, temperature, light intensity and duration, salinity, nitrogen content in the growth medium and the influence of competitor organisms in culture [12].

The selection of the extraction and solvent methods to be used would also affect the number of lipid yields of the microalgae *Botryococcus braunii*. The inappropriate selection of extraction and solvent method could increase the cost [16]. Since this study was conducted on a laboratory scale, the selection of the *soxhlet* extraction method was considered appropriate as it could save time as well as maximize the extraction process by repeated dilution.

The researcher also extraction with other methods such as maceration, percolation, autoclave, and osmotic by using 75 ml of n-hexane. The extraction results can be seen in Figure 4.

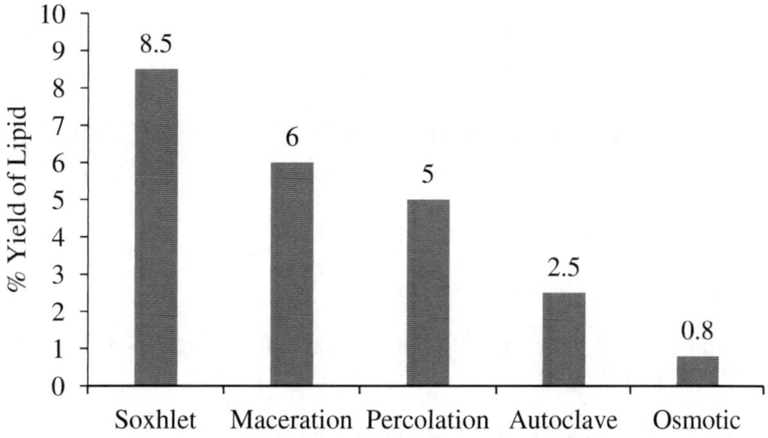

Figure 4. Graph of % Yield of Lipid Result with Various Extraction Methods

In Figure 4 it can be seen that the extraction results with *soxhlet* method yielded 8.5% lipid, maceration yielded 6% lipid, percolation yielded 5% lipid, autoclave yielded 2.5% lipid, and osmotic yielded 0.8% lipid. Then it can be concluded that the best extraction method is *soxhlet* which yields the highest % yield of lipid.

4. Conclusion
Based on research and literature study conducted, the best method of extraction is the soxhlet method because it had a dilution principle that occurs repeatedly. And based on literature studies, many researchers used the soxhlet extraction method to extract lipid from microalgae Botryococcus braunii. And n-hexane has also been extensively used throughout the world as because it is a non-polar organic solvent and this solvent can used for extracting vegetable oils such as lipid microalgae.

References
[1] Nigam P S and Singh A 2011 Progress in Energy and Combustion Science **37** 52.
[2] Alam F, Date A, Rasjidin R, Mobin S and Moria H 2012 Procedia Engineering **49** 221.
[3] Um B H and Kim Y S 2009 Journal of Industrial and Engineering Chemistry **15** 1.
[4] Schenk P, Thomas-Hall S, Stephens E, Marx U, Mussgnug J, Posten C, Kruse O, Hankamer B 2008 BioEnergy Research **1** 20..
[5] Chisti Y 2007 Biotechnology Advances **25** 294.
[6] Dragone G, Fernandes B, Vicente A A, and Teixeira J A 2010 Formatex 1355.
[7] Metzger P and Largeau C 2005 Applied Microbiology Biotechnology **66** 486.
[8] Banerjee A, Sharma R, Chisti Y and Banerjee U C Critical Reviews in Biotechnology **22** 245.
[9] Dayananda C, Sarada R, Gokare R, Kumar V 2007 Electronic Journal of Biotechnology 10 80.
[10] Kim J, Yoo G, Lee H, Lim J, Kim K, Kim C W, Park M S, Yang J W 2013 Biotechnology Advances **31** 862.
[11] Kalsum L, Sari I P and Silvia M 2016 Proceedings of the International Conference FIRST (Palembang) p B6
[12] Febriana E D, Mukti H dan Zullaikah S 2011 Jurnal Teknik POMITS 1–6
[13] Sari A M, Mayasari H E, Rachimoellah dan Zullaikah S 2013 Jurnal Teknik POMITS **2** 1–6.
[14] Nagaraja Y P, Biradar C, Manasa K S and Venkatesh H S 2014 International Journal of Current Microbiology and Applied Sciences (IJCMAS) **3** 851.
[15] Heinrich M, Barnes J, Gibbons S and Williamson E 2012 Fundamentals of Pharmacognosy and Phytotherapy 2nd Edition
[16] Mubarak M, Shaija A and Suchithra T V 2015 ALGAL Elsevier B.V. 7 117
[17] Balasubramanian S, Allen J D, Kanitkar A and Boldor D 2011 Bioresouce Technology **102** 3396

Partial Oxidation of Methane Over NiOx/Hierachichal ZSM-5 Catalyst

D A Nurani, Y K Krisnandi*andAkmal

Department of Chemistry, FMIPA Universitas Indonesia, Kampus UI Depok, Depok 16424, Indonesia

*Corresponding author's e-mail: yuni.krisnandi@sci.ui.ac.id

Abstract. Methane is a primary component of natural gas that is considered a greenhouse gas.In this study, methane will be catalytically oxidized to methanol and its derivatives using hierarchical ZSM-5. The products are potential to be used as fine chemicals and additives in gasoline. The catalyst ZSM-5 (Si/Al = 35) was successfullysynthesized by a double template method in which TPAOH is used as the MFI structure directing agent and PDDA-M is used as the secondary template. The as-synthesized hierarchical ZSM-5 was modified with metal oxide (M = Ni). Analysis with XRD indicates that ZSM-5 crystals have been formed with high crystallinity. SEM images showed hexagonal shape which is the characteristic of ZSM-5 crystals. Analysis with SAA indicated the presence of meso-sized pores by the presence of hysteresis loops. Surface area and pore volume have decreased after impregnation of Ni. Elemental analysis showed 2.73% nickel was loaded in the ZSM-5 framework. The catalytic reaction was carried out in batch reactor for 120 min. Two condition of gas feed was compared: CH_4: N_2(without Oxygen) and CH_4: N_2-O_2 (with oxygen)with identical pressure ratio 0.75: 2 bar. Analysis by GC-FID showed that the catalytic activity of catalyst ZSM-5with oxygen increase the methanol production.
Keywords : ***partial oxidation, ZSM-5 hierarchy, double template method, methane,methanol.***

1. Introduction

The greenhouse effect is a UV light radiation event that should be reflected back outside the earth in the form of IR radiation, but is reabsorbed by an existing greenhouse gas in the atmosphere and then re-emitted to the earth, so that the UV equivalent energy is trapped on the surface earth. Methane is one of the greenhouse gas contributors. When the methane content in the atmosphere increased, it becomes an insulator that holds more heat from the sun that is emitted to the earth and hard to escape from the earth's surface.

Methane is the main component of natural gas. One of the most widely conducted studies is to convert methane through partial oxidation reactions with catalysts to methanol and other derivative products, thus having economic value and reducing methane gas emissions released into the atmosphere. Although methane can be used as fuel, the constraints are in the packaging and distribution of methane. This is because at room pressure and temperature, methane is a gas, while methanol is a liquid. Therefore, many researchers prefer to do a partial oxidation reaction converting methane to methanol to make it easier on the packaging process and produce environmentally friendly energy [1].

Hierarchical zeolite possesses two levels of porosity i.e microporous and mesoporous structures. This modification is based on the dissatisfaction of the catalytic properties of ZSM-5 microporous because it is considered to have poor mass transport properties, making it difficult for the reactants to enter into the zeolite pore [2]. This modification can be done with two approaches; top-down approach and bottom-up approach. Top-down approach is a modification process in pore by adjusting Si / Al ratio. There are two types of treatments for this approach: dealumination and desilication. Dealumination can be done by heating, while desilication can be done with the addition of a base. The addition of a base can make the tetrahderal SiO_4 bind and loose from the zeolite frame leaving the gap, but this method is destructive for the zeolite framework. Bottom-up approach is a process of pore modification by using soft or hard templates and cationic surfactants [2].

2. Materials and Methods

2.1 Materials

Tetraethyl orthosilicate (TEOS, 98%, Sigma Aldrich), Poly(Diallyldimethylammonium Chloride) (PDDA, 35%, Aldrich),$NaAlO_2$(Sigma Aldrich), tetrapropylammonium hydroxide (TPAOH, 1 M, Sigma Aldrich),), nickel(II) nitrate ($Ni(NO_3)_2 \cdot 6H_2O$, 99.0% Merck), Ethanol (C_2H_5OH 99,9%), Methanol (CH_3OH 99,9%),formaldehyde (HCOH 37%), methane gas (99.99% UHP from BOC), N_2 gas (99.99 % UHP), O_2 gas (99.99% UHP) and deionized water.

2.2 NiOx/ZSM-5 Synthesis

Hierarchical ZSM-5 was synthesized by mixing0.2964 g $NaAlO_2$, 36.7159 g TPAOH, 85.991 g H_2O, and 24.6069 g TEOSat 373 K for 3 h. 9,003 g PDDA (10%wt) was added into the mixture and stirred for 15 h at room temperature. Then, the mixture was placed in an autoclave and hydrothermallyheated at 423 K for 144 h. The synthesized zeolite was dried at room temperature and followed by calcination at 823 K to remove the templates. Furthermore, Ni was impregnated into the zeolite by following method: 1 g of ZSM-5 was added to 1,7 mL larutan $Ni(NO_3)_2$ 0,2495 M and stirred for 24 h at room temperature followed by calcination at 823 K for 3 h. The synthesized zeolites were characterized by Brunauer-Emmett-Teller (BET), X-ray diffraction (XRD), Fourier Transform Infrared Spectroscopy (FTIR, Shimadzu Prestige 21), Scanning Electron Microscopy with Energy Dispersive Spectroscopy (SEM-EDX), and Atomic Absorption Spectroscopy (Shimadzu AA 13000).

2.3 Catalytic Test

The synthesized catalyst was activated by calcination at 823 K for 3 h prior to be used in catalytic test. 0.5 g catalyst was placed in batch reactor and flushed by N_2 gas for 5 min. Different composition of gas feed was compared i.e CH_4-N_2 and CH_4-O_2-N_2 with pressure ratio 0.75: 2 bar. The catalytic reaction was performed at 423 K for 120 min. The product was separated by extraction using ethanol as solvent and then analyzed by gas chromatography (GC Shimadzu 2010) with a Carbowax column and flame ionization detector (Shimadzu).

3. Results

3.1 Catalyst characterization

ZSM-5 was synthesized by double template method in which TPAOH as MFI structure directing agent (mesoporous zeolites) and PDDA as mesoporous template. The XRD pattern of as synthesized ZSM-5 can be seen in Figure 1. There are two peaks between $2\Theta = 7\text{-}10^o$ and three peaks between $2\Theta = 22\text{-}25^o$.This pattern shows typical peak of ZSM-5[3].

There are also peaks at 37.10 °, 44.40 °, 62.87 °, 76.50 °, and 79.22 ° which indicates the difractogram patterns of nickel-shaped cubic. Ni addition to the structure of ZSM-5 decreases the peak intensity.It means that the crystallinity in ZSM-5 zeolite decrease after NiO modification. The metal oxide NiO also appeared at 44° [4].

IOP Conf. Series: Journal of Physics: Conf. Series **1095** (2018) 012005 doi:10.1088/1742-6596/1095/1/012005

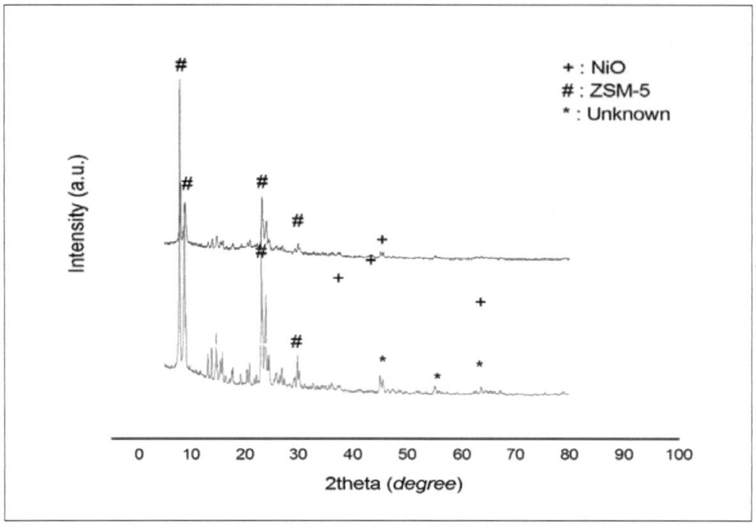

Figure 1. XRD patterns of ZSM-5 (red), NiOx/ZSM-5 (blue), and nickel oxide (yellow)

Figure 2 shows the FTIR spectra of ZSM-5 before calcination (red) and after calcination (blue). The peaks at 1375-1450 cm^{-1} (C-H stretching) and 2850-3000cm^{-1} (C-H bending) appeared in ZSM-5 before calcination but no peak observed at those wavenumber ranges in ZSM-5 after calcination. This indicates that both organic templates TPAOH and PDDA were removed from catalyst structure. The pores and channels of ZSM-5 were unoccupied after calcination.

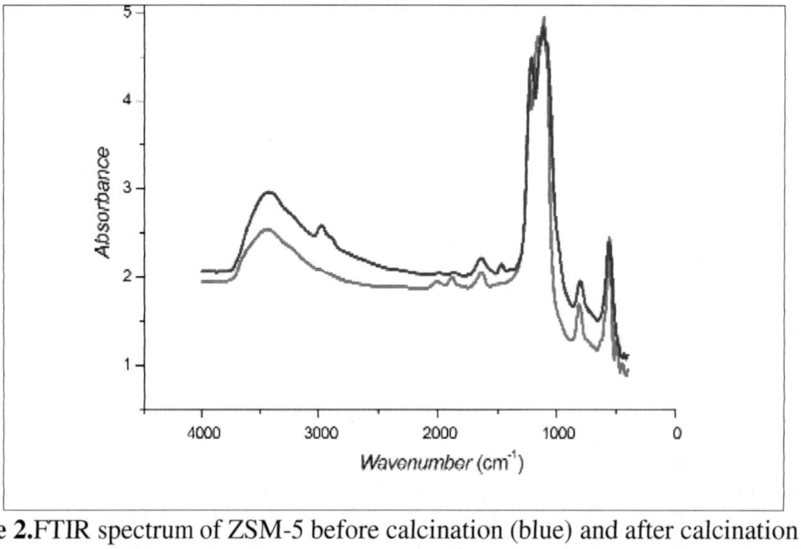

Figure 2.FTIR spectrum of ZSM-5 before calcination (blue) and after calcination (red)

The BET analysis shows that the average surface area of hierarchical ZSM-5 was 348.6 m^2/gr and after impregnated with NiOx was 330.3 m^2/gr. It can be seen in Figure 3a it can be seen that the catalyst shows H4 type hysteresis loop which indicates of mesoporous structure. In addition, the average pore diameter of catalyst was 1.8842 nm (micropore) and 2.741 nm which approximates the indication of pore formation with meso size (2 nm <d <50 nm).

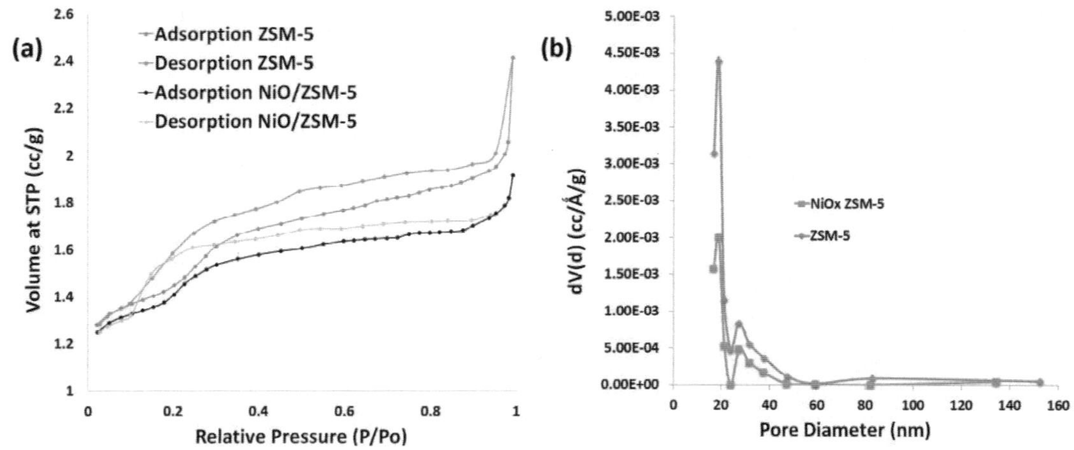

Figure 3. (a) Isotherm adsorption-desorption, and (b) pore size distribution of as-sythesized hierarchical ZSM-5 and NiOx/ZSM-5

Figure 4. SEM images of ZSM-5. Left: 20.000x; Right: 50.000x

In Figure 4, it is clear that the ZSM-5 zeolite synthesized in this experiment has hexagonal geometry and appears to form coffin-like. However, the characteristic of mesoporous ZSM-5 which should look rough surface is unclear. EDX analysis showed the mole ratio of Si / Al was 34 in the first analysis. It is still within range of Si/Al ratio of initial target (32).

3.2 Catalytic test: partial oxidation of methane

The products were analyzed with Gas Chromatography (GC) using carbowax column. The highest yield of methanol 22.9 % was formed in the reaction using ZSM-5 in which O_2 was added in gas feed whereas NiOx/ZSM-5 tends to form formic acid. This is a consequence of low crystallinity of zeolites which caused the dispersion of methane gas at the center of the activity is relatively unfavorable, this is crucial in the hydroxy formation stage. The low yield obtained in NiOx / ZSM-5 was influenced by oxygen-poor crystals (NiOx) in compared to metal with oxygen-rich crystals (CoOx) [5]. Furthermore, the presence of oxygen also affects the products yield.

IOP Conf. Series: Journal of Physics: Conf. Series **1095** (2018) 012005 doi:10.1088/1742-6596/1095/1/012005

4. Conclusions

The typical pores of hierarchical NiOx/ZrSM-5 (having both microporous and mesoporous) accelerate reactant diffusion into the catalyst pores. The low yield percentage with NiOx/ZSM-5 catalysts were assume because the lack of oxygen in nickel oxide crystal lattice, the molecular oxygen favour to undergo formic acid reaction mechanism.

Acknowledgement

This research was funded by Ministry of research technology and higher educationPUPT BOPTN Grant No. 2719/UN2.R3.1/HKP05.00/2017 and Osaka Gas Foundation 2016.

References

[1] Zhang, Qijian et al 2003 Recent Progress in Direct Partial Oxidation of Methane to Methanol (Journal of Natural Gas Chemistry) **12** pp 81–89
[2] Wang L, Zhang Z, Yin C, Shan Z, Xiao FS 2010 Hierarchical Mesoporous zeolites with controllable mesoporosity templeted from cationic polimers. (Microporous Mesoporous Mater) **131,** pp 58
[3] Treacy M M J 1986 Detection and Imaging of Supported Catalyst (USA: Material Research Soc) Vol **62** pp 367-378
[4] Mosayebi A 2014 Partial oxidation of butane to syngas using nano-structure Ni/zeolite catalysts Journal of Industrial and Engineering Chemistry **20** pp 1542–1548
[5] Krisnandi Y K Putra B A Bahtiar M Zahara Abdullah I Howe R F Partial Oxidation of Methane to Methanol over Heterogeneous Catalyst Co/ZSM-5. 2015, Procedia Chem. **14** 508

Free Solvent Amidation of Ursolic and Oleanolic Acids of *Fagraea Fragrans* Fruits: Their P-388 Antitumor Activity

D Basir[1], M. Hanafi[2], Julinar[1], A Saputra[1], and T Wati[1]

[1]Department of Chemistry, Faculty of Sciences, The University of Sriwijaya, Palembang, Indonesia, email: debasril_chem@yahoo.com, *Phone Number 0711-580927*
[2]Research Center for Chemistry, Indonesian Institute of Science, Kawasan PUSPIPTEK Serpong, Tangerang, Indonesia

E-mail email: debasril_chem@yahoo.com.

Abstract. This is an our laboratory effort in developing major compounds, ursolic acid (UA) and its isomer oleanolic acid (OA) of *Fagraea Fragrans* Roxb fruits, namely *buah tembesu*, *Loganiaceae* by free solvent amidation of the acids with *n*-buthylamine and phenylamine. The reactions have been conducted by using thionyl chloride as a chlorination reagent to make their acid chlorides respectively. In this reaction, the secondary hydroxyl group at C-3 position was preferably converted to be olefinic while the carboxyl group at C-28 position was normally amidated to give *N*-buthyl-urs-2,12-dien-28-amide [*N*-buthyl-olean-2,12-dien-28-amide] with yield 45.5% % and *N*-phenyl-urs-2,12-dien-28-amide [*N*-phenyl-olean-2,12-dien-28-amide] with yield 39%. Cytotoxic activity of amidation products toward P-388 murine leukemia cells was decreasing about half fold compared to starting material ursolic acid [oleanolic acid]. The IC_{50} values of N-butyl-urs-2,12-dien-28-amide [*N*-butyl-olean-2,12-dien-28-amide] and *N*-phenyl-urs-2,12-dien-28-amide [*N*-phenyl-olean-2,12-dien-28-amide] were 81, 4 μg/mL and 83.6 μg/mL respectively, whereas the IC_{50} value of ursolic acid [oleanolic acid] was 53.4 μg/mL.

Keywords ; Fagraea fragrans, fruits, oleanamide, ursolamide, P-388 marine leukemia cells

1. Introduction

This paper describes the thionyl chloride ractions on secondary alcohol and carboxyl groups in 3-hydroxyl-urs-12-en-28-oic acid and its isomer 3-hydroxyl-olean-12-en-28-oic acid isolating from the fruits of *Fagraea fragrans*, Loganiaceae [3,8]. In addition to develope those natural triterpenes to be their derivates of *N*-buthyl-urs-2,12-dien-28-amide [*N*-buthyl-olean-2,12-dien-28-amide] and *N*-phenyl-urs-2,12-dien-28-amide [*N*-phenyl-olean-2,12-dien-28-amide] and their cytotoxic activity toward P-388 murine leukemia cells. The amidation reactions were conducted between ursolic acid [oleanolic acid], thionyl chloride, aliphatic amine, and aromatic amine separately without using organic solvent.

It was realizing that in nature the triterpenes such as ursolic acid and oleanolic acid were found in many plant species but they were rarely found in form of amides. Therefore these works were capable to develope the natural acid triterpenes to be their amide derivatives, see structure 3[4] and 5[6] in Figure 1. These derivatives still need other biological activity test because in this experimental they have lower cytotoxic activity toward P-388 murine leukemia cells compared to their mother triterpenes or the synthetic design must be changed. Some of ursolamide derivatives were reported to have antimalarial agents [1].

As it has been shown in Figure 1 and referred to references of *Gnoatto et al* (2008) [1] and *Basir, D. et al* (2014) [2] dealing with biological activities, especially cytotoxic activity toward P-388 murine leukemia cells. These amidation reactions have involved eliminative reactions of hydroxyl group at C-3 position in both conditions of *n*-buthylamine and phenylamine eventhough those amines were different in base strength. As a conclusion, the hydroxyl group of carboxyl in C-28 was seem to have similar reactivty with secondary hydroxyl group at C-3 position toward thionyl chloride reagent and the heat of reaction (60 ^{o}C) might have been in powering the elimination reaction to give a double bond in C-2 and C-3 positions. In addition to those, *N*-buthyl-urs-2,12-dien-28-amide [*N*-buthyl-olean-2,12-dien-28-amide] and *N*-phenyl-urs-2,12-dien-28-amide [*N*-phenyl-olean-2,12-dien-28-amide] as amide derivatives of ursolic acid [its isomer oleanolic acid] were the first report in this paper [4,5,6,15], and this works could also give additional data to improve that there is no doubt; *Fagraea. Fragrans* fruits contain ursolic acid and its isomer oleanolic acid as major metabolites.

To maintain the secondary hydroxyl group in the starting materials and to amidate the carboxyl group selectively, the protecting group such as acetyl or oxo has to be inserted at C-3 position before the free solvent amidations[1,2]. Therefore, some of *N*-buthyl-3-acetyl-urs-12-en-28-amide [*N*-buthyl-3-acetyl-olean-12-en-28-amide] and *N*-phenyl-3-acetyl-urs-12-en-28-amide [*N*-phenyl-3-acetyl-olean-12-en-28-amide]; and *N*-buthyl-3-oxo-urs-12-en-28-amide [*N*-buthyl-3-oxo-olean-12-en-28-amide] and *N*-phenyl-3-oxo-urs-12-en-28-amide [*N*-phenyl-3-oxo-olean-12-en-28-amide] can be produced from the triterpenes of *Fagraea fragrans* fruits [2,9]. As a result, they are biological activities, especially toward P-388 murine leukemia cells, will be more potent then their mother triterpenes with IC_{50}= 53.4 μg/mL and 3-oxo-ursolic acid [its isomer 3-oxo-oleanolic acid] with IC_{50}= 18.3 μg/mL[2].

2. Materials and Methods

2.1 Materials

Fragraea fragrans fruits, technical methanol, activated carbon, *d*-pyridine, *d*-chloroform, *n*-buthylamin, anilin, thionyl chloride, silica gel G60 (70-230 mesh), Na_2SO_4 anhydrous, aquadest, 3-(4,5-dimethylthiazol-2yl)-2,5-diphenyl-tetrazolium bromide (MTT), fetal bovine serum (FBS), penicillin, streptomycin, P388 murine leukemia cell culture supplied by Department of Chemistry (ITB), dimethyl sulfoxide (DMSO), phosphor buffer solution (PBS), artonin E, sodium dodecyl sulphate, whatmann paper, ethyl acetate, n-hexane, acetic acid, formic acid, and methanol p.a. for LC MS-ESI positive ion.

2.2 Instrumentation

Melting point was determined using Fisher Johns apparatus. ^{13}C-NMR (*d*-$CHCl_3$) were determined on JEOL ECA500 spectrometer operating at 500 MHz (^{1}H), and 125 MHz (^{13}C), respectively. Mass spectra were obtained with a HITACHI L6200 LC-MS ESI positive ion spectrometer which was running on collumn C-18 (15mm x 1 mm), flow 0.1 mL/min, injection volume 2 μl, and 0.3% formic acid in MeOH as solvents.

2.3 LC-MS Analysis

LC-MS analysis was performed using an Mariner Biospectrometry equipped with a binary pump. The HPLC was interfaced with a Q-tof mass spectrometer fitted with an ESI source. Full-scan mode from *m/z* 100 to 1200 was performed with a source temperature of 140 °C. HPLC column (Phenomenex 5*μ* C18, 150 × 1 mm i.d.,) was used for the analysis. Solvent was methanol with 0.3% formic acid. Solvents were delivered at a total flow rate of 0.1 mL/min. The solvent running by isocratic elution.

2.4 Compouds 1[2]

The fresh *Fagraea fragrans* fruits were collected in Inderalaya swamp forest, Ogan Komring, South Sumatra on October 2014, see Figure 2b. The plant speciment was identified and deposited in the plant taxonomy laboratory of Deparment of Biology, Faculty of Science, Sriwijaya University. The fruits were then dried over two months at room temperature, and then milled to be powder. The powder were extracted with methanol according to the procedures of the isolation of the ursolic acid and its isomeric oleanolic acid from these fruits in our previous work, *Basir, D. et al.*, (2012 and 2014) [2, 3], and then they (m.p.=284-286 °C) were used as starting materials to make the compounds 3[4]. The spectroscopic elucidation data for these triterpenic acids as starting materials in this work has been given in those reference 2 as well.

2.5 Compounds 3 [4]

Amidation was conducted without using organic solvent according to the precedure of Kumagai *et al.*, (2010)[7]. Thionyl chloride 0.5 mL (6.9 mmol) was added to white crystals of ursolic [its isomer oleanolic] acids (154 mg, 0.34 mmol) and then the mixture was stirred for three hours in round flask at 60 ^{0}C. This later solution was allowed to be cooled at room temperature, thus the *n*-buthylamine 0.2 mL (2 mmol) was gradually dropped to the solution while it was stirred again for 2 hours. The reaction was stopped whenever it gave different retention time (r.f. = 0.8) from the retention time of starting material (r.f. = 0.4) respectively with 20% ethyl acetate in *n*-hexane. The amidation product was then dissolved in chloroform (30 mL), washed with water (30 mL). The chloroform phase was dried with anhydrous Na_2SO_4 and filtered with whatmann paper, and later on the chloroform was evaporated under reduced pressure to give dry residue (130mg). This residue was then subjected to silica gel G60 column with increasing solvent polarity from 1 to 5 % ethyl acetate in n-hexane to give 70 mg (45.5%) of compound 3 [4], and prepared to spectroscopy analysis including MS, ^{1}H NMR, ^{13}C NMR (DEPT 90 and 135), and tested for P-388 murine leukemia cells.

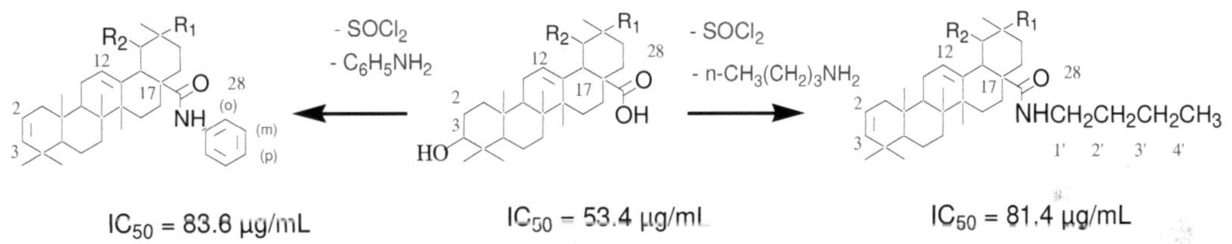

5 [6] 1 [2] 3 [4]

Figure 1. Free solvent amidation of ursolic acid (UA) and its isomer oleanolic acid (OA) isolated from *Fagraea fragrans* fruits.

2.6 Compounds 5 [6]

Amidation was also conducted without using organic solvent according to the precedure of Kumagai *et al.*, (2010)[7]. Thionyl chloride (0.3 mL, 3.5 mmol) was added to white crystals of ursolic [its isomer oleanolic] acids (100 mg, 2.2 mmol) and then the mixture was stirred for three hours in round flask at 60 ^{0}C. This later solution was allowed to be cooled at room temperature, thus the *N*-phenylamine 0.2 mL (3,5 mmol) was gradually dropped to the solution while it was stirred again for 2 hours. The reaction was stopped whenever it gave different retention time (r.f. = 0.7) from the retention time of starting material (r.f. = 0.4) respectively with 20% ethyl acetate in *n*-hexane. The amidation product was then dissolved in chloroform (50 mL), washed with water (50 mL). The chloroform phase was dried with anhydrous Na_2SO_4 and filtered with whatmann paper, and later on the chloroform was evaporated under reduced pressure to give dry residue (81.2). This residue was then subjected to silica gel G60 column with increasing solvent polarity from 1 to 5 % ethyl acetate in n-hexane to give 30.8 mg (39%) of compound 5 [6], and prepared to spectroscopy analysis including MS, ^{1}H NMR, ^{13}C NMR (DEPT 90 and 135), and tested for cytotoxic activity against P-388 murine leukemia cells.

Compound 5[6] were also made in solvent (benzene) condition. The result was still similar to free solvent amidation condition that took place via C-2 and C-3 elimination on ring-A and C-28 amidation.

2.7 Antiproliferation evaluation

P388 murine leukemia cancer cell cultures (3 x 10^3 cell/mL) were suspended into RPMI 1640 media having contained FBS (Fetal Bovine Serum), penicillin, and streptomycin. Cells were inoculated in microplate 96 well plate and incubated in CO_2 incubator for one day. At the second day, the DMSO (dimethylsulfoxide) solution of the crystals of 3[4] and 5[6] was diluted with PBS (phosphoric buffer solution, pH = 7.30-7.65) for variation concentrations of 100, 30, 10, 3, 1, 0.3, and 0.1 μg/mL media and then was dropped into those cells respectively. These last cells in microplate were incubated again in CO_2 incubator. The DMSO was used as negative control and artonin E (IC_{50} = 0.7 μg/mL) as positive control. After 48 h the incubation process, MTT reagent was added into the cells, incubated during 4 h, and SDS (Sodium Dodecyl Sulphate) was then added, shacked as well, and continuously incubated for other 24 h. The color change of MTT in viable mitochondria cells from yellow to purple could be quantified with spectrophotometer at λ = 550 nm. The values of OD and concentration (μg/mL) of tested compounds were reported as the mean three of replicates. The IC_{50} value was noted from antilog graphics based on the correlation of tested compound concentrations (μg/mL) and color intensity of cell viable solution, respectively [11-13].

3. Results and Discussion

Compounds 3[4] consisted of of *N*-buthyl-urs-2,12-dien-28-amide [*N*-buthyl-olean-2,12-dien-28-amide, They had LC chromatogram of 3[4] with r.t. = 13.6 minutes by LC column with 10% DMSO in methanol, while the compounds 5[6] consisted of *N*-phenyl-urs-2,12-dien-28-amide [*N*-phenyl-olean-2,12-dien-28-amide with LC chromatogram of 5[6] with r.t. = 34.6 minutes by LC column with 0.5% acetic acid in methanol.

Compounds 3[4] had some peaks of LCMS-ESI positive ion and methanol as eluent at m/z = 998.68 or 990.67 as protonated dimeric ions [2M + H^+], 516.84 as pseudomolecular ion [M + Na – H], and 494.85 as molecular ion peak $[M+H]^+$. while the compounds 5[6] gave some peaks at m/z = 1028.64 or 1050.66 as protonated dimeric ions [2M + H^+], 536.78 as pseudomolecular ion [M + Na – H], and 514.80 as molecular ion peak $[M+H]^+$.

Compounds 3[4] had some peaks of ^{1}H-NMR at δ (ppm) =5.99, 5.95, 5.93, 5.41, 5.40, 5.39, 5.38, 4.74, 4.70, 4.68, 4.44, 3.90, 3.88, 3.37, 3.35, 3.34, 3.04, 3.03, 3.02, 3.01, 2.51, 2.49, 2.27, 2.22, 1.66, 1.62, 1.45, 1.44, 1.43, 1.42, 1.34, 1.33, 1.32, 1.31, 1.25, 1.17, 0.98, 0.95, 0.94, 0.92, 0.90, 0.89, and 0.85. while the compounds 5[6] gave some main peaks at δ (ppm) = 7.75, 7.73, 7.47, 7.46, 7.30, 7.29, 7.26, 7.22, 6.74, 6.72, 6.58, 6.53, 5.62, 5.61, 5.60, 4.25, 4.23, 2.88, 2.68, 2.66, 2.37, 2.35, 2.07, 2.05, 1.71, 1.70, 1.41, 1.40, 1.37, 1.25, 1.22, 1.03, 0.95, 0.94, 0.93, and 0.74. The ^{13}C NMR of the compounds 3[4] was at δ (ppm) = 178.5 (C-28 of CONH for ursolamide, 178.3 [C-28 of CONH for its isomeric oleanamide]), 145.8 (C-13), 138.1 (C-2), 122.9 (C-12), 121.4 (C-3), 56.0, 52.1, 47.0, 46.5, 46.2, 42.5, 39.3, 34.4, 33.2, 32.6, 31.6, 31.0, 26.0, 24.1, 23.7, 20.4, 19.7, 14.0, while the compounds 5[6] was at δ (ppm) = 176.8 (C-28 of CONH- for ursolamide), 176.7 [C-28 of CONH- for its isomeric oleanamide]), 145.7 (C-13 for for ursolamide, 145.3 [C-28 for its isomeric oleanamide]), 138.3 (C-2 for for ursolamide, 138.1 [C-2 for for ursolamide]), 136.1 (C-3), 124.1 (C-12), 140.1, 129.1, 124.2, and 119.9 (carbons of phenylic), 46.2, 34.5, 34.1, 33.1, 32.6, 32.5, 30.9, 27.8, 25.6, 24.4, 23.7, 23.6, 19.7. Some of the chemical shifts of carbon peaks of compounds 5[6] could be clearly differentiated ursolamide peaks from its isomeric oleanamide peaks but they were not detected fully in the compounds 3[4], exception the carbonyl peaks of those amides at 178.5 (C-28 of CONH for ursolamide, 178.3 [C-28 of CONH for its isomeric oleanamide]) respectively.

3.1 Ursolic [Its Isomer Oleanolic] Acids, 1 ***[2]***

Isolation procedures, physical properties, and structure elucidation included δ_C, and δ_H values of triterpenes 1 [2] with m.p. = 284-286 °C,and molecular ion peak (FAB-MS) = 457 $(M + H)^+$ have already been reported in reference 3, Basir D. *et al.* (2012). In this paper, ^{13}C NMR data of 1 [2] in *d*-pyridine was given in Table 1.

3.2 N-buthyl-urs-2,12-dien-28-amide [N-buthyl-olean-2,12-dien-28-amide], 3 [4]

Compound 3 [4], 70 mg, 42 % counted from compound 1 [2], gave m.p. 141-143 ^{0}C . TLC (r.f.) was 0.37 in 20% ethyl acetate in *n*-hexane and LC chromatogram was 13.6 minutes with 0.5 % acetic acid in methanol. The LC-MS ESI gave molecular ion peak $(M + H)^+$ at *m/z* = 494.8, (M + Na) at *m/z* = 516.8, $(2M + H)^+$ at *m/z* = 988.7, and (2M + Na) at *m/z* = 1010.7. Molecular weight (MW) calculated was 493 which relatively correspond to the experimental (494.8 – H = 493.8) data with empirical formula, $C_{34}H_{55}O_1N_1$, double bond equivalent (DBE) = eight (8), concisted of five rings and three double bonds i.e. two of C=C and one of C=O. The chemical shifts of 3 [4] in ppm were given in Table 1. Table 1 indicated that ^{1}H NMR of 3 [4] at δH = 5.9 (br, 1H, *J* = 8.4 Hz) for its amide functional as reported in *N*-alkyl/aryl acetamide in reference 4 was coming from the loss of hydroxyl group of carboxyl at C-28 with δH =11.98 (s, 1H) in compound 1 [2]. Those were also supported by changing δC value of –COOH at 179.6 [180.1] to δC value of –CONH- at 178.3 [178.5]. The 3 [4] also has δH = 5.1 (1H at C-3) and 5.1 (1H at C-2] resulted in the elimination of H and OH coming from hydroxyl at C-2 and hydrogen at C-3 of 1 [2] with δH = 4.27 (t, 1H at C-3) and 2.98 (m, 1H at C-2) as well as δC of 3 [4] at 121.4 (C-3) and 138.1 (C-2) coming from δC of 1 [2] at 79.6 (C-3) and 23.6 (C-2).

3.3 N-phenyl-urs-2,12-dien-28-amide [N-phenyl-olean-2,12-dien-28-amide], 5 [6]

Compound 5 [6], 30.8 mg, 38 % counted from compound 1 [2], gave m.p. 122-123 ^{0}C . TLC (r.f.) was 0.4 in 20% ethyl acetate in *n*-hexane and LC chromatogram was 34.6 minutes with 10% DMSO in methanol[12], Figure 2b. The LC-MS ESI gave molecular ion peak $(M + H)^+$ at *m/z* = 514.8, (M + Na) at m/z = 536.8, $(2M + H)^+$ at *m/z* = 1028.6, and (2M + Na) at *m/z* = 1050.6. Molecular weight (MW) calculated was 513 which relatively correspond to the experimental (514.8 – H = 513.8) data with empirical formula, $C_{36}H_{51}O_1N_1$, double bond equivalent (DBE) = twelefth (12), concisted of six rings and three double bonds i.e. five rings in triterpenenoid skeleton, one ring of aromatic, two of C=C and one of C=O. In addition to the aboved free solvent condition, the LC-MS ESI of solvent (benzene) amidation condition gave protonated molecular ion peak $(M + H)^+$ at *m/z* = 514.83, and protonated dimeric ions $[2M + H+]^+$ = 1029.07, see Figure 2, with a single peak on retention time = 0.6 minutes (injection volume = 5 μL , flow 0.1 mL/min, and MeOH eluent). The chemical shifts of 5 [6] in ppm were given in Table 1. Table 1 indicated that ^{1}H NMR of 5 [6] at δH = 7.8 (s, br, 1H) for its amide functional was coming from the loss of hydroxyl group of carboxyl at C-28 with δH =11.98 (s, 1H) in compound 1 [2]. Those were also supported by changing δC value of –COOH at 179.6 [180.1] to δC value of –CONH- at 176.8 [176.7]. The 5 [6] also has δH = 5.6 (1H at C-3) and 5.6 (1H at C-2] resulted in the elimination of H and OH coming from hydroxyl at C-2 and hydrogen at C-3 of 1 [2] with δH = 4.27 (t, 1H at C-3) and 2.98 (m, 1H at C 2) as well as δC of 5 [6] at 136.1 (C-3) and 138.3 (C-2) coming from δC of 1 [2] at 79.6 (C-3) and 23.6 (C-2).

As it has been described in the introduction that the action of amine base in rate of elimition of H and OH from C-2 and C-3 positions of the starting mayerials 1[2] to be a double bond via the chlorinated intermadiate, see Figure 3a, might be affected significantly. Moreover aliphatic amine forced forming of the double bond faster then aromatic amines, thus they gave 3[4], 45.5% then 5[6], 39%. These are the *in situ* reaction for both of elimination and amidation reactions so that the presumable reaction mechanism of this works are written as below, see Figure 3a. As shown in Figure

3a, both of hydroxyl at C-3 and carbonyl at C-28 of 1[2] had been converted to be their chloronated compounds as well as those chlorinated intermadiates were converted to be 3[4] and 5[6] as elimination and amidation products respectively [10]. Amidation of 1[2] in both solvent (benzene) and free solvent conditions has given the same target compounds, 5[6]. This is indicated that secondary amine can not be made by means of solvent and free solvent conditions from secondary alkyl halide of ring-A. As a result, this elimination and amination products belong to new derivatives of ursolamide and oleanamide.

Table 1. ^{13}C-NMR and ^{1}H-NMR data of ursolic acid [its isomer oleanolic acid], 1 [2] in pyridine, *N*-buthyl-urs-2,12-dien-28-amide [its isomer *N*-buthyl-olean-2,12-dien-28-amide], 3 [4], and *N*-phenyl-urs-2,12-dien-28-amide [its isomer *N*-phenyl-olean-2,12-dien-28-amide], 5 [6] in $CDCl_3$

C. Number	1 [2], *δC*	1[2] , *δH*	3 [4], *δC*	3[4] , *δH*	5 [6], *δC*	5 [6], *δH*
1	38.8		39.2		40.9	
2	23.6	2.98 (m)	138.1	5.38 (m)	138.3	5.64 (d, 1H)
3	79.6	4.27 (t, 1H)	121.4	5.39 (m)	136.1	5.68 (d, 1H)
4	37.3		36.2		38.3	
5	55.7		55.8		52.1	
6	18.7		16.6		19.72	
7	33.1		33.1		32.6	
8	39.7		39.2		39.7	
9	46.6		46.4		46.2	
10	37.3		39.2		36.3	
11	23.7		23.7		24.4	
12	124.2	5.15 (t, 1H)	122.9	5.4 1(t)	124.1	5.62 (t, 1H)
13	144.8		145.8		145.3	
14	41.9		42.3		40.3	
15	28.3		30.8		28.1	
16	23.7		24.0		24.4	
17	48.0		46.4		49.5	
18	53.5		52.1		47.6	
19	39.3		39.2		42.9	
20	38.9		34.3		40.2	
21	30.4	1.27 (m)	31.5	1.3 (d)	31.1	1.25 (m)
22	34.2		34.3		34.4	
23	28.0		25.9		33.1	
24	16.5		25.8		16.8	
25	15.5		20.6		15.8	
26	15.6		16.7		15.9	
27	26.1		19.5		26.5	
28	179.6 (UA), 180.1 (OA)	11.98 (s, 1H)	178.3 (UA), 178.5 (OA)		176.8 (UA), 176.7 (OA)	
29	17.4		15.6		17.1	
30	21.4		16.6		21.5	
1'			40.1	3.2 (d,d *J* =6.5 Hz)	140.1	
2'			33.1	1.6 (m)	119.9, ortho (2' and 6')	7.1 (2H, dd, *J* =7.1Hz)
3'			20.4	1.3 (m)	129.1, meta (3' and 5')	7.5 (2H, m)
4'			13.8	1.0 (t)	124.2 (para)	7.3 (1H, d)
NH-				5.9 (br, *J* =8.4 Hz)		7.8 (s, br, 1H)

Not: UA = ursolic acid and OA = oleanolic acid

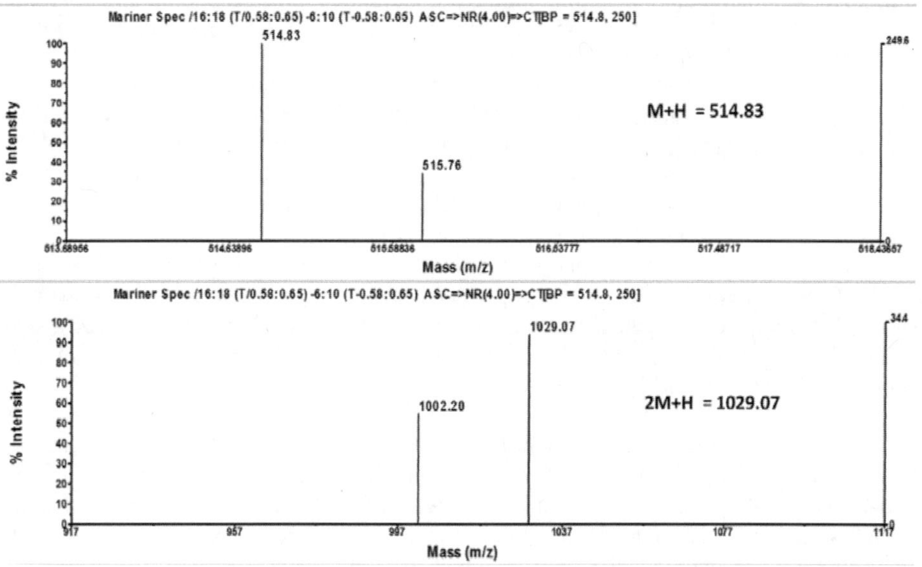

Figure 2. LC-MS ESI mass spectra of compound 5[6] made in the solvent (benzene) amidation condition of ursolic acid (UA) and its isomer oleanolic acid (OA) isolated from *Fagraea fragrans* fruits

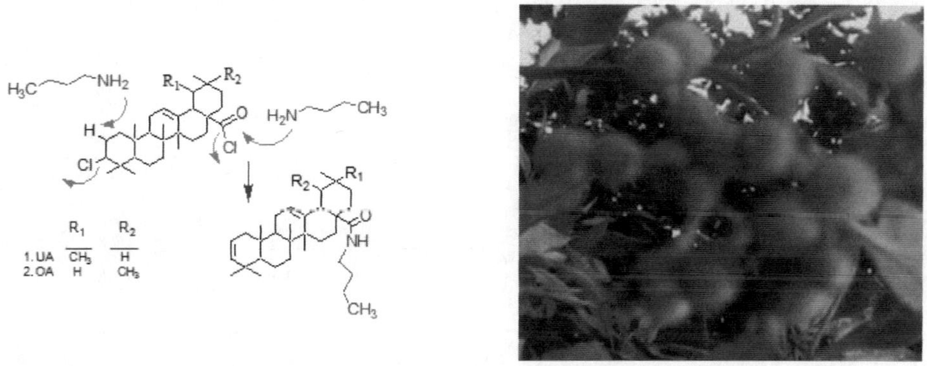

(a) (b)

Figure 3. **(a).** Free solvent elimination and amidation reaction mechanisms of the H and OH in ring-A and COOH at C-28 of 1[2] to give 3 (4) or 5 [6] via chlorinated intermadiates. **(b).** The *Fagraea fragrans* Roxb fruits as a potential resource of 3-hydroxyurs-12-en-28-oic acid [1] and its structural isomer 3-hydroxyolean-12-en-28-oic acid [2]

3.4 Antiproliferation Activities of 3[4] and 5[6]

Compound 1[2] as starting materials have IC_{50} values = 53.5 µg/mL, see reference 2, Basir D.et al. (2014) while the compound 3[4], and 5[6] as the products of C-28 amidation including C-2 and C-3 elimination respectively gave IC_{50} values of the cytotoxic againts P-388 murine leukemia cells: 81.4, and 83.6 µg/mL. The results of this *in vitro* test were given in Table 2. Artonin E and DMSO were used as positive and negative controls, while the absorbance (OD) was read at a wavelength of 550 nm. The IC_{50} values of compound 3[4] and 5[6] were respectively determined and the results were given in Table 2. Compound 3[4], *N*-buthyl-urs-2,12-dien-28-amide [*N*-buthyl-olean-2,12-dien-28-amide], and compound 5[6], *N*-phenyl-urs-2,12-dien-28-amide [*N*-phenyl-olean-2,12-dien-28-amide], gave IC_{50} = 81.4 µg/mL and IC_{50} = 83.6 µg/mL respectively as well as their semilog graphics of the tested compound concentration versus the P-388 murine leukemia cells absorption; 81.4335 (*x-axis*) as concentration (µg/mL) and 0.1735 (y-axis) as absorbance or optical density (OD) for 3[4] and and 83.6331 (x-axis) and 0.1750 (y-axis) as absorbance or optical density (OD) for 5[6], see Table 2. As a result, the effect of elimination of H and OH in ring-A of these amide products, 3[4] and 5[6] decreased half fold of the activities of P-388 murine leukemia cell antiproliferation compared to the their mother natural triterpenes 1[2] with IC_{50} =53,4 µg/mL [2,11,13]. Eventhough the *N*-buthyl-urs-2,12-dien-28-amide (3), *N*-buthyl-olean-2,12-dien-28-amide (4) as isomeric molecule of 3; and *N*-phenyl-urs-2,12-dien-28-amide (5), *N*-phenyl-olean-2,12-dien-28-amide (6) as isomeric of 5. They have similar P-388 murine leukemia cell antiproliferation activity.

As a conclusion, ursolic [its isomer oleanolic] acids (IC_{50} = 53.5 µg/mL) as inseparable white solid crystals coming from *Fragraea fragrans* fruits has been succesfully amidation to be inseparable *N*-buthyl-urs-2,12-dien-28-amide [*N*-buthyl-olean-2,12-dien-28-amide] (IC_{50} = 81.4 µg/mL), and *N*-phenyl-urs-2,12-dien-28-amide [*N*-phenyl-olean-2,12-dien-28-amide] (IC_{50} = 83.6 µg/mL). In addition to our previous works on *Fragraea fragrans* (called tembesu) fruits that are richly containing acid pentacyclic triterpenes [2,3], There is no doubt that those fruits are still the potential as resources of ursolic acid and its isomeric oleanolic acid to be developed to be other amide derivatives with more active anticancer compound by means of other amidation reaction routes eventhough in this works they gave lower antiproliferation activity against P388 leukemic cells. In short, the amide derivatives containing hydroxyl or amine group at C-3 are more prefered rather then alkene groups at C-2 and C-3 positions to be strong P-388 murine leukemia cell antiproliferation candidates.

Table 2. IC_{50} and absorbance (optical density) of compound 3[4], and 5[6] at a wave length of 550 nm.

	Concentration	Compound 3[4]		Compound 5[6]	
NO	(µg/ml)	Optical Density [OD]	IC_{50} (µg/ml)	Optical Density [OD]	IC_{50} (µg/ml)
1	100	-0.0163		-0.0106	
2	30	1.1822		1.2920	
3	10	1.9617	81.43	1.5203	83.63
4	3	1.3972		1.6270	
5	1	1.7417		1.6926	
6	0.3	1.7202		1.5060	
7	0.1	2.0812		1.2970	

OD of positive blank for 1[2] = 0.2943 and for 3[4] and 5[6] = 0,1778

Acknowledgment
This work was fully supported by funds from Indonesian Higher Education Department, and my Institution, the Sriwijaya University Research Center; *rancak* .

References

[1] Gnoatto SCB, Susplugas S, Vechia L D, Ferreira TB, Klimpt A D, Zimmer K R, Demailly C, Nascimento S D, Guillon J, Grellier P, Verli H, Gosmann G and Sonnet P 2008 *Bioorganic & Medicinal Chemistry,* **16** 771.
[2] Basir D, Julinar, Agustriana E and Untari B 2014 *Indonesian J. of Chemistry,* **14(3)** 269.
[3] Basir D and Julinar 2012 *Indonesian. J. Chem.,* **12(1)** 84
[4] Katke S A, Amrutkar S V, Bhor R J and Khainar M V 2011 *International J. of Pharma Sciences and Research (IJPSR),* **2(7)** 148.
[5] Chen H, Gao Y, Wang A, Zhou X, Zheng Y and Zhou J 2015 *European J. of Medicinal Chemistry,* **92**. 648.
[6] Meng Y, Song Y, Yan Z and Xia Y 2010 *Molecules,* **15** 4033.
[7] Kumagai, Takashi, Ebi T, Konishi A, MatsumotoK, Kurata H, Kubo T, Katsumoto K, Kitamura C and Kawase T, 2010 *Tetrahedron,* **66** 8968 .
[8] He X. and Liu R H 2007 *J. Agr.Food Chem.,* **55** 4366.
[9]. Mazumder K, Tanaka K and Fukase K 2013 *Molecules,* **18** 8929.
[10] Montalbetti A G, Christian N, and Falque V 2005 *Tetrahedon,* **61** 10827.
[11] Hanaf M, Salahuddin, and Haryanti 2013.*Indo. J. Chem,***13(2)** 166.
[12] Zou S., and Wu Chen 2008 *Journal Braz.Chem.Soc.,* **19(7)** 1429.
[13]. Batool A., Shahid K., and Muddasir M 2015 *Int. J. Pharm. Sci. Rev. Res.*, **30(2)** 25.
[14] Bi Y, Xu J X, Sun F, Wu X M, Ye W C, Sun Y J, Meng Q G, and Huang W W 2013 *Rec. Nat. Prod.* **7(4)** 254.
[15] Wang C, Lu Lu, Heya Na, Lifeng Cai, and Liu K 2014 *J.Med.Chem,* **57** 7342.
[16] Spom, M. B., Lybi, K. T., Yore, M. M., Fu, L., Lopchuk, J. M., and Gribble, G. W J. 2015 *Nat. Prod.* **74(3)** 537.

Fabrication and Characterization of Pt-Co/C Catalyst for Fuel Cell Electrode

Dedi Rohendi[1], Addy Rachmat[1], Nirwan Syarif[1]

[1]Department of Chemistry, Faculty of Mathematics and Natural Sciences, Universitas Sriwijaya, Jl. Palembang-Prabumulih Km. 32 Inderalaya Ogan Ilir, South Sumatera, Indonesia

Email: rohendi19@gmail.com / rohendi19@unsri.ac.id

Abstract. One of the most important components in fuel cell is electrode which hold as center for electrochemical reaction on converting fuel and oxygen into electricity. Special attention hence must be given to this component in order to achieve high performance especially on its catalytic activity and durability. To obtain such performance, some aspect must be take into consideration i.e. the content and type of catalyst as well as preparation method. Impregnation-reduction and electrodeposition method were evaluated on fabrication of fuel cell electrode from Pt-Co/C catalyst using $H_2PtCl_6.6H_2O$ and $CoCl_2.6H_2O$ in carbon Vulcan XC-72 substrate. Transition metal other than platinum as supporting catalyst was used not only to enhance catalytic activity but also to reduced cost and avoid poisoning which platinum suffers easily. Characterization of catalyst using Cyclic Voltammetry (CV) and Electrochemical Impedance Spectroscopy (EIS) confirmed that catalyst made from Pt-Co/C using ratio Pt : Co = 20:80 has good catalytic activity for fuel cell application. XRD diffractogram showed result of Pt reduction at 2-theta = 39.76° (111) and 46.23° (200) while the present of Co is not clear at 2-theta = 44.51° and 51.85°. Co was not detected due to it became an alloy with Pt and relatively small concentration (20%). Pt-Co/C catalyst prepared by impregnation-reduction and electrodeposition methods can be used as catalyst support in fuel cell electrode.

Keywords; Pt-Co/C; catalyst layer, fuel cell electrode.

1. Introduction

Fuel cell is an electrochemical equipment that can convert free energy available in fuel (chemical energy) directly into useful work in the form of electrical energy [1, 2]. Fuel cells with advantages on high conversion rates, low pollution levels, modular and diverse fuel sources are expected to replace existing conventional energy sources [3, 4]. Fuel Cells technology will surely be present in Indonesia due to the growing pressure to bring clean energy sources. Currently, the application of fuel cells in the world has entered to the commercialization stage for vehicles, while in Indonesia, fuel cell has started to be used as a source of backup power in Base Transceiver Station (BTS) without the support of independence in the mastery of technology.

Two type of fuel cell with the advantage of low operating temperatures, high conversion rates and diverse applications (portable equipment, vehicles and stationary energy sources) is the Proton Exchange Membrane Fuel Cell (PEMFC) and Direct Methanol Fuel Cell (DMFC)[1, 5].

One of the most important components in PEMFC/DMFC are the electrode contained in Membrane Electrode Assembly (MEA). Electrodes are the center of the electrochemical reaction of converting hydrogen and oxygen gas into electrical energy and water as effluents [6, 7]. Because the very important function, the PEMFC/DMFC electrode must get special attention in the effort of achieving high performance, especially in its catalytic activity and its durability [2]. MEA with high current density can significantly reduce PEMFC/DMFC stack size and decrease Pt catalyst content will decrease overall fuel cell price.

The catalysts used in fuel cell electrodes are generally platinum-based catalysts that are quite expensive and are easily poisoned [8-10]. In addition to a platinum single catalyst, using of other transition metal groups other than platinum is reportedly used as a platinum support catalyst [11-13]. It is intended to improve the catalytic function of the electrode and reduce the platinum content [10]. In this research, we used the alloys of Pt catalyst with Co in the carbon vulcan XC-72 as a substrate.

The aim of this research is to make and characterize Pt-Co / C catalyst for use in PEMFC electrode. The use of Pt catalyst alloys with Co is intended to reduce the use of Pt in the electrode and improve the catalytic function of the catalyst. The process of preparing of the catalyst is carried out by the reduction method based on a previous study [14].

2. Experimental Section

Material & Instrumentation

The material used in this study are $H_2PtCl_6.6H_2O$ (Sigma-Aldrich), $CoCl_2.6H_2O$ (Sigma-Aldrich). Electrodes are made from backing layer (Avcarb P75T), microporous layer and catalyst layer. The catalyst prepared in this study was a combination of platinum and cobalt catalysts in the XC-72 vulcanized carbon matrix in the form of Pt-Co / C by weight ratio of each composition 100:0, 80:20, 60:40, 40:60, 20:80 and 0:100 for comparison of Pt to Co. The CV and EIS analysis were performed using Autolab PGSTAT 302N and data processing was done with NOVA software version 1.8.14. The analysis was performed by using Pt electrode as counter electrode, Ag / AgCl / KCl electrode as reference electrode and electrode produced as working electrode with scan rate 50 mV / sec. Analysis of CV using electrolyte solution of H_2SO_4 0,3 M.

Procedure

Preparation of Pt-Co/C catalyst for spraying method was carried out by the method of impregnation and reduction of $H_2PtCl_6.6H_2O$ solution and $CoCl_2.6H_2O$ solution simultaneously on a vulcanized carbon powder matrix XC-72 with 1 M $NaBH_4$ solution as reducing agent. The resulting catalyst is then used to prepare the electrode with the spraying methods. The resulting catalyst was characterized using cyclic voltammetry (CV), Electrochemical Impedance Spectroscopy (EIS) and X-Ray diffraction (XRD) analysis. The catalytic activity of the catalyst is determined by calculating the electrochemical surface area (ECSA).

3. Results and Discussion

CV Analysis

CV analysis was performed to see the catalytic activity of the catalyst prepared. The CV analysis results are shown in Figure 1.

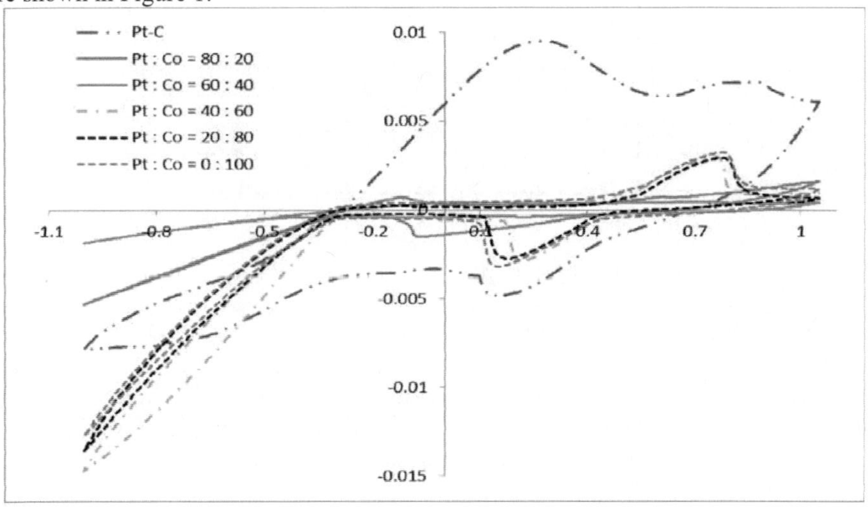

Figure 1. The CV curve for alloys of Pt / C and Pt-Co/C catalysts in various compositions

Based on Fig. 1 it can be seen that there are catalytic activity in almost all Pt-Co/C catalyst compositions, especially starting from Pt:Co = 40:60 composition showing reversible catalytic activity. In the catalyst composition Pt: Co = 0: 100, cathodic and anodic peaks are almost same like to the composition of 40:60 and 20:80. This indicates the expectation using Co catalyst as a support for Pt catalyst. Pt/C Catalyst have biggest area but not reversible. It can also be seen from ECSA calculations in which Pt / C catalyst has the highest ECSA value.

The ECSA values for each of the Pt-Co / C catalyst compositions are presented in Table 1.

Table 1. ECSA data for Pt / C and Pt-Co / C catalysts on various compositions

Catalyst	ECSA (m^2/g)
Pt/C 100 : 0	37,381
Pt-Co/C 80 : 20	18,700
Pt-Co/C 60 : 40	16,433
Pt-Co/C 40 : 60	17,285
Pt-Co/C 20 : 80	17,640
Pt-Co/C 0 : 100	25,284

The Pt/C catalyst (table 1) has the highest ECSA values. This is understandable because Pt is the main catalyst that used in PEMFC to convert hydrogen and oxygen gas into water. However, the catalytic activity of Co as a support of Pt catalyst is quite dominant, even in the Pt-Co/C 0:100 composition having a fairly high ECSA. This suggests that Co is a fairly responsive catalyst in catalyzing the reaction.

XRD Analysis

XRD observations for Pt-Co/C catalysts are shown in Fig. 2 and the combined results of all observations are shown in FIG. 3.

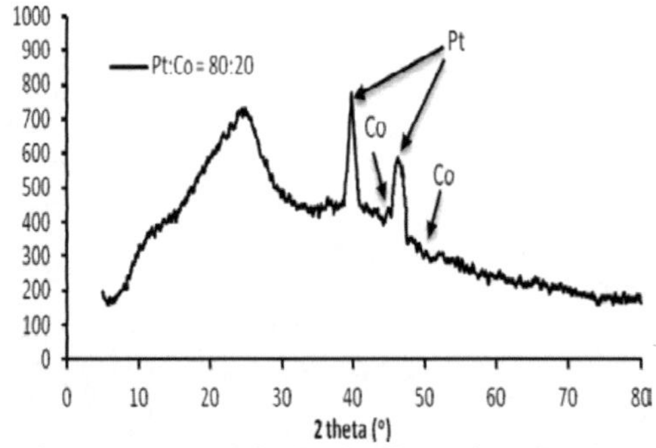

Figure 2. XRD pattern for Pt-Co/C catalyst (Pt:Co = 80:20)

Figure 2 shows the appearance of peak Pt at 2θ = 39.76° (111) and 46.23 ° (200) [14]. Co's presence is not so obvious at 2θ = 44.51° and 51.85 °. This is because Co particles are part of alloys with Pt with a relatively small concentration (20% b/b). The same results are also shown in previous studies [12, 14]. Meanwhile, the peak distribution between 2θ = 19 to 26 ° shows a typical peak for carbon. The higher of the Co content, the peak Co becomes more visible. This can be seen from Figure 3.

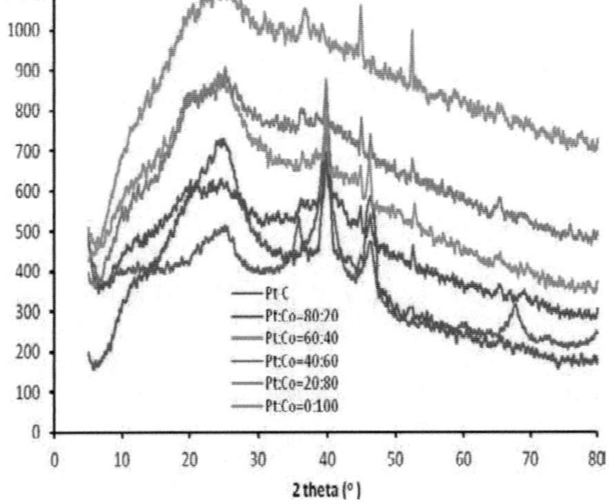

Figure 3. XRD pattern for combined Pt/C and Pt-Co/C catalysts in various compositions

Electrochemical Impedance Spectroscopy (EIS) Analysis

EIS is a perturbative characterization of the dynamics of an electrochemical process. EIS is widely used as a standard characterization technique for many material systems and applications (corrosion, plating, batteries, fuel cells, etc.). The EIS measurement data is shown in Figure 4.

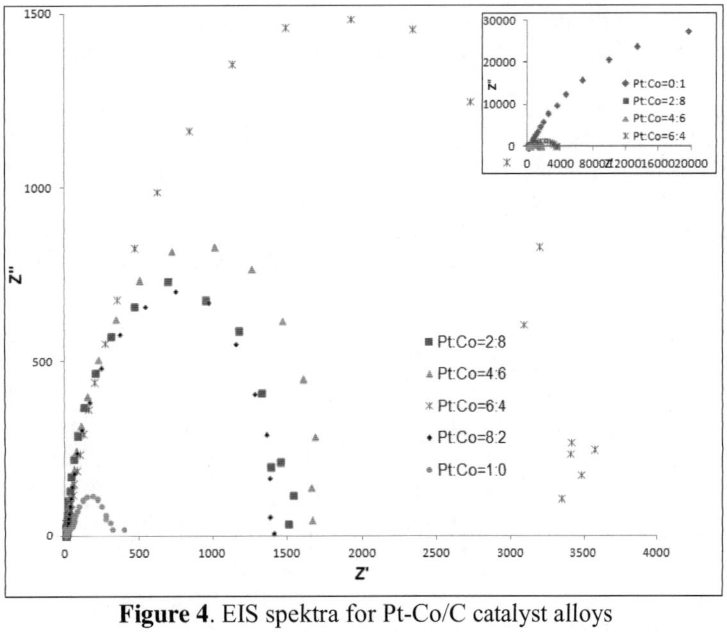

Figure 4. EIS spektra for Pt-Co/C catalyst alloys

Figure 4 shows the EIS curve of the Pt-Co/C catalyst. Based on the EIS results, the higher of the Pt level, the smaller the impedance that indicating the higher of the conductivity and catalyst activity. Although the performance of Co acting as a single catalyst (Pt: Co = 0: 1) looks good by CV and ECSA analysis, the EIS measurements show very large impedances (inzet images). Thus, the alloying composition of Pt catalyst with the lowest Co that can be used is Pt: Co = 20:80.

Conclusion

The reduction process of $H_2PtC_6.6H_2O$ and $CoCl_2.6H_2O$ in the carbon matrix by using $NaBH_4$ as the reducing agent successfully formed Pt-Co/C catalyst and alloy Pt and Co with composition Pt:Co = 20:80, showed good catalytic activity, as indicated by CV and EIS analysis.

Aknowledgment

This project was financed by a grant from Scheme "Hibah Kompetitif UNSRI 2018".

References

[1] Wang Z B, Zuo PJ, Chu YY, ShaoYY and Yin GP 2009 *International Journal of Hydrogen Energy,* **34(10)** 4387.

[2] Wang Y, Chen K S, Mishler J, Cho S C and Adroher X C 2011 *Applied Energy,* **88(4)** 981.

[3] Andr´e D, Taylor E Y K, Virgil P, Humes, Kizuka J, Levi T and Thompson 2007 *Journal of Power Sources* **171** 101.

[4] Daud W R W, Mohamad A B, Kadhum A A H, Chebbi R and Iyuke S E 2004) *Energy Conversion and Management,* **45** 3239.

[5] Su H N, Zeng Q, Liao SJ and Wu YN 2010 *International Journal of Hydrogen Energy,* **35(19)** 10430.

[6] Litster S and McLean G 2004 *Journal of Power Sources,* **130(1-2)** 61.

[7] Rohendi D, Majlan E.H, Mohamad A B, Wan Daud W R, Hassan Kadhum A A and Shyuan L K 2013 *International Journal of Hydrogen Energy,* **38(22)** 9431.

[8] Zhang S, Yuan XZ, Hin J N C, Wang H Friedrich K A and Schulze M 2009 *Journal of Power Sources,* **194(2)** 588.

[9] Sharma S and Pollet B G 2012 *Journal of Power Sources,* **208(0)** 96.

[10] Cho YH, Park HS, Cho YH, Jung DS, Park HY and Sung YE 2007 *Journal of Power Sources,* **172(1)** *89.*

[11] Xiong L, Kannan A M and Manthiram A 2002 *Electrochemistry Communications,* **4(11)** 898.

[12] Hsu SP, Liu CW, Chen HS, Chen TY, Lai CM, Lee CH, Lee JF, Chan TS, Tsai LD and Wang, KW 2013 *Electrochimica Acta,* **105(0)**. 180.

[13] Tasic G S, Miljanic S S, Marceta Kaninski M P, Saponjic D P and Nikolic V M 2009 *Electrochemistry Communications,* **11(11)** 2097.

[14] Yi L, Liu L, Liu X, Wang X, Yi W, He P and Wang X 2012 *International Journal of Hydrogen Energy,* **37(17)** 12650.

[15] Geng X, Zhang H, Ye W, Ma Y and Zhong H 2008) *Journal of Power Sources,* **185(2)** 627.

IOP Conf. Series: Journal of Physics: Conf. Series **1095** (2018) 012008 doi:10.1088/1742-6596/1095/1/012008

Extraction And Pre-Concentration of Zinc(Ii) Ion by Fatty Hydroxamic Acids Immobilized Onto Zeolit

Dedy Suhendra[1], Erin Ryantin Gunawan[1], Firman Ozaki[1]

[1]Department of Chemistry, Faculty of Mathematics and Science, University of Mataram ,Jl. Majapahit No. 62 Mataram, 83125, Indonesia

Email: dedysuhendra@unram.ac.id

Abstract. Mining in Indonesia contributes significantly to mineral supplies nationally, but illegal mining activities can cause natural destruction, habitat loss and loss of biodiversity. This is because only the main metal, which is the gold taken while the other metal is discharged into the drain as waste. One of the metal ions to be waste is the zinc(II) ion. A method of zinc(II) ion extraction from other ions by using a column containing fatty hydroxamic acids – loaded zeolit (FHA-Zeo) is described. Several factors, which affect the extraction efficiency such as pH, sample volume, and concentration of eluent and flow rate, have been investigated. The maximum sorption capacity of FHA-zeo obtained was 88.38 mg/g. A quantitative recovery of zinc(II) ion from FHA-zeo column was obtained using 10% HNO_3 solutions as eluent with a pre-concentration factor up to 60. A method for separation of zinc(II) from lead(II) is proposed. A rapid sample throughput, a clean separation, a high pre-concentration factor and simplicity are the main advantages in these analytical procedures.

Key words: Fatty Hydroxamic Acids, Zeolit, FHA-Zeo, Pre-concentration

1. Introduction

Zinc, one of heavy metals waste in unauthorized mining processing, is an essential trace element of great importance for humans, animals and plants. Zinc(II) ions have various biochemical functions in cell metabolism, including cell replication, protein synthesis, gene expression and cell division. However, as a heavy metal, zinc has hazard and harmful impacts on human health and the environment because they are toxic even at very low concentrations. Thus, accurate detection of zinc ions at trace concentration levels in environmental or water samples is very important in analytical chemistry and environmental pollution monitoring [1, 2]

Flame atomic absorption spectrometry (FAAS) has been a useful technique to detect trace heavy-metal ions in various environmental [3, 4, 5, 6], food and beverages [7, 8, 9, 10], and water samples [11, 12, 13] because of its high selectivity, speed, easy instrument usage and low costs. However, there are many difficulties to determine low concentration of heavy-metal ions due to in sufficient sensitivity of the instrument and high interfering influences of matrix ions on the signal of the analytes [14, 15]. Thus, a pre-concentration and/or separation step is necessary. Pre-concentration is necessary to bring the concentration of the analytes into the dynamic range of the flame atomic absorption detector used, while separation is employed to eliminate potential interfering matrix constituents.

Separation and pre-concentration techniques are of great importance owing to the limited sensitivity of an analytical instrument for trace analysis. Several technologies have been widely used to pre-concentrate and separate trace metals, among them are liquid–liquid extraction by using complexing agents [16, 17, 18], and solid phase extraction [19, 20, 21, 22]. Liquid-liquid solvent extraction of metal chelating complexes has been used as a separation method for a long time. However, this method leads to many disadvantages. Among them are, usually, time consuming and labor intensive, use of large amounts of organic solvents for extraction and subsequent disposal of the solvent, which creates another environmental problem, and cannot be combines with FAAS or inductive couple plasma–atomic emission spectrometry (ICP-AES) due to the difficulties of introducing organic solvents into nebulizer of AAS or plasma of ICP-AES. Meanwhile, the solid phase eaxtraction has low selectivity for metal ion adsorption [23].

A very efficient system for separation and pre-concentration of trace metal ions is immobilization of chelating agent on various supports. This method has several advantages over liquid-liquid extraction and solid phase extraction. The advantage includes pre-concentration from a larger sample volume and establishing higher concentration factors. In this paper, we report the results of our investigations on fatty hydroxamic acids (FHA), synthesized from ketapang (*Terminalia cattapa* L.) kernel oil, that immobilized onto zeolit to be used as an extractant for zinc(II) ions.

2. Materials and Methods

2.1. Materials

All of the chemicals used in this work were of analytical grade and were obtained from Merck (Darmstadt, Germany) or Fluka (Bunch., Switzerland) and bidistilled/ deionized water was used in all experiments. Hydroxylamine hydrochloride, sodium phosphate, sodium hydroxide, phosphoric acid, vanadium pentoxide and nitrate salts of copper, zinc, cobalt and nickel were of the highest purity available and used without any further purification. The standard stock solutions of metal ions (1000 mg L^{-1}) were prepared by dissolving nitrate salts of the metals in 0.5 mol L^{-1} HNO_3. The standard solutions used for calibration were prepared by diluting the metal stock solutions with 0.5 mol L^{-1} HNO_3 before use. Non-edible and non-commercial oil, that is ketapang kernel oil was extracted from ketapang fruit kernel. The oil purified by drain off into chromatography column and eluted using a mixture of hexane-diethyl ether (87:13, v/v). Atomic absorption spectrophotometer (AAnalist 400 Perkin Elmer) with an air-acetylene flame was used for the metal ion analysis.

2.2. Procedure

2.2.1. Synthesis and characterization of FHA. The FHA was prepared as described in our previous publication with a little modification. The reaction was carried out by reacting a required amount of ketapang kernel oil with hydroxylamine in the present of Lypozyme in a 100 mL stoppered flask. The mixture was incubated in water shaker batch at 100 rpm. The FHA formed in water-hexane interface was then separated from the water and the lipase by filtration. To obtain the solid FHA, the hexane fraction was cooled in a refrigerator (<-5°C) for 5 h and then filtered and washed by hexane for several times and dried in vacuum desiccator over phosphorous pentoxide for 24 h [24]

Qualitative analysis of hydroxamic acid groups on FHA was carried out by observes the colored complex formed after the methanolic solution of the FHA was reacted with iron (III), copper (II) and vanadium (V) in a dilute hydrochloric acid solution. The amount of the hydroxamic acid group was estimated based on nitrogen content of the dry FHA by using Kjeldahl method. FTIR spectra of FHA were recorded with Fourier Transform Infrared Spectrophotometer (Perkin Elmer FTIR-Frontier, USA) using KBr disk

2.2.2. *Preparation of the FHA loaded zeolit (FHA-Zeo).* The zeolite was activated by heating in a furnace at 450 for 2 h. The activated zeolite (0.5 g) was mixed and shaken with 2000 ppm of FHA solution in methanol at 35 rpm for 12 h. The mixture was filtered and then kept in refrigerator

2.2.3. *Effect of pH on metal ions uptake by FHA-Zeo.* The effect of pH on metal ions uptake by FHA-Zeo was determined by batch equilibrium techniques. The metal ion (20 mL, 0.001 M) was shaken with 100 mg of FHA-Zeo for 12 h to ensure the equilibrium is attained in all cases. The pH of the metal solutions was adjusted to 1-6 with 0.01 M phospate buffer solutions before shaken. After the equilibration, the concentration of ion in solution was determined by AAS for the calculation of sorption capacity

2.2.4. *General extraction and desorption procedures on zinc(II) ion uptake by column technique.* FHA-Zeo (4g) was packed into a column with an internal diameter 1 cm to 10 cm height. A 1000 mL solution containing the desired concentration of zinc(II) ion at pH 5, adjusted with 0.01 M of phosphate buffer was passed through the column. The concentration of zinc(II) ion before and after passing the column was determined using FAAS for the calculation of percentage sorption of zinc(II) ion. The columnar stripping of copper (II) ion was performed using nitric acid solution as eluent. The acid was passed through the column containing FHA-Zeo-Ni. The successive fractions of effluent (3 mL) were collected. The concentration of copper after stripping was determined using FAAS for the calculation of percentage recovery.

3. Results and Discussion

3.1. Characterization of the FHA

Complexes of the FHA with vanadium (V), iron (III) and copper (II) are purple, dark red and green, respectively. These are the common color of the complexes observed when these metal ions are reacted with hydroxamic acids [24]. FTIR spectrum of the FHA (Figure 1b) shows the characteristic absorption bands of monosubstituent amides at 3050 - 3500 cm (medium peak), 1650 - 1750 cm^{-1} (sharp peak) and 1450 – 1600 cm^{-1}(medium peak), which indicate the presence of O-H and N-H stretching and bending modes. Absorption bands that appear at 2800 – 3000 cm^{-1}(sharp and medium peaks) and 1024 $cm^{-1}$are due to C-H stretching and bending respectively. Both qualitative tests above support that the FHA is successfully produced from the ketapang kernel oil.

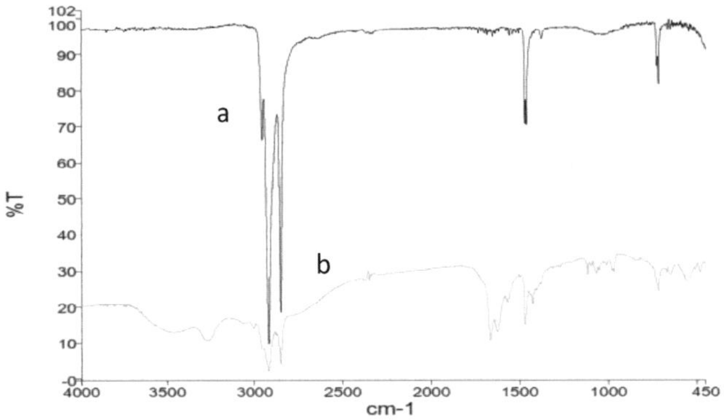

Figure 1. FTIR spectra of Ketapang kernel oil (a) and FHA (b)

The nitrogen content of the dry FHA analyzed by Kjeldahl method was found to be 4.19%. This indicates that there is 2.99 mmol/g of hydroxamic acid groups in one gram of the FHA. Meanwhile, the maximum sorption capacity obtained of the FHA immobilization onto zeolite was 4.31 mg/g.

3.2. Effect of pH on sorption of Zn(II), Cu(II) and Pb(II) at FHA-Zeo

The sorption behavior of Zn(II), Cu(II) and Pb(II) on FHA-Zeo at different pH values was examined by using batch method. The results show that all metal ions are absorbed into FHA-Zeo at the variuos pH used (Table 1). However, the highest capacity shows by FHA-Zeo-Zn(II) complex. The capacity sorption of the FHA-Zeo is an important factor to determine the amount of FHA-Zeo that required to quantitatively remove a specific metal ion from a solution.

Table 1. Metal ions sorption of FHA-Zeo at various pH

Metal Ion	Sorption Capacity (mg/g)					
	pH					
	1	2	3	4	5	6
Zn (II)	0.00	0.00	1.1	5.4	17.2	9.5
Cu (II)	0.00	0.00	0.01	2.14	3.74	5.08
Pb (II)	0.00	0.03	0.53	3.24	6.53	8.25

3.3 Pre-concentration and separation of Zn(II) from Cu(II) and Pb(II)

The pre-concentration and separation studies of zinc(II) ion was carried out by extracting 1000 mL solution containing 100 ppm zinc(II) ion at pH 5. In addition to zinc(II) ion, the solution also containing 100 ppm of copper(II) and lead(II) ion, respectively. The sorption of metal ions was performed at eluent flow rate of 0.3 mL min^{-1}. The results (Table 2) show that almost 100% of the zinc is absorbed into the FHA-Zeo. Meanwhile copper and lead are absorbed 38 and 57%, respectively. This indicates that complex of zinc(II)-FHA-Zeo on the resin is more stable than complexes of copper(II)- and lead(II)-FHA-Zeo. The difference of sorption capacity will affect to the retention of metal ions during desorption from the FHA-Zeo.

Table 2. Metal ions sorption on FHA-Zeo at pH 5

Metal Ion	Concentration (ppm)	Sorption (%)
Zn	100	96.3
Cu	100	38.9
Pb	100	56.7

Desorption of the metal ions from FHA-Zeo was performed using 10% of nitric acid as eluent, at flow rate of 0.5 mL min^{-1}. The results show that the percentage recoveries for all pre-concentration factors are more than 99%. This is indicating that the FHA-Zeo can be applied for the pre-concentration of zinc(II) ion in aqueous media. Figure 1 shows the results of desorption of metal ions from the FHA-Zeo at the column. It can be seen that zinc is separated well from both copper and lead, while copper and lead we're not separated. This is probably because the stability of these two complexes is very similar.

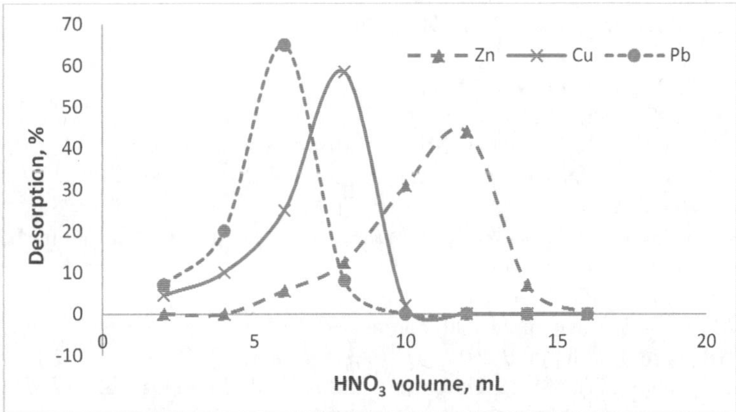

Figure 2. Separation of zinc(II) from copper(II) and lead(II). zinc extracted 96.3 ppm, total desorption 99.87%. copper extracted 3.89 ppm, total desorption 98.94%. lead extracted 56.7 ppm, total desorption 99.93%

4. Conclusion

The present studies highlight that immobilization of FHA onto zeolite is an effective method for extraction and pre-concentration of zinc(II) ion in water sample. A pre-concentration process has been proposed for the determination of zinc(II) ion in water samples, which contain trace

IOP Conf. Series: Journal of Physics: Conf. Series **1095** (2018) 012008 doi:10.1088/1742-6596/1095/1/012008

concentrations of zinc that they cannot be measured directly by FAAS. A quantitative recovery of zinc(II) ion from FHA-zeo column was obtained using 10% HNO_3 solutions as eluent with a pre-concentration factor up to 60. Extraction data obtained indicate clean separations of zinc(II) from copper(II) and lead(II) ions with reasonably good separation factors.

References

[1] Li Z., Xiang, Y., Tong A. 2008, *Analytica Chimica Acta*. **619**, 75.
[2] M. Shamsipur, Zahedi, M.M., De Filippo, G., Lippolis, V 2011. *Talanta,* **85**, 687.
[3] Sahan S., Sacmacı S., Sahin, U., Ülgen, A., Kartal S. 2010 *Talanta*, **80**, 2127.
[4] Shokrollahi A.and Gohari, M. 2016 *Beni-Suef University Journal of Basic and Applied Sciences*, **5 (4)**, 350.
[5] Gondim T.A., Guedes, J.A.C , Ribeiro LPD, Lopes GS and Matos W.O 2017 *Marine Pollution Bulletin,* **114 (2)** 786
[6] North A.E, Kumankomah S., Bellavie A.R., White W.M and Gailer J 2017. *Journal of Environmental Sciences,* **57** 249-257.
[7] Tokay F.and Bagdat, S 2016. *Food Chemistry*, **197** 445–449
[8] F.A.C. Amorim, Costa, V.C., da Silva, E.G.P., de Castro Lima, D., de Jesus, R.M., de Almeida and Bezerra, M. 2017, *Food Chemistry,* **227** 41.
[9] Gürkan, R. and Altunay, N 2015. *Food Chemistry* **177** 102.
[10] Daşbaş T, Saçmacı Ş, Çankaya N and Soykan, C 2016,. *Food Chemistry*, **203** 283.
[11] Dos Santos W N L, Costa, Rennan J L.O, Araujo G.O, de Jesus D.S and Costa 2006 A.C.S. *Journal of Hazardous Materials*, **137 (3)** 1357.
[12] Knap M., Kilian K and Pyrzynska, K 2007 *Talanta* **71** (1) 406.
[13] Meira L.A and de Souza Dias F 2017 *Microchemical Journal* **130** 56..
[14] Srogi K 2008 *Analytical Letter* **41** 677.
[15] Gouda A A 2016 *Talanta*, **146** 435
[16] Stasiak M, Regel-Rosocka M, Borowiak and Resterna A 2016. *Hydrometallurgy* **162** 57
[17] Daryabor M, Ahmadi A., Zilouei H 2017 *Ultrasonics Sonochemistry* **34** 931
[18] Ding K, Liu,Y, Tang, J, Zhou Y, Hu J, Lin X., Wu W 2017. *Separation and Purification Technology*. **186** 264
[19] Suhendra D., Yeen K P, Haron M J, Silong S, Basri, M. And Wan Y, Wan M. Z. 2005 *Solvent Extraction and Ion Exchange*. **23** 13.
[20] Gouda A.A 2014 *Int. J. Environ. Anal. Chem* **94** 1210
[21] Roushani M., Abbasi S. Khani H. and Sahraei, R 2015 *Food Chemistry,* **173** 266.
[22] Singh R, Mahandra H, Gupta B 2017. *Hydrometallurgy,* **172** 11.
[23] Suhendra D., Wan Y, Wan.M.Z., Haron, M. J., Basri, M., Silong, S 2006 . *Indo. J. Chem.,* **6** (2) 165
[24] Suhendra D, Wan Y, Wan.M.Z., Haron, M. J., Basri, M., Silong, S 2005. *J. Oleo Sci*, **54 (1)** 33

Nickel Slag Coated by Titanium Dioxide for Degradation of Methylene Blue

Destri Muliastri[1], A B Widyartha[1],Wasiara[1], R. Saputra[1], Muhammad Nurdin[1], L O A N Ramadhan[2]

[1]Department of Chemistry, Faculty of Mathematics and Natural Sciences, Universitas Halu Oleo, Kendari 93232 – Southeast Sulawesi, Indonesia

E-mail: ramadhan305@gmail.com

Abstract. The utilization of nickel slag waste coated titanium dioxide (TiO_2) as material for organic contaminant degradation was carried out. The aim of this study is to influence of degradation time and nickel slag mass for contaminant degradation in photolysis and photocatalysis system. TiO_2 was synthesized by usingsol-gel method. Material characterizations were performed using X-Ray Diffraction (XRD), X-Ray Fluorescence (XRF), and Optical Microscope. XRD Analysis of slag coated TiO_2(Slag-TiO_2) showed that TiO_2 in anatase phase at $2\theta = 68.30°$ (230).XRF analysis showed that dominant compound in nickel slag was silicon. Furthermore, optical microscope analysis showed that surface morphology of nickel slag was porous and solid. The result of research reveals that degradation test of methylene blue was conducted in photolysis and photocatalysis system for 4 hours irradiation with time interval for 30 minutes.The percentage degradation of photolysis and photocatalysis were 48% and 87.50%, respectively. Meanwhile, a mass variety of slag used in methylene blue degradation were 5.46 gram; 7.43 gram, and 9.32 gramfor photolysis and 5.49 gram; 8.02 gram, and 10.02 gram for photocatalysis. Degradation percentage was increased along with the increase of nickel slag mass.

Keywords: Nickel slag, TiO_2, photocatalysis, photolysis, methylene blue

1. Introduction

Nickel can be produced either by reduction-smelting process from laterite ores [1] or roasting-smelting process from sulfide ores [2]. Both of these processes produce nickel slag, which will take up a lot of space and pollution to the environment. The nickel slag has a porous structure and a solid form. In addition the nickel slag can be utilized as the medium for overcoming the dye pollution by coating method that should be addressed [3].

Meanwhile, Methylene Blue (methylthionine chloride) is a heterocyclic aromatic chemical compoundwith the chemical name [3,7-bis (Dimethylamino)-phenazathionium chloride Tetramethylthionine chloride] under the molecular formula ($C_{16}H_{18}ClN_3S.3H_2O$) [4]. It is cationic and considered a significant threat to human and biota, as carcinogenic and mutagenic. Methylene blue is used for coloring paper, wool, and cotton staining [5]. In addition, the aromatic structure of methylene blue provides the stability of the chemical and physical attacks, so it needs serious treatment as a water pollution in environment.

Photocatalyst can be used to reduce and damage the organic or inorganic compounds through a photocatalytic process [6].Titanium (IV) oxide (TiO_2) is a heterogeneous photocatalyst most effective to degrade the organic compound pollutant. This is due to the photocatalytic activity of TiO_2 has a stable and high-photoactivity [7]. Photocatalyst with TiO_2 heterogeneous underultraviolet (UV) light is

an emerging technology to be applied in the removal or changing of waste through the mechanism of reduction and oxidation (redox) [8]. The addition of the metal surface of TiO_2 is the object of research that is important to examine the photocatalytic activity of TiO_2.
The addition of the metal widely increases the area of the spectral response of TiO_2 to the UV band gap changes effects [9]. In TiO_2 photoexcitation process, electrons can be transferred from the conduction band to metal particles on the TiO_2 surface. Dopant ions can create trap electrons which will minimize the recombinant process of electron and hole. Metals such as Ni and Zn have been reported to increase the catalytic activity of the photocatalyst [10]. Porous materials are frequently adopted as an adsorbent and still investigated in the design of adsorption assisted photocatalytic reaction systems [11].
Nickel slag porous structure and containing a metal oxide when coated with TiO_2 semiconductor was potentially helping the performance of TiO_2 in degrading organic pollutants. This study performs the coating with TiO_2 slag through a dip-coating method, which is expected to oxidize pollutants organic substances such as methylene blue and turn it into simpler compounds and does not harm the environment. The purpose of this study is to obtain the effect of TiO_2-nickel slag as materialto degrade organic pollutants in photolysis and photocatalytic process.

2. Material and Methods

2.1 Materials
The materials used in this study were nickel slag, methylene blue, Titanium tetra-isopropoxide (TTIP) 97% (Aldrich), Nitric acid (HNO_3 65% E. Merck),Ethanol (C_2H_5OH E. Merck) and distilled water.

2.2 Preparation of Sol-Gel TiO_2
Synthesis of TiO_2 sol-gel was adopted according to the method described by Kim et al. [12] TiO_2 sol-gel made using two solutions, namely solution I (250 mL of distilled water is added by nitric acid to a pH of 1.5 and then stored at cold temperatures) and solution II (Titanium tetra-isopropoxide (TTIP) 1.25 mL, then added with 25 mL of ethanol). The solution I and solution II were mixed using a magnetic stirrer for 30 minutes. Subsequently, it was stirred gently at 4°C for 12 hours and stored in the refrigerator.

2.3 Coating Process of Slag Nickel with TiO_2
Nickel slag with mass variation were coated with TiO_2 sol-gel by dip-coating technique based on the modified method described by Li et al. [13]. The dip-coating was performed for 10 minutes and calcined for 120 minutes at a temperature of 500°C, thus producing of slag coated TiO_2 (slag-TiO_2).

2.4 Preparation of Reactor
Reactor of wood was coated by aluminum foil on the insidecube with dimension of 30cm× 30cm× 30cm. Lights were placed in the form of two UV LEDs mounted on the top surface of the cube wall. The reactor equipped with a switch to turn off and turn on the UV lightsthat can be seen in Figure 1.

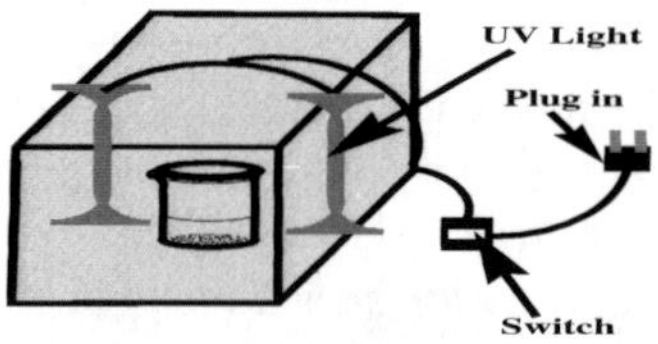

Figure 1. Illustration of photodegradation reactor

2.5 Characterization of Slag and Slag-TiO_2

Morphology and diffraction pattern analysis. Morphology of TiO_2 slag and slag-known were observed by optical microscopy imaging tools. This analysis is done at 35 times magnification. Meanwhile, analysis of the diffraction pattern-TiO_2 slag and slag was carried out using X-Ray Diffraction (XRD) Shimadzu 6000.

Analysis of element compositions. Analysis of the elemental composition of TiO_2 slag and slag were performed using Portable X-Ray Fluorescence (XRF) with the type of Niton XL3t GOLDD. All samples were analyzed in the form of press powder. Measured data in the form of intensity (I) and elemental energy which is then converted in the form of numbers so that the data generated in the form of a percentage elemental composition of a sample.

2.6 Degradation Process of Methylene Blue

Variations in exposure time. Methylene blue dye degradation was carried out in a glass reactor with a volume of 50 mL, with a concentration of 4 mg.L-1. Then put the catalyst that had been made previously, namely nickel slag coated TiO_2. The reactor was irradiatedby UV light time variation of 30, 60, 90, 120, 150, 180, 210 and 240 minutes. After irradiation for 4 hours, the absorbance of degradation was analyzed by UV-Vis spectrophotometer at a maximum wavelength of 664 nm. The analysis results werecompared to the concentration of methylene blue in photolysis and photocatalytic degradation. The percentage degradation has determined by equation (1) [14]:

$$(\%D) = \frac{C_0 - C_t}{C_0} \times 100\% \qquad \textbf{(1)}$$

Where C_0 is the initial concentration of the dye before irradiation, C_t is the concentration of the dye on t.

Nickel slag mass variation. Degradation of methylene blue is carried out by using a mass variation. Mass variations were 5.49, 8.02, 10.02 grams. Then each beaker was put 50 mL of methylene blue and inserted nickel slag coated TiO_2. The reactor was irradiated with UV light with time variation of 30, 60, 90, 120, 150, 180, 210 and 240 minutes. After UV irradiation for 4 hours, degradation results absorbance were analyzed by spectrophotometer at a maximum wavelength of 664 nm. The results of the analysis were compared to the concentration of dye in photocatalytic degradation results determined the percentage of degradation of dyes by equation (1). Furthermore, the same treatments were performed in the photolysis of the slag mass variation of 5.46 gram, 7.43 gram, and 9.32 gram.

3. Results and Discussion

3.1. Process of Nickel Slag coated TiO_2

Nickel slag has been successfully coated TiO_2 as indicated by the white appearance on the surface of the slag. Slag mass after being coated successively were 5.49, 8.02, and 10.2 grams based on the slag mass original sample, so that the mass of TiO_2 deposited were 0.03; 0.59 and 0.88 grams. The TiO_2 synthesis are expected to have a crystal structure of anatase for calcined at 500°C for 2 hours. Anatase crystals have a largeband gap energy (3.2 eV).so that it has also larger surface area that makes the high-performance for photoactivity [15].

3.2. Characterization of Nickel Slag

Analysis of optical microscope. Slag surface morphology are shown in Figure 2. Before coating the entire slag with different masses has pores and dark gray as the color of iron in general. It is certainly not out of the composition of the slag, but it is mostly in the form of iron.

a b c

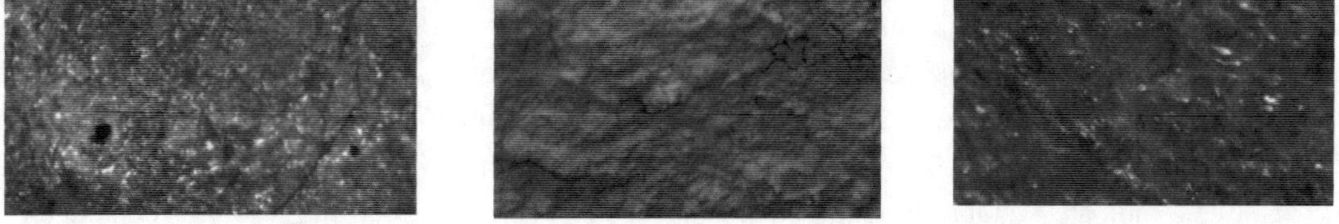

d f

IOP Conf. Series: Journal of Physics: Conf. Series **1095** (2018) 012009 doi:10.1088/1742-6596/1095/1/012009

e

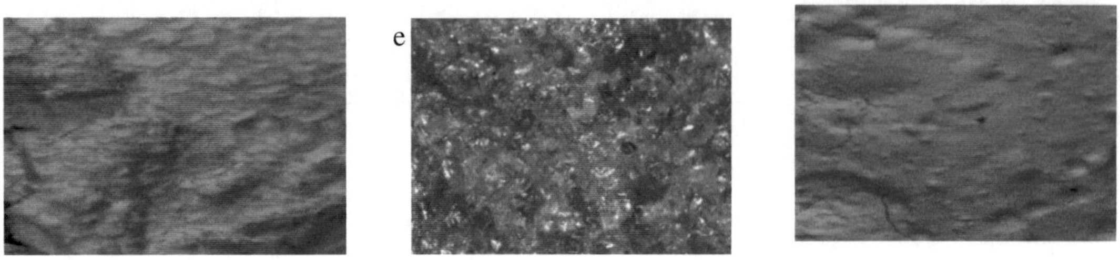

Figure 2. Slag surfaces original samples (a) before TiO_2 coated 5.46 gram (b) after TiO_2 coated of 5.46 gram(c) 7.43 grams (d) after TiO_2 coated of 7.43 gram (e) 9.32 gram mass (f) after TiO_2 coated of 9.32 gram

X-Ray Diffraction Analysis (XRD). XRD analysis aims to interpret the dominant presence of the compounds contained in nickel slag and determine the crystal phase of TiO2 that has been synthesized by sol-gel method. Figure 3 shows the results of XRD spectrum on Pomalaa nickel slag and slag-TiO2. Slag diffractogram shows the pure slag where there are peaks at the position of 2θ = 28.57°; 62.98°; and 63.16°which indicate the presence of a compound $CaMg(Si_2O_6)$. The analysis showed that the compound is a dominant compound in the slag with a composition of 50%. Other minerals with relatively large amounts of the $Ni_2(SiO_4)$ and $Fe_2(SiO_4)$ were detected at 2θ = 42.4° and 2θ = 25.57° with the composition of their respective 24% and 19%. These results demonstrate conformity with the research report by Tangahu et al. [16].

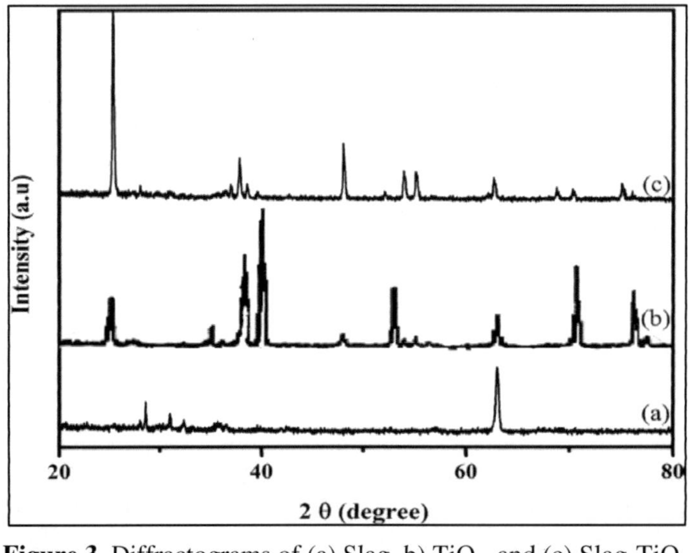

Figure 3. Diffractograms of (a) Slag, b) TiO_2, and (c) Slag-TiO_2

In addition to these three minerals, there is also diffraction at 2θ = 25.31°. Peaks are formed in very small or in other words have a very low intensity due to the minimal amount of slag-TiO_2. This confirms the validity of the research by Rusmini showed a small amount of TiO_2 in slag through XRD analysis [17].

Slag-TiO_2 diffraction spectrum show that the peaks are formed in common with pure slag where there are peaks of $CaMg(Si_2O_6)$, $Ni_2(SiO_4)$, and $Fe_2(SiO_4)$ minerals with similar intensity. However, the coating of slag-TiO_2 indicate a very high peak intensity at the position of 2θ = 25.30° and moderate intensity at 2θ = 53.98°. The existence of this second peak indicates the presence of TiO_2 anatase phase corresponding by JCPDS No.21-1272. In previous studies, Salmin and Al-Shamali [18] have reported that the structure of TiO_2 can be obtained by calcination at a temperature of 500 °C so that the results of this study indicate suitability. Compounds of TiO_2 anatase has been successfully coated on the nickel slag surface of Pomalaa expected to provide efficiency degradation of organic pollutants.

X-Ray Fluorescence (XRF). The result of XRF as an elemental analysis technique can be seen in Table1.

Table 1. Measurement of nickel slag composition (%) used XRF

No.	Elements/ compounds	Composition	No.	Elements/ compounds	Composition
1	Si	29.50 %	9	Zr	0.02 %
2	Fe	8.75 %	10	Zn	10 %
3	Ni	0.13%	11	Ca	42.3 %
4	Zn	0.11 %	12	Mg	5.40 %
5	Cu	0.8 %	13	Nb	1.02 %
6	Co	0.6 %	14	C	0.1 %
7	Mn	0.2 %	15	F	0.3 %
8	Cr	0.4 %	16	S	0.16 %
Total Composition				100%	

3.3. Degradation of Methylene Blue Dye

Photolysis Degradation. Methylene blue degradation by photolysis was performed on two conditions, namely degradation by using nickel slag with a mass variation of 5.46 gram, 7.43 gram, and 9.32 gram, without the use of TiO_2 catalyst, under UV light irradiation and degradation without using nickel slag (only UV light) for 4 hours. The relationship between the irradiation time by photolysis, and only with UV light, is shown in Figure 4 a, and the relationship between the irradiation time and slag variation mass is shown in Figure 4 b.

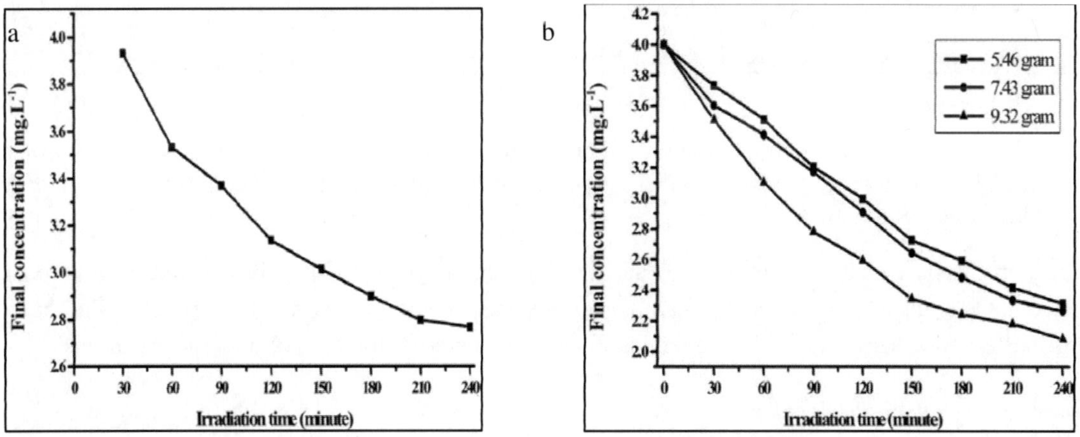

Figure 4. (a) Graph of time related to the concentration of methylene blue in the photolysis (Only Assistance UV rays). (b) Graph of irradiation time relationship and slag mass variation toward the concentration of methylene blue in the photolysis

Figure 4 (a) shows the degradation of methylene blue under UV light without nickel slag for 4 hours in intervals of 30 minutes. The results showed that in intervals of 30 minutes the methylene blue dye has decreased of concentration. But a decrease in the concentration likely to be small. Decreasing the concentration of methylene blue to the irradiation time of 240 minutes was 2.7 $mg.L^{-1}$, with the percent degradation reaches 31%. Figure 4 (b) shows the results of degradation of methylene blue 4

$mg.L^{-1}$ by using nickel slag mass 5.46 gram, 7.43 gram, and 9.32 gram and the time of irradiation for 4 hours in intervals of 30 minutes. The optimum time degradation is determined by observing at the most significant decrease in concentration, so the optimum degradation time is 60 minutes.

Based on Figure 4 (b) shows that the irradiation was greatly affects the decreasing of the concentration of methylene blue, the longer of exposure time was obtained the greater decrease in the concentration of methylene blue. In the higher percent degradation increases, the energy of photons produced by UV rays to generate OH radicals. OH radicals will be formed exponentially with time. Decreasing the concentration of methylene blue which tends to stay constant because in the longer OH radicals are formed H_2O_2 [19]. Slag mass difference in the degradation of methylene blue also affects the efficiency of degradation. The greater the mass of slag is used, the greater the decrease in the concentration of methylene blue. Figure 4 (b) shows that the mass of slag 9.32 gram, can degrade methylene blue, it is quite large compared to the mass of 5.46 gram and 7.43 gram.

Photocatalytic degradation. The photocatalytic process is one of the degradation processes that use catalyst and photons. Photocatalytic degradation of methylene blue is done in two conditions, namely the degradation of methylene blue only with a catalyst of TiO_2 and UV light, as well as the degradation of methylene blue with nickel catalyst coated by TiO_2 slag and UV light irradiation time of 240 minutes as can be seen in Figure 5.

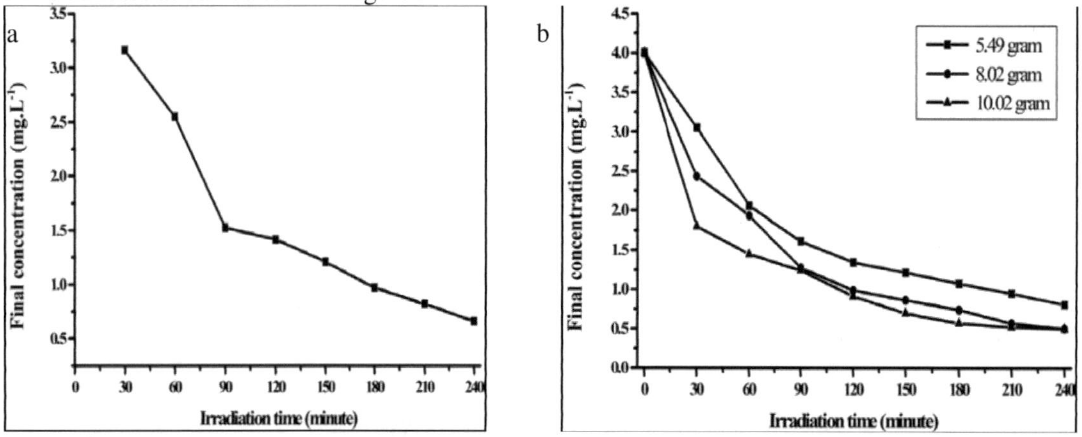

Figure 5. (a) Graph of irradiation time relationship toward the final concentration and (b) graph of irradiation time relationship and mass variation toward the final concentration of methylene blue in photocatalytic

Based on the data obtained can be seenthat the degradation of methylene blue using a nickel slag coated TiO_2 is much more effective than using TiO_2 suspension, this is due to the nickel slag has a large surface, and the metal oxide is a semiconductor that helps the process of photodegradation.

3.4. The Degradation Efficiency

Based on Figure 6 the photocatalytic processes was provided better results than the photolysis process because the presence of radical species reproduce TiO_2 catalyst to degrade the methylene blue which would accelerate the degradation reaction. Methods of using photon photolysis only produce radical species that tend to be a bit because it does not happen generating pairs of electron/hole by semiconductor TiO_2.

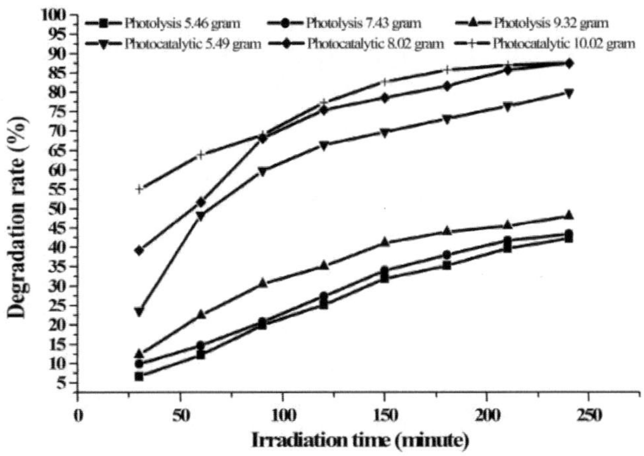

Figure 6. Comparison of degradation results of methylene blue

4. Conclusion

Based on the results of research and discussion that has been described, it can be concluded that the percentage of degradation of methylene blue dye solution using nickel coated slag as photolysis and photocatalytic TiO_2 under UV light irradiation for 4 hours by 48% and 87.50%, respectively.The increase in mass of slag used is directly proportional to the increase in the percentage of degradation of methylene blue dye both photolysis and photocatalytic. Meanwhile, the optimum time of degradation is determined by observing at the most significant decrease in concentration so that the optimum time of degradation on the photolysis and photocatalytic process is 60 minutes.

Acknowledgment

The authors would like to thank Ministry of Research, Technology and Higher Education for Student Creativity Program Research Grant.

References

[1] Yan W C, Yin F, Chen Y Q, Wang Z, and Wang J 2008 *The Chinese Journal of Nonferrous Metals* **18** 1.
[2] Fu C, Liu H X, Zhong D L and Zeng D W 1999 *The Chinese Journal of Nonferrous Metals*, **9(4)** 805.
[3] Saptahari 2005 Jurnal Infrastruktur dan Lingkungan Binaan **1(1)** 2.
[4] Wiklund L, Basu S and Miclescu A 2007 *Ann. N.Y. Acad. Sci.* **1122** 231
[5] Zhang D, Guisheng L, Hexing L and Yunfeng L 2013 *Engineering Chem Asian J* **8** 26.
[6] Sarici-Ozdemir C 2014 *Particul. Sci. Technol.* **32(3)** 31.
[7] Idris A, Misran E, Hassan N, Jalil A A and Seng C E 2012 *J. Hazard. Mater.* **227–228** 309.
[8] Dian D, Rahmayeni and Syukri A 2013 *Jurnal Kimia Unand* **2(3)** 58.
[9] Litter M I 1999 *Appl. Catal.* B **23** 89
[10] Hermawan P, Pranowo H D and Indriana K 2011 *Indo Journal of Chemistry* **11(2)** 135.
[11] El-Bahy Z M, Ismail A A and Mohamed R M 2008 *J. Hazard. Mater.* **11(1)** 138.
[12] Kim T K, Lee M N, Lee S H, Park Y C, Jung C K and Boo J H 2005 Development Of Surface Coating Technology Of TiO_2 Powder And Improvement Of Photocatalytic Activity By Surface Modification Thin Solid Films **475** 171.
[13] Li J H, You Y X and Li-ping Zhu 2009 Fabrication and Characterization of a Novel TiO_2 Nano Particle Self- Assembly Membrane With improved Fouling Resistance, Institute of Polymer Science Key Laboratory of Macromolecule Synthesis and Functionalization Ministry of Education Zhejiang University Hangzhou 310027 People's Republic of China **326** 659.
[14] Raheem Z and Ahmed M H 2015 IJBAS **4(1)** 81.
[15] Carp O, Huisman C L and Reller A 2004 Photoinduced Reactivity of Titanium Dioxide. Solid State Chemistry, **32** 33.
[16] Tangahu, Bieby V, Idaa W, Dian S and Lily P 2015 *Indonesia ACES* **5(5)** 408.

[17] Rusmini 2010 *Jurnal Teknologi Technoscientia* **ISSN:1979-8415** 56
[18] Salmin S and Al-Shamali 2013 AJBAS **7(4)** 172.
[19] Xu M W, Bao S J and Zhang X G 2005 *Materials Letters* **59** 2194

Fatty Acids From Microalgae *Botryococcus braunii* For Raw Material of Biodiesel

P Dilia[1]**, K Leila**[2]**, Rusdianasari**[2]

[1] Applied Master of Renewable Energy Engineering, Politeknik Negeri Sriwijaya, Jl. Srijaya Negara, 30139, Palembang, Indonesia
[2] Renewable Energy Engineering Study Program, Politeknik Negeri Sriwijaya, Jl. Srijaya Negara, 30139, Palembang, Indonesia

E-mail : puspadilia1602@gmail.com

Abstract. One of the alternative sources of renewable energy that can overcome the environmental crises and the current fossil energy crisis is microalgae. Microalgae contain fatty acids that can be converted into biofuel products, such as biodiesel. This study was a preliminary research which was conducted to determine the amount of fatty acids contained in microalgae and its potentiality to be converted into biodiesel. The microalgae used in this study were *Botryococcus braunii*. Data collection of fatty acids of *Botryococcus braunii* were obtained by the direct analysis in the laboratory using an instrument in the form of gas chromatography mass spectrometry (GC-MS) and also obtained from literature study in several research of previous researcher. Based on that, it was found out that *Botryococcus braunii* contained with oleic acid ranged from 15.65 – 35.85%, linoleic acid 3.05 – 23.15%, palmitic acid 4.03 – 22.13 % and stearic acid 0.48 – 15.41%. The profile and fatty acid content of this preliminary study could be one of the decisive parameters in the selection of microalgae species. Based on results concluded that *Botryococcus braunii* had good potentiality to be converted into biodiesel because the ratio of saturated and unsaturated fatty acids in this species is not much different.

1. Introduction

Over the last decade, the issue of environmental crises and the fossil energy crisis has become one of the global issues that motivates researchers in the world to conduct research concerning alternative and renewable energy. Oil-based biofuel derived from plants is one of the many solutions developed due to the availability of abundant raw materials and fairly simple production technology [1]. One research on biofuels that is being developed to be an alternative is to use photosynthetic microorganisms in the form of microalgae that are converted into biodiesel [2].

Microalgae, according to Mata et al. (2010), is a prokaryotic or eukaryotic photosynthetic microorganism whose growth is very productive and can outperform other land plants and is considered efficient enough to be used as biofuel due to the high fat or oil content and product of the photosynthesis from microalgae is also very high around 3 – 8%, while for land plants it only reaches 0.5% [3-4]. The potential comparison of some types of biodiesel raw materials can be seen in Table 1.

Table 1. Comparison of the Potential of Some Biodiesel Raw Materials [5]

Raw Material	Produced Lipid Yield (L/Ha)	Required Area Width (M ha)
Corn	172	1,540
Soybean	446	594
Canola flowers	1,190	223
Jatropha curcas	1,892	140
Coconut	2,689	99
Oil palm	5,950	45
Microalgae[a]	136,900	2
Microalgae[b]	58,700	4.5

[a]70% lipid assumption in biomass (% dry weight)

[b]30% lipid assumption in biomass (% dry weight)

One type of microalgae that can be utilized to biodiesel is Botryococcus braunii. Botryococcus braunii is a single green cell microalgae and is commonly found in the lakes, ponds, or brackish waters and the sea with a chlorophyll content of ± 1.5 – 2.8%. Botryococcus braunii has a cell nucleus with a size of ± 15 – 20 μm and lives in colonies. It is non-motile and each movement is strongly influenced by the flow of water. Botryococcus braunii is a potential source of renewable energy raw materials because this type of microalgae contains many hydrocarbon compounds. According to Samori et al. (2010), in the hydro cracking process of this type of microalgae, distillate can produce 67% gasoline, 15% aviation fuel, 15% diesel fuel and 3% residual oil. Fat content (fatty acids) of this type of microalgae is also quite high, ranging from 40 – 85%. In the process of photosynthesis, Botryococcus braunii also contributes significantly in reducing CO2 emissions, namely 1.5×10^5 tons y^{-1} per 8.4×10^3 in the microlaga cultivation area [6-8]. The compounds composition contained in Botryococcus braunii is shown in Table 2.

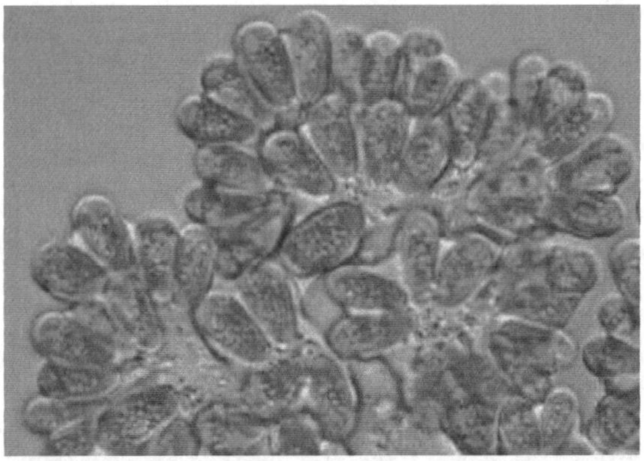

Figure 1. Microscopic of Botryococcus braunii [9]

Table 2. Compounds Composition in *Botryococcus braunii* [10]

Compound Type	Composition (%)
Chlorophyll a	0.4
Carotenoids	1.2
Protein	17.8
Carbohydrates	18.9
Lipid	61.4

Lipid are insoluble compounds in water of organic solvents such as chloroform hexane, toluene, and acetone, and lipids can also form combinations with other simple compounds such as triglycerides, and phospholipids [11]. Microalgae have a number of lipids with the same composition and even tend to be more than the other plants. The largest form of fat in microalgae cells is the triglycerides that can account for 80% of the total fat. This fat is the most important of the microalgae because it is the kind of oil that is good for producing biodiesel. Microalgae fats are usually esters composed of glycerol and fatty acids with a long chain of C14 to C22. The fatty acid composition of microalgae is monounsaturated fatty acids (MUFAs) and polyunsaturated fatty acids (PUFAs), such as palmitic acid (C16:0), palmitoelic acid (C16:1), stearic acid (C18:0), oleic acid (C18:1), linoleic acid (C18:2), as well as several other types of acids [12].
The purpose of this study is as a preliminary research to determine the profile and composition of fatty acids contained in microalgae *Botryococcus braunii* and the relationship between the identification of fatty acids and the potential of this type of microalgae as a source of renewable energy, i.e. biodiesel. Beside the direct research in laboratorium, researcher also adding another data as reference materials, some data obtained from some literature that discusses fatty acids *Botryococcus braunii* were implemented.

2. Material and Methods

The study was started with cultivation, extraction and transesterification process until biodiesel were obtained. The data for the cultivation and harvesting process of microalgae were obtained from direct observation conducted for ± 2 months in the laboratory of Process Unit of Chemical Engineering Department of Sriwijaya State Polytechnic, and the data of fatty acids were obtained by doing GC-MS analysis done in chemistry laboratory of materials processing of Bandung Institute of Technology. As supporting data of the main data obtained from the results of the study, the researcher used some literature study from some previous researchers. For the literature used as supporting data in this paper can be seen in Table 3.

Table 3. Literature studies on Fatty Acids Profile of *Botryococcus braunii*

Ref.	Lead Author	Description
[8]	Rao	Focused on the cultivation of *B. braunii* in two different media: raceway ponds and circular ponds from which both cultivation mediums obtained different composition of *B. raunii* fatty acids.
[13]	Ramaraj	Evaluated the feasibility of biodiesel production directly from *B. Braunii* biomass at laboratory scale achieved through direct transesterification process.
[14]	Diraman	Analyzed the fatty acid profiles of two types of microalgae namely *B.braunii* and *P.cruentum* from which it was known that *B. braunii* had potentiality as a source of biodiesel feedstock and *P.cruentum* was essential for fish feed.
[15]	Custódio	Analyzed the profile and fatty acid composition of *B.braunii* and *N.oculata* from which it was found out that the fatty acid composition which was a biodiesel fatty acid in *B.braunii* was higher than *N.oculata*

The study phase was begun with the cultivation of *Botryococcus braunii* microalgae seedlings obtained from the Balai Budidaya Perikanan dan Air Payau (BBPAP) in the area of Situbondo, East Java as much as 100 ml.

This microalgae cultivation was carried out in a culture bottle attached to the aerator with a duration of 10 days cultivation, and the cultivation medium used in the form of seawater with 25 – 35 ppt salinity of 500 ml. The medium of nutrition used in this cultivation is the walne's medium as much as 1 ml on each cultivation. Lighting was done using a 20 watt TL lamp with a bright dark method with a 12 hour timeframe (12 light hours:12 hours of dark). Further harvesting and cell density calculations were performed using a hemocytometer on which the amount of the cell was calculated using equation (1) [16].

$$\text{Number of cells per box} = \left(\frac{\text{number of calculated cells}}{\text{number of boxes}}\right) \times 25 \times 10^4 \quad (1)$$

The second stage after cultivation was the process of taking lipids from microalgae that have been cultivated or known as the extraction process. Extraction method used was soxhlet method, that is extraction method which is done by heating and using solvent. The solvent used was 175 ml of n-hexane to extract 10 grams of *Botryococcus braunii*. This extraction process lasted for 4 hours in order to produce optimal lipids [16]. Furthermore, the obtained lipid was weighed and calculated in quantity by using equation (2).

$$\%\text{yield} = \frac{\text{mass of extracted lipid}}{\text{initial sample mass}} \times 100\% \quad (2)$$

The last step conducted in this study was the making of biodiesel or FAME *(Fatty Acid Methyl Ester)* by using transeterification process with solvent in the form of methanol. The process was carried out for 2 hours until the two layers of triglycerides were formed on the bottom layer and FAME on the top layer. After that, several analyses of FAME were produced, one of which was fatty acid component analysis in FAME made from *Botryococcus braunii* by using GC-MS (Gas Chromatography-Mass Spectrometer) tool which was done in chemistry laboratory of material processing of Bandung Institute of Technology.

3. Results and Discussion

One of the parameters that could be considered as the reason in selecting the species of microalgae as the raw material of renewable energy, particularly biodiesel, is the lipid quantity resulted from the extraction process and the fatty acids composition owned by those microalgae. For this reason, this preliminary study was conducted in order to know the potential of *Botryococcus braunii* species of microalgae to be converted into biodiesel.

Results

This study was conducted through 3 stages, namely cultivation stage, extraction stage and FAME production stage. Then, several analyses and calculations were also done, the calculation of cell density amount, the calculation of resulted-lipid quantity, and the analysis of fatty acids. For the calculation of the amount of cell density and lipid quantity, the data were directly obtained from the conducted study. Meanwhile, for the analysis of fatty acids contained in *Botryococcus braunii*, the data were obtained from two sources which were the data of the study and the companion data from the existing literature (Table 3).

In the cultivation process done by the researchers, the value of cell density was observed by using microscope and hemocytometer chamber. The calculation of the amount of cell density was done in order to know the amount of microalgae cell developing during the growth period. By the duration of cultivation for 10 days, the highest amount of cell density was obtained on the 7th day of cultivation which was 5.208×10^4 cells/ml. The calculated data of the amount of the cultivated *Botryococcus braunii* cell density could be seen in Figure 2.

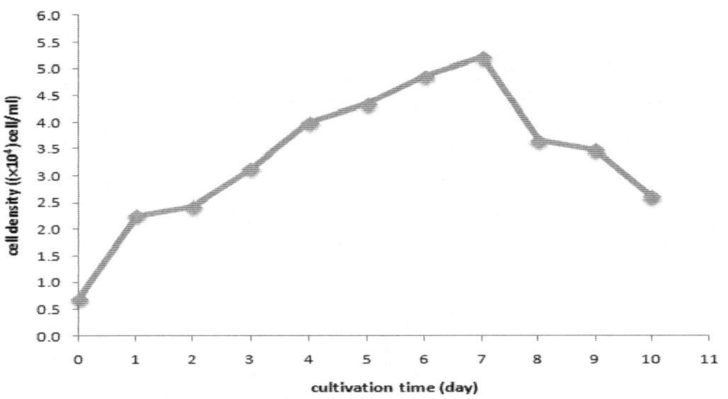

Figure 2. The Graph of *Botryococcus braunii* Cell Density

In the stage of extraction, the quantity of lipid rendemen, the extraction result, was calculated or % yield. % yield is a comparison between product mass and initial raw material mass [16]. The lipid rendemen resulted from extraction is colored as bright yellow and not too smelly and slightly thick. The lipid % yield resulted from soxhlet extraction was 24% with 175 ml of n-hexane solvent.

After the extraction process was complete, the third stage, producting FAME, was done using transesterification method. The resulted FAME was bright yellow with a volume of ± 10 ml. From the resulted FAME, the composition of fatty acids from microalgae used as raw material that was *Botryococcus braunii* could be analyzed by using GC-MS instrument. Based on the results of the analysis, 3 types of fatty acid compounds in FAME produced, which was a compound contained in biodiesel, palmitic acid, stearic acid, oleic acid and linoleic acid, could be detected [12]. The graph of GC-MS analysis result from FAME and the composition of each fatty acid detected in the GC-MS analysis could be seen in Figure 3 and Table 4.

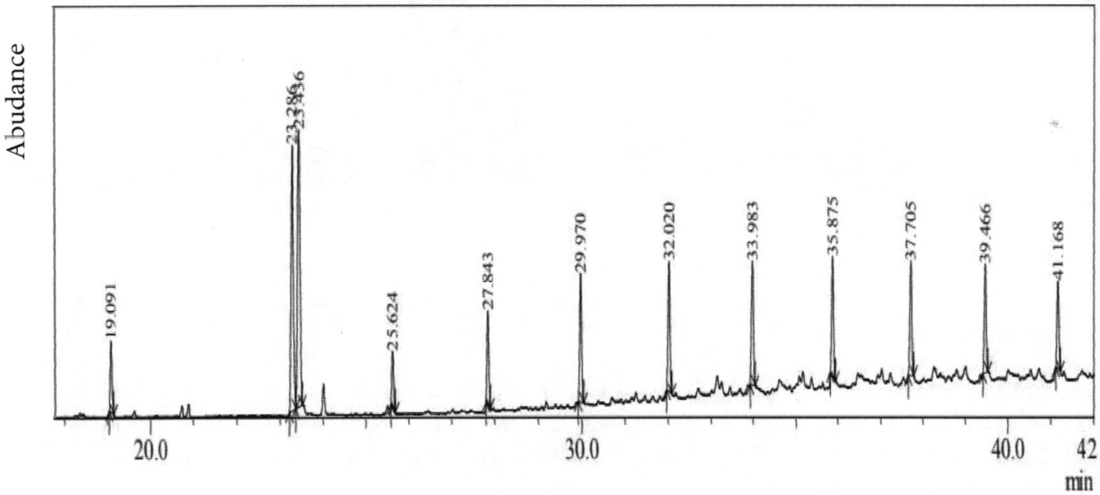

Figure 3. The Graph of GC-MS analysis from FAME in Direct Study

Table 4. Profile of *Botryococcus braunii* Fatty Acids in Direct Study

Fatty Acid	Retention Time	Area (%)	Height (%)
Palmitic acid (C16:0)	19.091	3.59	4.03
Oleic acid (C18:1)	23.286	16.91	15.65
Linoleic acid (C18:2)	23.436	20.66	15.01

As a companion data, the researchers also conducted a literature study on *Botryococcus braunii* fatty acids from the previous studies. There were 4 literatures used as the companion data (Table 3). The research data explained that there were 4 types of fatty acids contained in *Botryococcus braunii* that were biodiesel content. The composition of those four types of fatty acids in the studies could be seen in Table 5 and the comparison of fatty acid data from the results of direct study and previous study could be seen in Figure 4.

Table 5. Fatty Acids of *Botryococcus braunii* in Several Previous Studies

Fatty Acid	Composition (%)				
	Ramaraj	Diraman	Rao[a]	Rao[b]	Custódio
C16:0	11.66	12.55	16.52	22.13	18.05
C18:0	5.85	15.41	8.21	5.19	0.84
C18:1	45.4	35.47	34.23	28.35	35.85
C18:2	3.05	4.32	12.26	16.24	23.15

[a]cultivation in raceway ponds
[b]cultivation in circular ponds

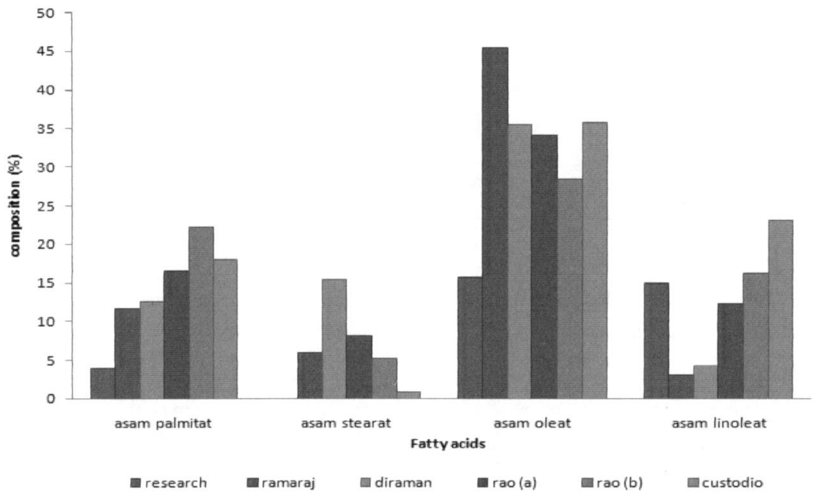

Figure 4. The Graph of *Botryococcus braunii* Fatty Acids Composition from Direct Study and Literature Study

Based on the existing fatty acids data, it was known that the range of fatty acid composition of *Botryococcus braunii*, especially for 4 types of fatty acid which was the biodiesel content was 4.03 – 22.13% for palmitic acid, 0.84 – 15.41% was stearic acid, then for oleic acid ranged between 15.65 – 45.4% and linoleic acid ranged from 3.05 – 23.15%.

Discussion

The growth phase of microalgae is generally divided into five phases of growth, namely lag phase, exponential phase, phase of growth rate lowering, stationary phase and phase of death [17]. From the research results, it was known that in the lag phase, the number of *Botryococcus braunii* cells were 2.257×10^4 cells/ml. Lag phase is an early phase of growth or adaptation of microalgae to the culture medium and nutrient medium used.

The highest growth phase of *Botryococcus braunii* occurred on the 7th day of cultivation with a cell number 5.028×10^4 cells/ml. This highest growth phase is called the exponential phase in which the cells divide rapidly and there is an increase in the number of cells. Based on the growth graph of *Botryococcus braunii* (Figure 2), it could be seen that the increase of *Botryococcus braunii* cells number occured from the 2nd day until the 6th day of cultivation, yet the number of cells peaked on the 7th day. This happened because the microalgae cells cultivated have been in stable culture conditions.

After the peak increase or exponential phase, the microalgae cells would have the lowering of growth rate until there would be a stop of growth and the microalgae would undergo the phase with the lowest number of cells on the 10th day of cultivation. This phase is called as the death phase. The phase of death occurs as the stock of nutrients in the culture medium has decreased or discharged that causes microalgae cells in culture medium can no longer develop and slowly die until the number of cells in the culture medium becomes constant. The number of cells on the 10th day of cultivation was 2.604×10^4 cells/ml.

After the cultivation, microalgae were harvested and dried for later being extracted using the soxhlet method. Soxhlet extraction is usually used to extract the compound with limited solubility in a solvent and the impurities are insoluble in that solvent [18]. In this study, the compound of n-hexane was used as the solvent. The selection of this solvent was due to the nature of fat which is water-insoluble or polar solvent but soluble only in non-polar solvents and was also because the type of microalgae used in this study would be easily extracted by using n-hexane.

The resulted lipid rendemen from soxhlet extraction using n-hexane of 175 ml had a yield % of 24%. The resulted rendemen was quite high considering that the sample of *Botryococcus braunii* extracted was only 10 grams. The volume of solvent used also affected the lipid produced with 175 ml of solvent that 45 solvent cycles were occured. Once the solvent is added then there will be more cycle happened and the resulted lipid rendemen will also increase.

Furthermore, the obtained lipid was converted to FAME, then the obtained FAME was analyzed to find out the profile and composition of the fatty acids contained there that whether the fatty acids of *Botryococcus braunii* was a component of biodiesel fatty acid or not could be identified. The analysis was performed using GC-MS instrument. From the analysis conducted directly, 3 types of fatty acids which were the components of biodiesel fatty acids, namely palmitic acid, oleic acid and linoleic acid were obtained, while the stearic acid was not detected by the instrument. This may be due to excessive sample storage time and poor storage mode allowing for the occurrence of oxidation in the sample before the sample arrived at the analysis site.

In the study conducted by Rao et al., (2012), two cultivations with different culture media, raceway ponds and circular ponds, in outdoor conditions were done. The total lipid produced by *Botryococcus braunii* was about 24% (w/w) with the dominant component of palmitic acid and oleic acid, both in cultivation using raceway and circular ponds [8].

The fatty acid profile contained in the microalgae oil was related to the biodiesel produced. Unsaturated fatty acids had a lower liquid point compared to saturated fatty acids that had good flowing ability at low temperatures. The opposite occured with saturated fatty acids that had a high melting point so that at low temperatures, it tends to not form a liquid or become a gel. According to Hu et al. (2008) and Chinnasamy et al. (2010), saturated fatty acids will tend to be at low temperatures but produce biodiesel with high oxidative stability [18-19].

According to Ramaraj et al. (2016), *Botryococcus braunii* is included to the microalgae which have good potential once being converted into biodiesel. It can be known from the high amount of lipid produced by *Botryococcus* and fatty acid profile owned by *Botryococcus braunii* was a component of biodiesel fatty acids as well. In the study conducted, it could be seen that the components of fatty acids dominantly owned by *Botryococcus braunii* were oleic acid, followed by palmitic acid, linoleic acid and stearic acid [13].

Lipids with high oleic acid content are known to have a good balance to the characteristic of fuel. The higher oleic acid in the fuel can increase oxidative stability for longer storage. Besides, *Botryococcus* has saturated and saturated fatty acids that are well balanced so that the biodiesel produced can still flow at low temperatures but have high oxidation values as well [13-14].

In a study conducted by Custódio et al. (2014) on fatty acids in *Botryococcus braunii* and *Nannochloropsis oculata,* it was also known that the fatty acid composition which was the content of biodiesel fatty acids in *Botryococcus braunii* was higher than the one in *Nannochloropsis oculata* and *Botryococcus braunii* was more potential to be converted to biodiesel compared to *Nannochloropsis oculata* for fatty acids in *Nannochloropsis oculata* were more dominated by saturated fatty acids [15].

4. Conclusion

Based on the study and literature study conducted, it could be known that *Botryococcus braunii* had a growth with a pretty high amount of cell density and could result plentiful lipid rendemen that had a good potential once being converted to biodiesel. Besides, some fatty acids components contained in *Botryococcus braunii* were the fatty acids components of biodiesel and the content of saturated and unsaturated fatty acids of *Botryococcus braunii* which were balanced that could be considered to be the raw material of biodiesel. The high composition of oleic acid in *Botryococcus braunii* could also be the basis in getting *Botryococcus braunii* as the raw material of biodiesel.

Acknowledgments
Thanks to all people who give me support for writing this paper, also thanks to reviewers who make this paper will be better and accept this paper

References

[1] Lardon L, Helias A, Sialve B, Steyer J P and Bernard O 2009 Atmospheric Chemistry and Physics Discussions 43 6475.
[2] Crutzen P J, Mosier A R, Smith K A and Winiwarter W 2007 Atmospheric Chemistry and Physics Discussions 7 11191.
[3] Mata T M, Martins A A and Caetano N S 2010 Renewable and Sustainable Energy Reviews 14 217.
[4] Huntley M E and Redalje D G 2007. Mitigation and Adaptation Strategies for Global Change 12 573.
[5] Chisti Y 2007 Biotechnology Advances 26 126.
[6] Samori C, Torr C, Samori G, Fabbri D, Galletti P, Guerrini F, Pistocchi R and Tagliavini E 2010 Bioresource Technology 101 3274.
[7] Weiss T L , Chun H J, Okada S, Vitha S, Holzenburg A, Laane J and Devarenne T P 2010 J. Biol. Chem 285 32458.
[8] Rao A R, Ravishankar G A and Sarada R 2012 Bioresource Technology. Elsevier Ltd 123 528
[9] Dayananda C, Sarada R, Komar V and Ravishankar G A 2007 Electronic Journal of Biotechnology 10 80.
[10] Saputro B R, Kusdiyantini E and Pancasakti H 2015 Jurnal Sains dan Matematika 23 94.
[11] Sween Mc, Harry Y, Richardson JR SM and Uhle ME 2003 Geochemistry: Pathways and Processes 2nd ed. (New York: Colombia University Press)
[12] Katili V R A 2012 Komposisi Asam Lemak Mikroalga Jenis Skeletonema costatum, Thalassiosira sp., dan Chaetoceros gracilis (Skripsi) (Bogor-IPB).
[13] Ramaraj R, Kawaree R and Unpaprom Y 2016 Emer Life Sci Res 2 (1) 7
[14] Diraman H and Koru E 2009 Fatty acid profiles of some microalgae, Porphyridium cruentum and Botryococcus braunii from Turkey (Turkey)
[15] Custódio L, Soares F, Pereira H, Rodrigues M J, Barreira L, Rauter A P, Albericio F and Varela J 2015 Journal of Applied Phycology 27 839.
[16] Kalsum L, Sari I P and Silvia M 2016 Proceedings of the International Conference FIRST (Palembang) p B6
[17] Rini I S 2008 Pengaruh Konsentrasi Limbah Cair Tahu Terhadap Pertumbuhan dan Kadar Lipid Chlorella sp. (Skripsi) (Malang – UIN Maulana Malik Ibrahim)
[18] Melwita E, Fatmawati and Oktaviani S 2014 Jurnal Teknik Kimia 20 20
[19] Hu Q, Sommerfeld M, Jarvis E, Ghirardi M, Posewitz M, Seibert M and Darzins A 2008 Plant Journal 54 621
[20] Chinnasamy, Ashish, Bhatnagar, Ryan, Hunt W and Das K C 2010 J. Biortech. 101 3097

Textile Dyes Removal by ZSM-5 from Bangka Kaolin

Iryani, A[1] and Hartanto, D[2]

[1]Chemistry Departement, FMIPA UNPAK Bogor, Indonesia
[2]Chemistry Departement, FIA ITS Surabaya, Indonesia

[2]E-mail : djokohar023@gmail.com; djokohar@its.ac.id

Abstract. The ZSM-5- adsorption capacity, which is synthesized directly from Bangka kaolin, has been thoroughly to remove the basic dye in a textile waste solution. The Langmuir and Freundlich isotherm models were used to analyze and measure the results of experimental adsorption data from the red adsorption of Congo Red (CR), Auramin (Au), Methylen Blue (MB), Azin (Az), Rodamin B (RB) on ZSM-5 at a temperature of 40 ^{0}C. The ZSM-5 adsorption capacity of Bangka kaolin showed dye absorption : Congo Red, Auramin, Methylen Blue, Azin and Rodamin B respectively were 129,157, 134, 205 and 210 mg/g. Therefore, ZM-5 of Bangka kaolin is respectively good adsorbent to remove the textile base dye waste.

Keyword : Bangka-kaolin-ZSM-5, Textile-basic-dyes, Adsorption, Removal-dyes-wastewater

1. Introduction

Organic dyes and pigments are widely used in industry to dye their production materials, and produce waste [1]. The industry not only produces dye waste but is mixed with other materials such as dispersants, salts, emulsifiers and heavy metals. This mixture makes the dyestuff waste stable in aqous waste [2] and is generally considered to be highly toxic to aquatic systems or ponds. Even some of these dyestuff wastes are reported to cause a fatal effect on living and human bodies that intentionally or unintentionally consume [3].

Many dyes used in the textile industry, some of which are often used in are Congo Red for red dye, Auramin for yellow dye, methylen blue for blue, Azin for orange, Rodamin B for blue color. This dyes can be grouped dyes of diazo and xanthat compounds, this group dye is a water soluble compound with very high solubility (above 3.5 g / 100 mL). These dye compounds as crystals are stable in the form of acid or anionic and cationic salts [4]. The dye also causes significant toxicity to humans and animals that can cause burning eyes, irritation of the skin, gastrointestinal tract and respiratory tract. Based on the above analysis, the dye waste treatment is very important before being discharged to water bodies such as rivers or ponds [5], this is due to high solubility and its stability so that it is difficult to be processed in a simple way.

The usual treatment of the above textile dyestuffs is difficult to treat in order to obtain acceptable concentrations according to the existing rules for dye waste allowed to be fed into rivers or lakes. Special treatment should be used, usually by physics and chemistry methods, it is still very expensive and complicated to be widely used primarily for developing countries. Several commonly used processes are sonochemically degradation [6], photochemistry [7], electrochemistry [8], coagulation and flocculation [9], membrane separation [10], adsorption [11] and oxidation or ozonation [12, 13]. Some of these processes are an option which needs to find an appropriate economic process for removing the textile dyestuff. The adsorption process is a common and appropriate process to be developed into textile dyestuff processing and this process of recent years has attracted great attention. The application of several applications of cheap adsorbents for the processing of dyes and the removal of metal ions has been reviewed and offered [14]. Interestingly, the adsorption method has been commercially proven to be more economically efficient and feasible to remove textile dyes from wastewater.

In this study, ZSM-5 synthesis was prepared using Kaolin Bangka raw material without activation method with calcination which means the direct process of kaolin without using the template, the result is used as adsorbent for dyestuff waste, previously the result of ZSM-5 is characterized by scanning electron

IOP Conf. Series: Journal of Physics: Conf. Series **1095** (2018) 012011 doi:10.1088/1742-6596/1095/1/012011

microscopy (SEM), FTIR spectroscopy, BET surface area and XRD and further study of adsorption, kinetics and thermodynamic.

2. Material and Methods

2.1 Adsorbent preparation
ZSM-5 was synthesized directly from Bangka kaolin without pretreatment and without using organic template. The synthesis procedure of ZSM-5 was adopted from Hartanto, D., et al [15, 16], taking the recipe for Si/Al molar ratio was 60.

2.2 Adsorbate preparation
The dyes materials were used in this experiments are The dye Congo red [C.I name = Direct Red 28, chemical formula weight = 696.65, λmax-498nm]; Auramin; Methylen Blue; Azin and Rhodamine-B [C.I. name = Basic violet 10, chemical formula weight = 479.02, λmax-543nm]. The stock dyes solution have concentration are 500 mg/L, that was prepared by dissolving 0.5 g of dye powder in 1000 ml of distilled water. Different dilutions ware made any ranging from 60 to 250 mg/L, were then prepared using distilled water.

2.3 Adsorption Isotherm
The relationship between the amount of adsorbate absorbed by the concentration of adsorbate in solution at equilibrium and temperature conditions can be expressed by adsorption isotherm. In isothermal adsorption, in that circumstance the amount of adsorbate (molecule adsorbed) is a function of pressure (if its molecule in the form of a gas) or as a function of concentration at a constant temperature state.

The isotherm adsorption process can be understood through isotherms produced between the contact time (20-180 minutes) and the amount of adsorbate absorbed by the adsorbent mass unit at constant temperature (298 K). In the results of this study used experimental data analysis using Langmuir isotherm adsorption model, Freundlich,. The applicability and suitability of the isotherm equation to the equili-brium data were compared using the values of the correlation coefficients, R^2.

3. Results and Discussion

3.1 ZSM-5 from Bangka Kaolin
ZSM-5 synthesized from Bangka Kaolin and It were characterized using X-ray diffraction (XRD) with a Philips X'Pert Powder Cu Kα radiation source (λ = 1.54 Å) in the 2θ between 5-40 °[17]. Characteristic of XRD is show at fig 1, that representative samples of kaolin (a) and sample ZSM-5 result of synthesis with Bangka kaolin (b). Figure 1(a) shows the peak at 2θ about 12, 20-27 and 35-40 °. Base on Fig 1a. Kaolinite is the main content of the kaolin with it have height peak XRD pattern [18]. Fig 1.b did not indicate kaolin content again and that only have ZSM-5 crystal phase, it show difractogram peaks only XRD patter of ZSM-5 synthesized And the height 2θ pattern showed sharp peaks with high intensity. Therefore, crystallinity of ZSM-5 have height crystallinity.

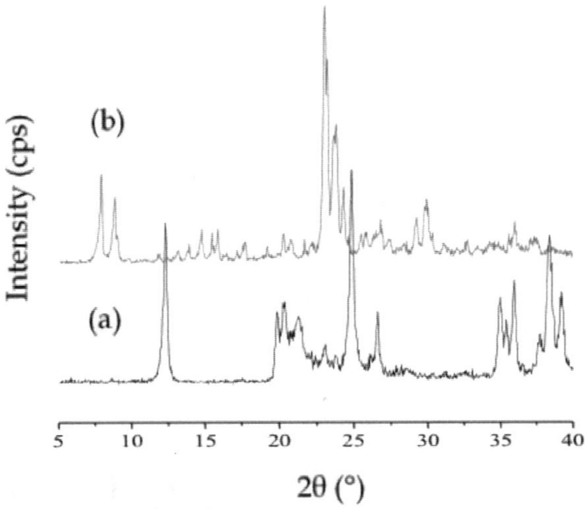

Figure 1. Difraktogram (a) kaolin and (b). ZSM-5 synthesis from Bangka

The surface area of ZSM-5 synthetic from Bangka kaolin calculation uses the BET (Brunauer, Emmet and Teller) calculation model with the assumption that the gas forms an infinite number of layers above a surface [19, 20]. Based on calculations using the BET model, the surface area of the catalyst sample is obtained. The result of isothermal adsorption of nitrogen gas at temperature 77K for determination of surface area of one zeolite with ZSM-5 Si/Al 50 shown that surface area a 333 m2/g and a pore diameter of 5.72 A and 20-28 A Figure 2.

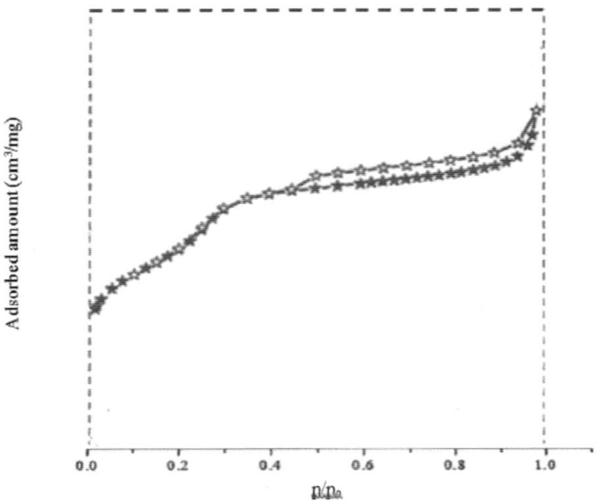

Figure 2. Pore and surface area ZSM-5 synthesis from Bangka kaolin

The surface area of ZSM-5 calculation uses the BET (Brunauer, Emmet and Teller) calculation model with the assumption that the gas forms an infinite number of layers above a surface [21, 22]. Based on calculations using the BET model, the surface area of the catalyst sample is obtained. The result of isothermal adsorption of nitrogen gas at temperature 77K for determination of surface area of one zeolite with ZSM-5 Si/Al 50 shown that surface area a 333 m2/g and a pore diameter of 5.72 A and 20-23 A figure 2.

3.2 Batch Adsorption Studies

Effect of Initial dye concentration, loading catalyst and effect time contact. In order to determine the optimum initial concentration of the target dye, its value was varied (for both dye under study) in the range 50-150 mg/L, while keeping the temperature, adsorbent mass and the contact time constant at 303 K, 0.010 g and 180 min respectively. The effect of initial concentration on the removal of CR, Au, MB, Az and RB by the adsorbent is indicated in Table 1. It is evident from the table that percentage of CR, Au, MB, Az and RB removal decreases with increase in dye concentrations; however actual amount of the dye adsorbed is increased. This is due to increase in dyes concentration, surface area and active sites of the adsorbent were saturated and hence percentage removal decreases [23, 24].

The effect of contact time on the removal of CR Au, MB, Az and RB by the adsorbent is indicated in Figure 3. It is indicated that uptake of the dye is rapid in the beginning and then it becomes constant. The adsorption curves are single, smooth, and continuous leading to saturation and indicate the possible monolayer coverage on the surface of adsorbents by the molecules. In fig3. It give figure of height capacity of absorption adsorpt varying dyes diluted in water [25, 26, 27].

Table 1. The effect of initial concentration and loading of adsorbent

Parameter	Absorption Efficiency by intial concentration (%)			Amount of Absorbent (mg/g)			Absorption Capacity mg/g
Initial Concentration	50 mg/l	100 mg/l	150 mg/l	20 mg	30 mg	40 mg	Average
CR	98.2	84.0	65.3	194.5	120.0	194.5	129
Au	78.6	82.5	84.0	125	215	193	157
MB	57.0	59.1	59.5	85.0	235	285	134
Az	56.0	55.2	58.1	183	235	221	205
RB	50.0	59.5	58.0	150.0	197.5	435	210

IOP Conf. Series: Journal of Physics: Conf. Series **1095** (2018) 012011 doi:10.1088/1742-6596/1095/1/012011

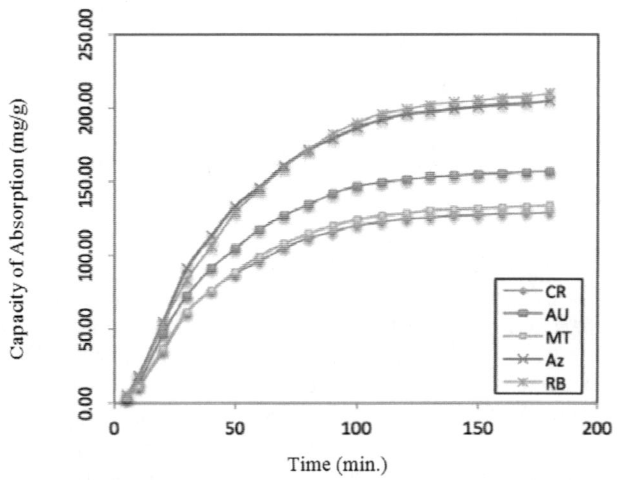

Figure 3. The Influence Time Absorption and Absorption Capacity

Table 2. Different Adsorption Isotherm model parameters for the adsorption of CR, Au, MB, Az and RB on ZSM-5

Methods	Parameters	CR	Au	MB	Az	RB
Langmuir Isotherm	R^2	0.993	0.9971	0.9944	0.9984	0.9982
	K_L	0.009	0.007	0.008	0.005	0.004
	Q_m(mg/g)	129.201	157.21	134.21	205.33	210.21
Freundlich Isotherm	R^2	0.949	0.973	0,954	0.9873	0.9883
	K_f	13.6×10^4	4.9×10^6	8.1×10^4	8.1×10^4	4.6×10^6
	n	0.48	0.56	0.54	0.54	0.54

Thermodynamic parameters are important in adsorption studies; they provide a better understanding of the effect of temperature on the adsorption process. The thermodynamic parameters are calculation and the negative values of ΔH° obtained in the adsorption of both CR, Au, MB, Az and RB dye onto ZSM-5 signify that the adsorption process was exothermic in nature. It can be proven from the results of research conducted showed CR, Au, MB, Az and RB adsorption process using ZSM-5 decreased with increasing temperature adsorption [28, 29].

The ΔH° value was calculated to be -188,201 kJ/mol (CR), -132 kJ/mol (Au), -144kJ/mol (MB), -161kJ/mol (Az) and -161.414 kJ/mol (RB). In the other hand, the positive values of ΔS° indicate that increase in randomness occurred at solid-solution interface during the adsorption process. This indirectly shows the affinity of adsorbent toward dye molecules [30,31]. ΔG° values were negative at all temperatures studied, inferring that the adsorption was spontaneous in nature. While a higher negative value indicates that the process is energetically preferred over other adsorption mechanisms. Since the values of ΔG° decreased with increasing temperature, it suggests that at higher temperature, the driving force was less, resulting in lower adsorption uptake [32, 33].

4. Conclusion

The ZSM-5 adsorption capacity of Bangka kaolin showed dye absorption: Congo Red, Auramin, Methylen Blue, Azin and Rodamin B respectively were 129, 157, 134, 205 and 210 mg/g, the adsorption phenomena are allow Langmuir model. Therefore, ZSM-5 of Bangka kaolin is a relatively good adsorbent to remove the textile base dye waste.

5. Acknowledgments
The E-mail acknowledges This work was funded by the Ministry of Research, Technology and Higher Education. The authors would like to thank Professor Hadi Nur which support this research and instrumental analysis.

References
[1] Bagane M and Guiza S 2000 *Ann. Chim. Sci. Mater.* **25** 615.
[2] Imamura K, Ikeda E, Nagayasu T, Sakiyama T and Nakanishi K 2002 *J. Colloid Interface Sci.* **245** 50
[3] El-Geundi M 1997 *Adsorp. Sci. Technol.* **15** 777
[4] McKay G 1992 *J. Chem. Technol. Biotechnol.* **32** 759.
[5] Dai M 1998 *J. Colloid Interface Sci.* **198** 6.
[6] Gemeay A H, El-Sherbiny A S and Zaki A 2002 *J. Colloid Interface Sci.* **245** 116.
[7] Abbasi M N and Asl R 2008 *J. Hazard. Mater.* **153** 942.
[8] Gupta V.K., Mittal A., Jain R., Mathur M., Sikarwar S., 2007 *J. Colloid. Interface Sci.* **309** 460.
[9] Fan L, Zhou Y, Yang W, Chen G, Yang F 2008 Electrochemical degradation of aqueous solution of Amaranth azo dye on ACF under potentiostatic model, Dyes Pigments, **76** 440.
[10] Zonoozi M.H., Moghaddam M.R.A., Arami M. 2009 *Water Sci. Technol.* **59** 1343.
[11] Sachdeva S and Kumar A 2009 *Membr J. Sci.* **329** 2.
[12] Tan I A W, Ahmad A L and Hameed B H 2008 *J. Hazard. Mater.* **154** 337.
[13] Malik P K and Saha S K 2003 *Sep. Purif. Technol.* **31** 241.
[14] Holzheu, S. and Hoffmann, H. 2002 *J. Colloid Interface Sci.* **245** 16
[15] Prasetyoko D, Ayunanda N, Fansuri H and Hartanto D 2012 J. Math. Fund. Sci., 44 A (3) 250.
[16] Hartanto D, Saputro O, Utomo WP, Rosyidah A, Sugiarso D, Ersam T, Nur H and Prasetyoko D 2016 *Chemistry - An Asian Journal*, **28(1)** 211.
[17] Hartanto D, Yuan LS, Sari SM, Sugiarso D, Murwani IK and Ersam T 2016, *Malaysian Journal of Fundamental and Applied Sciences* **12** **(2)** 85
[18] Prasetyoko D, Ayunanda N, Fansuri H, Hartanto D and Ramli Z 2016 *Journal of Mathematical and Fundamental Sciences* **44** **(3)** 250.
[19] Ojedoku, Adedamola Titi n and Bello O S 2017Journal of Dispersion Science and Technology, **38(9)** 1285.
[20] Langmuir I 1918 *J. Am. Chem. Soc.* **40** 1361.
[21] Freundlich H M F 1906 *Uber die adsorption in losungen.Zeitschrift fur Physikalische Chemie* (Leipzig) **57A** 385.
[22] Dada A O, Olalekan A P, Olatunya A M and Dada O 2012. IOSR *J. Appl. Chem.* **3** 38.
[23] Aharoni C and Ungarish M 1977 *J. Chem. Soc., Faraday Trans* **73**. 456.
[24] Nisaa S 2011 *Adsorpsi biru metilena pada kaolin dan nanokomposit kaolin TiO2 serta uji sifat fotokatalis. [skripsi].* (Bogor - FMIPA IPB)
[25] Armaroli T, Simon LJ, Digne M, Montanari T, Bevilacqua M, Valtchev V, Patarin J and Busca G 2006. *Appl. Catal A Gen* **306** 78.
[26] Mohamed M, Fouad O A, Ismail A A and Ibrahim I A 2005 *In Materials Letters* **59** 27.
[27] Somani O G,Anant L Choudhari, Rao B S and Mirajkar S P 2006 *Journal of Materials Chemistry and Physics* **82(3)** 538.
[28] Adamson A.W 1990 *Physical Chemistry of Surface 5 th ed* (New York. John Willey &. Sons)
[29] Yahya S A, Musa I E, Amjad H E, and Gavin M W (2008). *Effect of solution pH, ionic strength, and temperature on adsorption behavior of reactive dyes on activated carbon Dyes Pigm* **77** 16.
[30] Gercel O, Ozcan A, Ozcan A S, and Gercel H F 2007 *Appl. Surf. Sci.* **253** 843.
[31] Wu F C, Tseng R I, and Jung R S 2001 **Water Res. 35** 613.
[32] Harni, Rudy M, Iryani A and Affandi H 2013 *Pemanfaatan serbuk gergaji kayu jati (tectona grandis l.f.) Sebagai adsorben logam timbal (pb).*
[33] Venkata S, Mohan N, Rao C, and Karthikeyan J 2002 *J. Hazard. Mater.* **90** 189.

X-ray Diffraction Analysis on Effect of Time Reaction and Alkali Concentration in Merlinoite

Eddy Heraldy[1,2], Syaiful Ahmad Nur Cahyo[1], Nurul Apri Indri[1], Fitria Rahmawati[1,2]

[1]Department of Chemistry, Mathematics and Natural Sciences Faculty, Sebelas Maret University, Jl.Ir. Sutami 36 A Kentingan, Surakarta, Indonesia 57126
[2]Solid State Chemistry & Catalysis Research Group, Chemistry Department, Sebelas Maret University, Jl.Ir. Sutami 36 A Kentingan, Surakarta, Indonesia 57126

E-mail : eheraldy@mipa.uns.ac.id

Abstract. This research objectives is to determine the effect of time reaction and alkali concentration on the synthesis of merlinoite. The synthesis of merlinoite was performed by hydrothermal method with variation of reaction time in 8, 12, 24 hours and various concentration at 1, 3, 5M. The quantitative X-Ray Diffraction (QXRD) was employed on product using Le Bail refinement method. Characterization XRD shows the highest crystallinity of the product was obtained at 3M of alkali concentration in 8 hours reaction time. The refinement result revealed three other phases such as quartz, mullite, and tobermorite.

Keywords : Fly ash, Le Bail refinement, Merlinoite, X-ray Diffraction

1. Introduction

Merlinoite is one type of zeolite that has small pores and its presence is very rare in nature [1,2,3,4]. According to Sherman [5], merlinoite has a good ability in ion exchange and adsorption capacity. Merlinoite, also known as zeolite W [6,7,8] is a potential material because it has been widely used as an adsorbent removal of pollutant in water [9], ion exchange, and slow-release potassium fertilizer [10]. In addition, merlinoite also can be used as catalyst through a process of dehydration of methanol [11].

Some researchers have developed a synthesis of merlinoite from various sources (raw material). Collela et al. [12] used natural glass for synthesis merlinoite. Strohmaier and Robson [13] reported that merlinoite have been made from alumina, colloidal silica, and potassium hydroxide. While Medina et al. [9] had succesfully converted fly ash to merlinoite. Moreover, Schubert and Husing [14] suggested the parameters that affect the synthesis of merlinoite include in time, alkali concentration, temperature, and molar ratio. Alkali concentration and reaction time has very influent in the formation of merlinoite and its cristallinity [11].

Determine of quantitative phase analysis from merlinoite can be used lattice parameter from standart X-ray diffraction (XRD). Quantitative phase analysis using XRD data could be further by approached with Rietveld refinement [15,16] and Le Bail method [17]. The aim of this paper is to discuss the crystal structure with the Le Bail method on the effect of time reaction and KOH concentration on synthesis of merlinoite from fly ash.

2. Material and methods

2.1 Material

Raw material that used in this synthesis is coal fly ash (CFA) which was taken from Tanjung Jati B CFPP, Jepara, Indonesia. The reagents KOH, H_2SO_4 98%, and aquadest was used as solvent.

2.2 Synthesis of merlinoite

Synthesis merlinoite was performed with alkaline hydrothermal method. To remove the unburned carbon and some impurity compounds, CFA was purified with H_2SO_4 solution. The mixture was soaked for 8 h at 60° C and then filtered, washed with distilled water and dried. After CFA was dried, then it mixed with 3M KOH solution into 150 mL Teflon liner of stainless-steel autoclave in comparison of composition between KOH : fly ash is 2:1. Furthermore, the activation is done by the time variation of 8, 12, and 24 h at 150°C in the oven. After a varying of time was completed, the autoclave was cooled to the room temperature and the solid phase is separated with their liquid phase. Solid product is washed with distilled water until pH 10-11. The solid was dried at 60°C for one night. The dried solids were characterized using X-Ray Diffraction (XRD) in order to note the optimum condition of the reaction time variations. The variation of alkali concentration can be conducted with the same step of the time reaction variation.

The crystallographic structure of merlinoite was studied using XRD Brucker AXS D8 Advance diffractometer with Cu Kα radiation. The diffraction intensity was determined between 10° and 60°. The crystal size of merlinoite was computed using the Scherrer formula $D = 0.9.\lambda/(\beta.\cos\theta)$. the product was proven by powder XRD (ICSD#81895 for merlinoite). Le Bail refinement was performed using the refinement program, i.e. Rietica. Experimental parameter refined were the instrument zero, scale factor, peak shape parameters u, v, w, γ_0 and γ_1 [15,16].

3. Result and Discussion

The characterization of fly ash was characterized by X-ray fluorescence (XRF) to determine the composition of chemical constituents in the form of metal oxides. SiO_2 and Al_2O_3 in this material can be use as source to form the framework of merlinoite (Table 1).

Table 1. Chemical composition (%) of fly ash by XRF

Compound	Composition (%)
Si	49.86
Al	21.71
Ca	6.76
Mg	5.83
K	2.52
Fe	2.34
Ti	1.33

Table 1 shown that Silica and Alumina (Si and Al) are the main component of CFA sample which is shows the similarity with zeolite. Thus it possible to converted CFA into merlinoite due to the CFA components were able to serve as a source of Al and Si.

Crystallinity's character from merlinoite were analyzed using XRD with a range of 2θ 10-60°. Diffractogram pattern from merlinoite at the time variation of 8, 12, and 24 hours is present in Fig. 1.

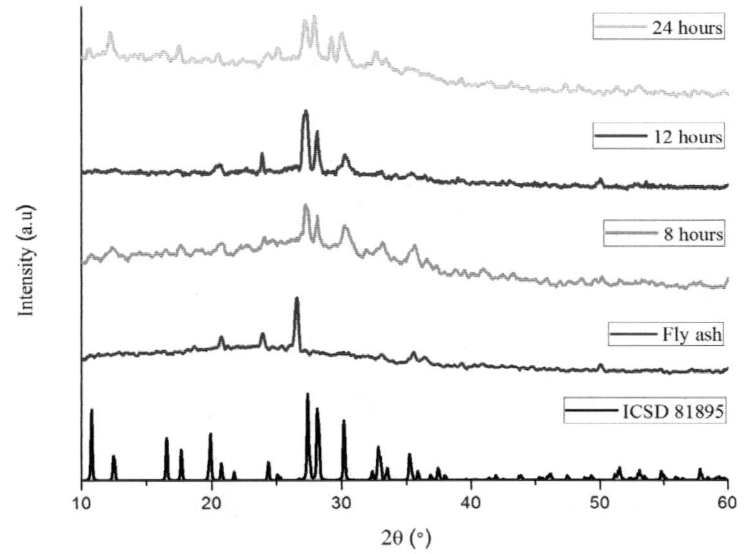

Figure 1. XRD pattern of merlinoite synthesized with variety reaction of time

Figure 1 showed the pattern of merlinoite synthesized from variety of time. As shown in Figure 1, the XRD pattern of merlinoite has sharp peak and typical in the range of 2θ = 27° (*013*); 28° (*240*); and 30° (*123*) [7]. However, the diffractogram showed that still quite amorphous and crystalline yet entirely. The differences in high or low intensity was expected any differences in the crystal growth of merlinoite.

Furthermore, to determine the conformity of XRD diffraction with the crystal structure based on comparison with ICSD, followed by refinement process using Le Bail method. Result of refinement process showed the conformity with the standard data of merlinoite (ICSD#81895). In addition, it found other phases such as quartz (ICSD#280364), mullite (ICSD#23726). Even, at 24 hours in time variation appear new phase like tobermorite (ICSD#100405). There is still any quartz because that phase is stable. Therefore, it was quite difficult to bind with KOH and affecting the formation of merlinoite [18].

Something that is not different showed in Figure 2 for XRD pattern of merlinoite in the variation of alkali concentration (KOH). In Figure 2 was shown that the higher concentration of alkali emerging a new phase like tobermorite (ICSD#100405). It phase was as same as phase at variation of reaction time in 24 hour.

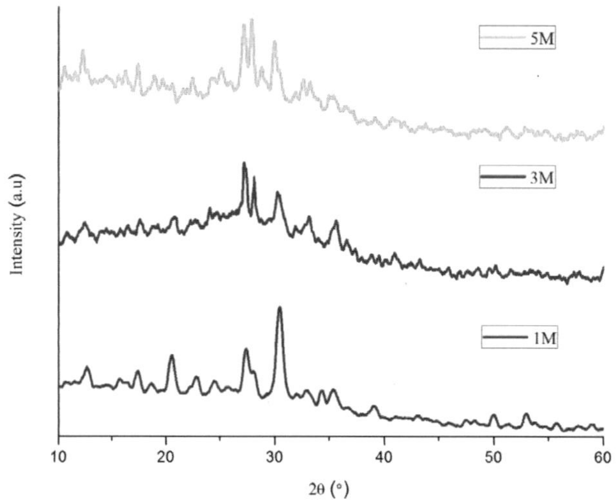

Figure 2. XRD pattern of merlinoite synthesized with variety alkali concentration

The refinement had the following order zero, lattice parameter, histogram parameter and instrumental peak shape. For every step were used 30 cycles. Table 2 presented the detail about result of refinement for merlinoite with variety treatment. The reliability index parameters are the pattern factor (*Rp*), the weighted pattern factor (*Rwp*), and the goodness of fit (χ^2), which is shown the quality of refinement in this study. The ideal value for χ^2 is 1.0 [19] or 3.0, and all the refinement were acceptable quality [19]. In Table 2, merlinoite with variation in 8h and 3M were shown smallest *Rp* and *Rwp* than other variations. It is indicated that optimum condition in synthesis of merlinoite. On the other hand, the higher reaction of time and alkali concentration would make the possibility to emerge a new phase [10].

Table 2. The result of Le Bail refinement with various phase

Variation	Phase composition				Value		
	merlinoite	quartz	mullite	tobermorite	*Rp*	*Rwp*	χ^2
8h	76.85	7.89	15.26		0.79	1.43	0.0573
12h	73.67	12.66	13.67		1.28	2.17	0.1423
24h	49.72		4.16	46.13	1.91	2.73	0.1847
1M	68.34	17.43	14.23		1.94	2.46	0.1828
3M	81.73	7.90	10.37		0.79	1.43	0.0573
5M	48.58	2.68	4.01	44.72	1.71	2.11	0.0955

Table 3 showed that the treatment of variation in synthesize of merlinoite made some difference in crystallite size (*D*). The diminished in crystallite size with the increasing of various concentration or time reaction may be explained that a merlinoite structure lead to form a small piece. Furthermore, it can be explained that there was an incomplete dissolution of early formed amorphous of merlinoite which contribute to the aggregation of that material.

Table 3. The crystallite size (*D*) of merlinoite with Debye-Scherer method

Variation	Scherer method (nm)
8h	5.324
12h	8.761
24h	3.882
1M	4.766
3M	5.372
5M	3.394

4. Conclusion

The end of result from this investigation proved that merlinoite was successfully synthesized from fly ash. The optimum conditions in synthesis are in 8h of reaction time and 3M of alkali concentration (KOH). The higher reaction time and alkali concentration has decreased the crystallite size of merlinoite. It is showed by crystal growth occurred in slow formation. The higher reaction time and alkali concentration also provided a possibility to emerge a new phase. On X-ray diffraction has got another phase beside merlinoite, such as quartz, mullite, and tobermorite, which is confirmed by analysis of refinement Le Bail.

References

[1] Passaglia E, Pongiluppi D, Rinaldi R and Merlinoite 1977 *N. Jb. Miner* 355.
[2] Alberti A 1979 *J. Appl. Crystallogr.***9**, 373.
[3] Khomyakov A.P., Kurova, T.A., Muravitskaya, T.N., 1981 *Doklady Akademiy Nauk SSSR*, **1** 172.
[4] Donahoe RJ and Liou JG 1984 *Clay. Clay Miner.* **32** 433.
[5] Sherman J D 1978 American Chemical Society, Washington, **40** 30.
[6] Sherman D 1977 *Amer. Chem. Sot.*, Symposium Series 40 30.
[7] Bieniok A, Bornhold K and Brendel U 1996 *J. Mater. Chem.* **6** (2) 271.
[8] Skofteland, B.M., Ellestad, O.H., Lillerud, K.P., 2001 *Micropor. Mesopor. Mat.*, **43**, 61.
[9] Medina A, Gamero P, Almanza J.M, Vargas A, Montoya A, Vargas GM and.Izquierdo M 2010 *J. Hazard. Mater.*, **181** 91.
[10] Jing L, Zhuang X, Font O, Moreno N, Vallejo VR, Querol X and Tobias A 2014 *J. Hazard. Mater.* **265** 242.
[11] Hou J, Yuan J and Shang R 2012 *Powder Technol.* **226** 222.
[12] Colella C, Caputo D, de Gennaro B and Torracca E 2004 *Stud. Surf. Sci. Catal.*, **154** (2) 1920.
[13] Strohmaier K. and Robson H 2001 *Verified Synthesis of Zeolitic Material*, second ed., (Netherlands – Elsevier)
[14] Schubert and Husing, 2000 *Micropor. Mesopor. Mat.*, **34** 301.
[15] Peterson V.K 2005 *Powder Diffr* **20** 14.
[16] Heraldy E, Nugrahaningtyas K D and Heriyanto 2016 *IOP Conf. Ser. Mater. Sci. Eng.*, **176**,
[17] Sanches E.A, Carolino ADS, Dos Santos A.L, Fernandes EGR, Triches DM and Mascarenhas YP 2015 *Adv. Mater. Sci. Eng.*
[18] Wu D 2008 *Fuel*, **87** 2194.
[19] Yakubovich, O V , Massa W , Pekov I V , Kucherinenko T V, 1999 *Crystallogr. Rep.*, **44** (5) 776.

The Efficacy of Noni Fruit Methanol Extract (Morinda citrofolia) to Brain Derived Neurotrophic Factor (BDNF) on Male Swiss Webster Mice Induced By Immobilization Stress

Eka Febri Zulissetiana [1] and Susilawati[2]

[1] Department of Physical Medicine and Rehabilitation, Faculty of Medicine, Sriwijaya University, Palembang, Indonesia.
[2] Department of Clinical Pathology, Faculty of Medicine, Sriwijaya University, Palembang, Indonesia.

E-mail : susilwt78@gmail.com

Abstract. Aging process is associated with declines in certain cognitive abilities such as learning and memory ability and impact on high risk of dementia, physical disability and death. Oxidative stress is believed as basic mechanism of aging process. *Morinda citrofolia* (Noni fruit) has long been used as a traditional plant in worldwide and was proven empirically in traditional medicine as antioxidant. This study was to determine the effect of noni fruit methanol extract on BDNF levels in white mice of Swiss Webster strain induced by immobilization stress. This research is experimental study with post test only control group design. Male Swiss Webster Mice were induced by immobilization stress and randomized into seven groups (5 mice/groups). The first group was negative control group; Group 2,3,4,5 and 6 was given treatment with varied concentrations of *Morinda citrofolia* (50, 100, 200, 400 and 800 mg/kgBW). Vitamin E (70 mg/kgBW), standart antioxidant was used as positive control (Group 7). Brain BDNF level of the white mice were measured by using ELISA method. Experiment result showed that treatment with *Morinda citrofolia* extract (50, 100, 200,400 mg/kgBW) and Vitamin E 70 mg/kg BW showed a sigificant differences in BDNF level compare with the negative control rats. There is no significant differences between 200 mg/kgBW Metanol Extract of *Morinda citrofolia* group and vitamin E (p=0,301;p>0,05). Metanol Extract of *Morinda citrofolia* could prevent the decrement of BDNF level due to stress.

Keywords : *Morinda citrofolia*, Noni fruit, BDNF, stress

1. Introduction

Aging process is a gradual and irreversible decrease in the physiological capacity of the body system to the extent of failure and death. Aging is a major cause of decreased brain function such as learning and memory processes (1,2). An increase of the elderly population along with the high life expectancy around the world causing cognitive decline caused by aging has become one of the serious problem in health care system. The decrease of cognitive function will have an impact on the high risk of dementia, physical disability and death (3,4). Aging process is believed to be mediated by oxidative stress. Oxidative stress is a redox state caused by an imbalance between the production and detoxification of reactive oxygen species (ROS) that can cause oxidative damage to lipids and proteins and decrease natural glutathione and antioxidant levels in the synapse of nerve cells (1,5,6).

Brain-derived neurotrophic factor (BDNF) is a neurotropin family that plays a role in the proliferation, differentiation and survival of nerve cells (7). BDNF also protects nerve cells from oxidative stress caused by hypoglycaemia, ischemia, hypoxia and ethanol toxicity (8). Stress, both acute and chronic, affects the synthesis of BDNF in the brain. Chronic stress decreases BDNF protein expression (9,10) whereas acute stress is quite the opposite (11,12).

Previous studies have shown that *Morinda citrofolia* could increase the levels of antioxidant enzymes (13), as well as inhibit the formation of malondialdehyde (MDA) and Nitrit Oxide (NO) (14). The extract of methanol and ethyl acetate of noni fruit can prevent oxidation of Low Density Lipoprotein (LDL) induced by Copper (15) and inhibit oxidation process due to lipid hyperxidation or superoxide anion (16).

Until now, many studies have been conducted on the antioxidative and antiinflammatory capabilities of noni fruit, so the pharmacological effects of noni fruit as antioxidant and antiinflammatory are also thought to be able to protect the nerve damage caused by oxidative stress so that the need to be studied and examined how the effect of giving extract methanol of *Morinda citrofolia*against BDNF levels in white mice of swiss strains of male Webster induced chronic immobilized stress.

2. Materials and Methods

2.1 Preparation of Extract

Morinda citrofolia fruits were collected from Tebedak, Ogan Ilir, South Sumatera Indonesia during the month of October 2016. The fruits were washed, blotted, dried and sliced into 100 mesh. Five hundred grams of dried fruit powder were macerated with 1000 ml of methanol for 24 hours, stirring regularly, filtered. The residual residue is macerated again 3 times so that all the substances contained in the mengkudu fruit is extracted. All the filtrate is then collected and evaporated in rotary evaporator.

2.2 Animal and Experimental Design

The research subjects were 35 of Mus Musculus, 10 to 12-week-old male Swiss Webster strain with the weight ranging from 25 to 35 grams, conserved in the Animal House of Medicine Faculty of Sriwijaya University. Each cage housed eight mice fed by food and drinks every day ad libitum. The conservation room was well ventilated with the maintained room temperature of 25 to 30 ^{o}C, humidity of 50 % to 60% and the cycle ratio of darkness to brightness of 12:12 hours.

Seven days after acclimatization, the subjects were then randomly divided into four groups with five mice for each. Group 1 was negative control group; group 2,3,4,5 and 6 was given treatment with varied concentrations of *Morinda citrofolia* (50, 100, 200, 400 and 800 mg/kgBW. Vitamin E (70 mg/kgBW), standart antioxidant was used as positive control (Group 7). All mice were put in a particular plastic canister, specially designed to ensure the mice would stay in dorsal recumbent position without access to food and drink. Such immobilized stress would be applied for two hours (10 am to 12 pm) for 21 consecutive days.

2.3 ELISA Assay

Mice had to be sacrificed 24 hours after the last training session by decapitation to obtain fresh tissues by decapitation procedure without anesthesia as outlined by Institutional Animal Care and Use Committee (IACUC). Brain was separated from the skull and brain stem manually with a scalpel blade and a micro tweezer, then was restored at -80^{0}C. The brain tissue sample was homogenized with blade homogenizer after being added with PBS. Next, the sample was centrifuged with the velocity of 3,000 rpm for 20 minutes in 4°C temperature. BDNF levels of brain tissue were measured by the ELISA method as described in manufacturer's instructions of ELISA kit

3. Results and Discussion

3.1 Result

The mean BDNF levels for all groups were between (174.8 ± 6.81) and (470.93 ± 10.9) pg/ml.

Table 1 The Efficacy of *Morinda citrofolia* Extract on BDNF Level (pg/ml)

Vitamin E	**464,53± 6,53**
Aquadest	174,8± 6,81[b]
***Morinda citrofolia* extract 50mg/kgBW**	411,40± 8,24[b]
***Morinda citrofolia* extract 100mg/kgBW**	413,67± 6,57[b]
***Morinda citrofolia* extract 200mg/kgBW**	470,93± 10,9[a]
***Morinda citrofolia* extract 400mg/kgBW**	339,67± 8,57[b]
***Morinda citrofolia* extract 800mg/kgBW**	234,73± 15,94[b]

Independent T Test, a : p > 0,05 vs vitamin A. b : p < 0,05 vs vitamin A; Significance level was determined by one way ANOVA followed by LSD pos-hoc test.

Table 1 showed BDNF Level as negatif control and subjected to *Morinda citrofolia* extract (50,100,200,400,800 mg/kgBW) and Vitamin E. Treatment either extract and Vitamin E significantly increased BDNF Level (p < 0,05) compare with negative control. But, there was no differences between Vitamin E group and *Morinda citrofolia* extract 200mg/kgBW groups in BDNF Level after treatment (p = 0,301) p > 0,05).

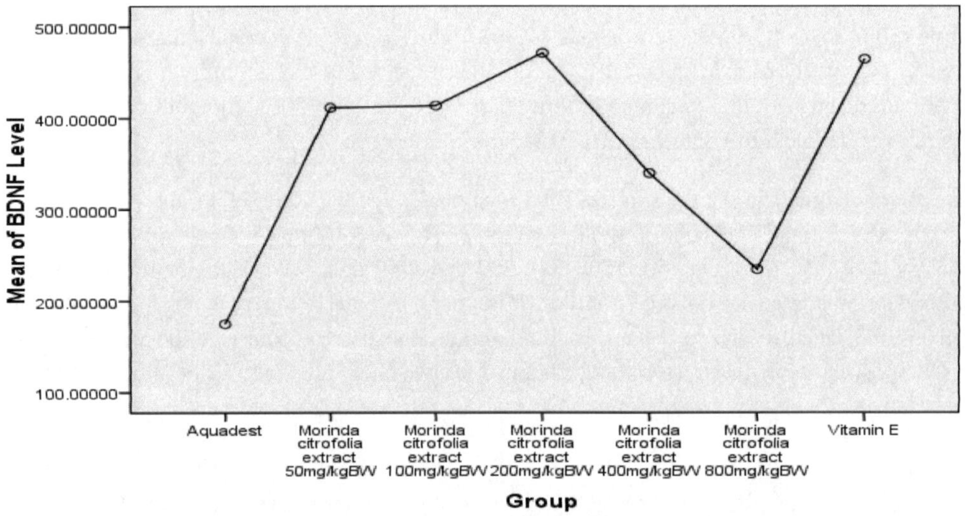

Figure 1 Average Graph of BDNF Level (pg/ml)

3.2 Disscussion

As the main organ that responds to stress, brain undergoes structural and chemical changes (17). It is a form of brain adaptation (allostasis) in maintaining homeostasis. Although the brain is able to adapt, extreme and repetitive stress exposure can cause brain plasticity abnormalities, while lowering the ability to control stress (allostasis overload). Some forms of brain plasticity include neurogenesis (18), dendrite remodeling (19,20) and synaptic reinforcement (21).

The glucocorticoid hormone is the last effector of the HPA axis that plays an important role in stress management (22). The interactions of glucocorticoids hormones and their receptors affect the expression of some of the genes involved in cell metabolism, the structure and transmission of synapses, such as enzymes, neuropeptides, growth factors, and cell adhesion molecules (23). One candidate who plays an important role in brain plasticity is the brain-derived neurotrophic factor (BDNF) (24). Glucocorticoid hormone has been shown to influence the expression of the BDNF gene (25).

BDNF expression in the brain is strongly influenced by the development of phylogenetic brain, type, intensity and duration of stress, as well as the rhythm of the glucocorticoid hormone in the blood circulation (25). In addition, the role of epigenetic factors, such as CREB as transcription factor, methyl-CpG-binding protein 2 (MeCP2) as transcriptional repressor, Ca2 + ions and other intracellular signaling cascades also affect BDNF gene expression (26,27).

In this study, the cerebroprotective effect of of Noni fruit methanol extract on BDNF was clearly demonstrated using stress immobilization model mice. The results of this study showed that the extract of 200mg / kgBW dose of Noni fruit methanol extract showed the highest increase of BDNF, but the increase of dose of methanol extract to 400 mg/kgBW or 800mg/kgBW did not show better improvement than the 200mg/kgBW dose. It was concluded that the effect of Noni fruit methanol extract of 200 mg / kg BW can significantly increase BDNF levels compared to vitamin E.and methanol extract of noni fruit on BDNF concentration in this research is flat dose.

The efficacy of Noni fruit (*Morinda citrofolia*) as a neuroprotector against cell damage caused by oxidative stress has long been proven. It is mediated by the active chemicals contained in noni fruit, including enzymatic antioxidant (Superoxide dismutase, catalase, peroxidase, glutathione reductase, glutathione-s-transferase) and non enzymatic antioxidant (ascorbic acid, alpha tocopherol, vitamine A, flavonoids, phenols and tannins (28). Studies have also reported that Noni fruit extract could also pass the blood brain barrier (29).

Research conducted by Muralidharan *et al.* (2010) mentioned that giving ethyl acetate fraction of noni fruit in mice with beta-amyloid induced cognitive dysfunction model at dose of 400 mg/kg/day could increased serotonin, dopamine and antioxidant enzyme serum levels (30). According to Pauchauri *et al.* (2012) giving ethanol extract of noni fruit and ethyl acetate fraction in mice enhanced memory and brain blood flow and decrease oxidative stress and acetylcholinesterase activity (31). The same result is also conveyed by Hussin *et al.* (2007) who mentioned that giving ethanol extract of noni fruit could inhibit lipid peroxidation where it is characterized by decrease serum MDA level and increased activity of antioxidant enzyme (32).

4. Conclusion

Noni fruit extract of 200mg/kgBW dose was effective in increasing BDNF levels in white mice of Swiss Webster strain induced by immobilized stress which the effect of noni fruits extract on BDNF levels was flat dose.

Acknowledgements
This study was supported by the research grant No 01714/IT2.11/PN.08/2017 funded by the Faculty of Medicine, Sriwijaya University, Palembang, Indonesia

References :

[1] ReferenHansakul P 2010 *Thammasat Medical Journal*, **10(3)** 311.
[2] Muto I, Oda T and Sudo N 2012 *Macroeconomic impact of population aging in japan: A perspective from an overlapping generations model.*
[3] Williams KN and Kemper S 2010 *Interventions to reduce cognitive decline in aging. Journal of psychosocial nursing and mental health services.* **48(5)** 42.
[4] Deary I J, Corley J, Gow A J, Harris S E, Houlihan L, Marioni RE, Penke L, Rafnsson S B and Starr JM, 2009. *British medical bulletin*, **92(1)** 135.
[5] Azman A, Koller J and Plesnicar B 1979 *Journal of the American Chemical Society*, **101(5)** 1107
[6] Wang X, Wan, W, Li L, Perry G, Lee HG and Zhu X 2014. B*iochimica et Biophysica Acta (BBA)-Molecular Basis of Disease* **1842(8)** 1240.
[7] Bekinschtein P, Cammarota M, Katche C, Slipczuk L, Rossato J I, Goldin A, Izquierdo I and Medina JH, 2008. *Proceedings of the National Academy of Sciences* **105(7)** 2711.
[8] Sun M, Kong L, Wang X, Lu XG, Gao Q and Geller A I 2005 *Brain research* **1052(2)** 119.
[9] Grønli J, Bramham C, Murison R, Kanhema T, Fiske E, Bjorvatn B, Ursin R and Portas CM 2006. *Pharmacology Biochemistry and Behavior* **85(4)** 842
[10] Shi S S, Shao S H, Yuan B P, Pan F and Li Z L 2010. *Yonsei medical journal*, **51(5)** 661.
[11] Molteni R, Calabrese F, Bedogni F, Tongiorgi E, Fumagalli F, Racagni G. and Riva MA, 2006 *The The International Journal of Neuropsychopharmacology* **9(3)** 307.
[12] Bland S T, Schmid MJ, Der-Avakian A, Watkins L R., Spencer R L and Maier S F 2005. *Brain Research* **1051(1)** 90.
[13] Kumar KT, Panda DS, Nanda UN and Khuntia S 2010 *Int. J. PharmTech Res*, *2*(**2)** 1030
[14] Hussain S, Tamizhselvi R, George L and Manickam V 2016 *International journal of stem cells*, **9(2)** 22.
[15] Kamiya K, Tanaka Y, Endang H, Umar M and Satake T 2004 *Journal of agricultural and food chemistry* **52(19)** 5843
[16] Wang X and Michaelis E K 2010. *Frontiers in aging neuroscience*, *2.*
[17] Radley J J and Morrison J H 2005. *Ageing research reviews*, **4(2)** 271.
[18] Joëls M, Karst H, Krugers H J and Lucassen P J 2007 *Frontiers in neuroendocrinology* **28(2)** 72.
[19] Radley J J, Sisti H M, Hao J, Rocher A, McCall T, Hof P R, McEwen B S and Morrison J H 2004. *Neuroscience*, **125(1)** 1
[20] Goldwater D S, Pavlides C, Hunter RG, Bloss E B, Hof P R, McEwen B S and Morrison JH 2009. *Neuroscience*, **164(2)** 798.
[21] Sapolsky R M 1999 *Experimental gerontology* **34(6)** 721
[22] Tsigos C and Chrousos G P 2002 *Journal of psychosomatic research* **53(4)** 865.
[23] De Kloet E R, Joëls M and Holsboer F 2005. *Nature reviews. Neuroscience*, **6(6)** 463.
[24] Santos AR, Comprido D and Duarte CB 2010. *Progress in neurobiology*, **92(4)** 505-.
[25] Suri D and Vaidya VA 2013. *Neuroscience* **239** 196.

IOP Conf. Series: Journal of Physics: Conf. Series **1095** (2018) 012013 doi:10.1088/1742-6596/1095/1/012013

[26] Roth TL, Zoladz PR, Sweatt JD and Diamond DM 2011 *Journal of psychiatric research*, **45(7)** 919.
[27] 27. Karpova NN, 2014 *Neuropharmacology* **76** 709.
[28] 28. Srinivasahan V and Durairaj, B.R.I.N.D.H.A 2014 *International Journal of Pharmacy and Pharmaceutical Sciences*, **6(4)** 55.
[29] Vanisree AJ 2012, Technical Bulletin - 09, *World Noni Research Foundation, Chennai and University of Madras, Chennai, Tamil Nadu, India.* **36**
[30] Muralidharan P and Srikanth 2010 *Journal of Chemistry* **7(2)** 612.
[31] Pachauri SD, Tota S, Khandelwal K, Verma PR, Nath C, Hanif K, Shukla R, Saxena JK, Dwivedi AK 2012. *Journal of ethnopharmacology*. **139(1)** 34.
[32] Hussin, M., Abdul-Hamid, A., *Food Chemistry*, **100(2)**, 535-541.

IOP Conf. Series: Journal of Physics: Conf. Series **1095** (2018) 012014 doi:10.1088/1742-6596/1095/1/012014

Optimization Of The Enzymatic Ammonolysis Of Alkanolamide From Ketapang Kernel Oil

Erin Ryantin Gunawan[1], Dedy Suhendra[1], Trisnasari1[1], Lely Kurniawati[1]

[1]Department of Chemistry, Faculty of Mathematics and Science, University of Mataram ,Jl. Majapahit No. 62 Mataram, 83125, Indonesia

Email: erinryantin@unram.ac.id

Abstract. The enzymatic ammonolysis reaction of alkanolamide using *Terminalia catappa* L. (local language: Ketapang) kernel oil and etanolamine as substrates has been conducted. It's commonly synthesized from derivative of petroleum and natural gas. In this research, the raw material used was ketapang seed oil, the renewable and non-edible oil. The effects of various reaction parameters such as reaction time, temperature, amount of enzyme, and molar ratio of substrates were investigated. The optimum reaction conditions obtained by a one-step lipase catalyzed reaction were ratio of ketapang seed oil and etanolamine of 1 : 12.5, amount of lipase (Lipozyme) of 0.1 g, temperature at 40°C for 2 h, and hexane as a solvent. The percentage yields of alkanolamide obtained at these optimum ammonolysis reaction conditions was 60.07 %.

Key words: alkanolamide, ketapang kernel oil, enzymatic ammonolysis

1. Introduction

Alkanolamide or fatty acid alkanolamide are widely used as a surfactant, formulations on cosmetic and pharmaceutical ingredients. Commonly, fatty acid alkanolamides can be produced by transamidation of the fatty methyl ester with monoethanolamide at high temperature and pressure. However, it can also lead side reactions.[1,2] Recent studies show that the fatty acid alkanolamides can be produced enzymatically by lipase-catalyzed transesterification.[2,3] The enzymatic synthesis offers mild reaction conditions and environmentally friendly process.[4] Usually, the basic material to synthesize alkanolamide is pure fatty acids (uneconomist price) or edible oils (unsability competition) [2,3,5,6]. In this research, natural non-edible oils (usability noncompetition) used to sustainable source of fatty acids, which represent raw materials that can replace pure fatty acids or edible oils. The natural non-edible oil such as *Terminalia catappa L.* Kernel Oil or ketapang kernel oil (local language) has been used to synthesize alkanolamide via enzymatic ammonolysis reaction.

Previous research showed that the yield of crude oil from ketapang kernel is 50,40-58.40%. Fatty acid content of Ketapang kernel oil is palmitic acid (25-30%), oleic acid (35-55%), linoleic acid (12-16%) and stearic acid (4-7%). [7,8] Thus, this show that economically, ketapang kernel oil can be as a source of fatty acid and it can converted into alkanolamide. For that, it required optimization step to produce the optimal product. The effects of various parameters on ammonolysis reaction were investigated.

2. **Materials and Methods**

2.1 Materials

Materials that used are ketapang kernel oil, lipozyme TL.IM was produced by Novo Nordisk (Denmark), Ethanolamine, fatty acids alkanolamide standards and n-hexane were obtained from Sigma Aldrich (USA). All other chemicals were analytical grade. The reactions were analyzed by a gas chromatograph (Shimadzu GCMS-QP2010) RTx-65TG capillary column (30 m x 0,25 mm). Helium was used as the carrier gas at a flow rate 30 ml/min. The temperature was programmed at 1 min at 160°C, 10°C/min to 280°C was held for 25 min.[5] Detector used was FID (Flame Ionization Detector) at 300°C. FT-IR Spectrophotometer from Perkin Elmer Model Frontier.

2.2 Methods

2.2.1 Synthesis and purification of alkanolamide. Ketapang kernel oil that has been extracted from the fruit is used as a source of fatty acids. The reaction mixture consisted of ketapang kernel oil, ethanolamine and lipase/lipozyme and hexane as a solvent. The reaction was incubated in a horizontal water bath shaker (150 rpm) at different variables. The product was isolated and then purified by crystallization method at 4°C. The product was collected by filtration, washed with amount volume of hexane and dried in a vacuum desicator until constant. The pure product of alkanolamide was quantitated gravimetrically.

2.2.2 *Effect of Reaction Time.* A series of reactions catalyzed by 0.1 g Lipozyme IM, incubated at 40°C at 150 rpm at various reactions periods (1,2,3,4 and 5 h) were studied to determine the effect of reaction time. After completion of each assigned period, the alkanolamide were isolated and purified. The product was quantitated as described earlier.

2.2.3 Effect of Ratio Substrate. This effect was carried out using different mole ratios of the reactants amount of the lipase and other parameters were kept constant. The amount of ketapang kernel oil at 2 g and the amount of ethanolamine were varied from 5 to 30.0 mmol. The reaction mixtures were incubated at 40°C for 5 h at 150 rpm. The amount of the enzyme used was 0.1 g. The product was quantitated as described earlier.

2.2.4 Effect of Amount of Enzyme. The reaction mixtures containing different quantities (0.05 to 0.25 g) of enzyme were incubated at 40°C for 5 h at 150 rpm. The amount of ketapang kernel oil (2 g) and ethanolamine (10 mmol) were kept constant. The product was quantitated as described earlier.

2.2.5 Effect of Temperature. To study the effect of reaction temperature, another series of samples were incubated in a water bath shaker at 150 rpm for 5 h and different reaction temperatures (30, 40, 50, 60 and 70°C). The amount of enzyme used was 0.1 g. The product was quantitated as described earlier.

2.2.6 Analysis of the product. The products of the reactions were examined by thin layer chromatography (TLC) on precoated silica gels plate (60F, Merck, Darmstadt, Germany) and developed in hexane/diethyleter (87:13, v/v). Identification was made by comparison with known standards. The analysis of standards and products formed were analysed by a gas chromatograph (Shimadzu) RTx-65TG capillary column (30 m x 0,25 mm). Helium was used as the carrier gas at a flow rate 30 ml/min. The temperature was programmed at 1 min at 160°C, 10°C/min to 280°C was held for 25 min.[5] Detector used was FID (Flame Ionization Detector) at 300°C. Infrared (IR) analysis of the products was carried out using Perkin Elmer Fourier-transform Infrared spectrophotometer. Amount of total nitrogen in alkanolamide was determined by Kjedahl method.

3. Results and Discussion

3.1 Effect of Reaction Time

The results (Fig. 1) show that enzymatic reaction, the reaction proceeded in 1 h. Although the maximum yield is obtained at the reaction period at around 2 h, then slows down up to 5 h. The decrease in the reaction rate may be due to some mass-transfer limitations, which inevitably arise in a reaction mixture containing a high proportion of solid product and the reaction reaches the equilibrium state where the rate of forward reaction was equal to the backward reaction.[9,10] However, this result is very faster compared with the result of Rahman et al,[2]which shows that the synthesis of alkanolamide from palm kernel oil using free enzyme for *Candida rugose* and the maximum conversion at 36 h. This may be because the enzymes used in this study are immobilized enzymes. Immobilized enzymes typically have greater thermal and operational stability than free enzyme.[11]

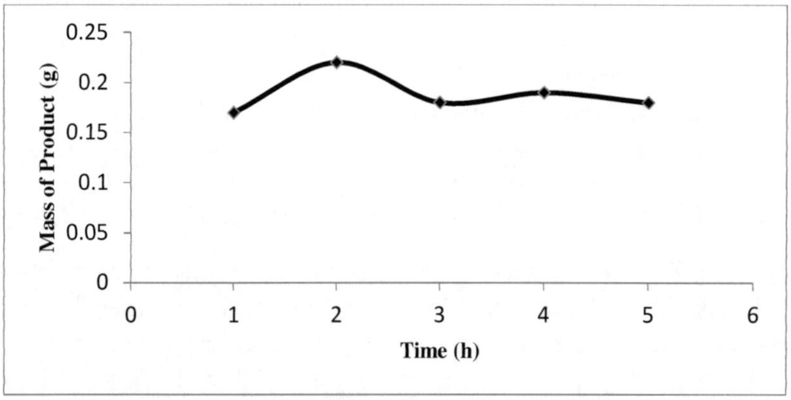

Figure 1 Effect of Reaction Time

3.2. Effect of Ratio Substrate

The effect of ratio substrate on the amonolysis reaction indicated competitive nature of ethanolamine and fatty acids (in ketapang kernel oil) binding. Fig. 2 shows that the highest result when the ratio of the ketapang kernel oil and the ethanolamine was at 2 and 25 mmol or 1:12.5 mmol. This result is in agreement with Wang et al[6] that shows the reaction yield increases with the increase in the concentration of ethanolamine. The high concentratios of ethanolamine will facilitate the formation bond of carbon and nitrogen.[12]

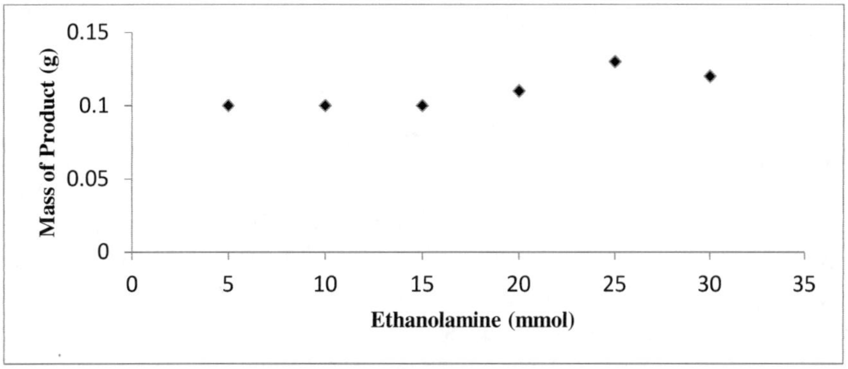

Figure 2 Effect of Ratio Substrate

3.3 Effect of Enzyme Amount

Fig. 3 depict the result of using different amount of enzyme. The weight of alkanolamide was increased with increasing amount of enzyme from 0,05 to 0.2 g. Further increase of lipase to substrate ratio does not significantly increase the yield. This behavior of leveling off of the reaction at higher enzyme concentrations can be explained by considering that the active sites of the enzyme molecules present in excess would not be exposed to the substrates and remain inside the bulk of enzyme particles without contributing significantly to the reaction.[13]

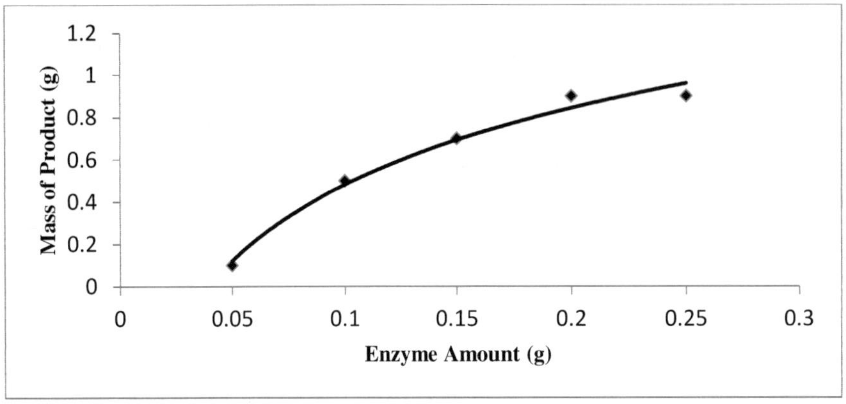

Figure 3. Effect of Enzyme Amount

3.4. Effect of Temperature

The ammonolysis reactions showed an increment of the product as the temperature was increased from 30°C to 40°C as shown in Fig. 4. At temperature above 40°C, the product was decreased and slightly decreased again which may cause by denaturing of the lipase. Similar result was reported by some research.[2,3,6] Garcia et. al.[14] reported when temperature above 50°C, the activity of energy for enzymatic reaction was low.

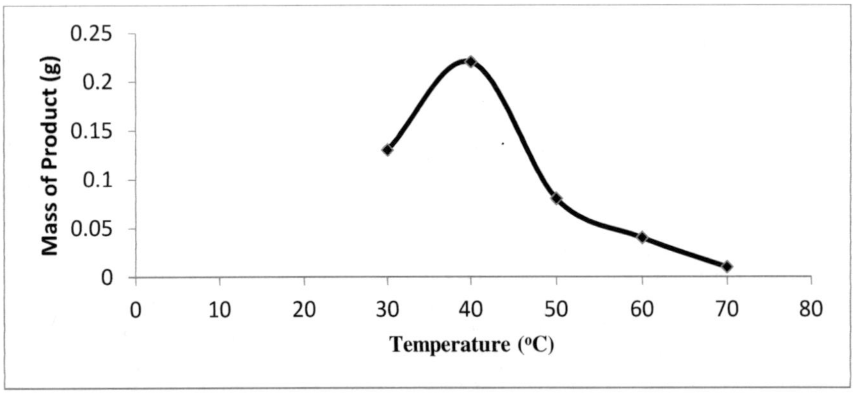

Figure 4. Effect of Temperature

3.5 Characterization of Products

The enzymatic ammonolysis reaction of alkanolamide was ascertained by TLC. Rf value of the standar and the product are 0.97 and 0.95 respectively with eluen hexane/diethyleter (87:13). The characteristic peaks of IR spectrum bands at 3190 cm^{-1} and 1751-1715 cm^{-1} for stretching N-H and C=O. Similar result of characteristic peaks was reported by Adewuyi *et. al.*[15] Characterization of the fatty acids alkanolamide using GC-MS showed similar resemblance of the compound,

Palmitoylethanolamide and oleylethanolamide with respective retention time of the compounds was 11.9821 and 12.565 minutes.

3.6 Ammonolysis Reaction at Optimum Conditions

The enzymatic ammonolysis reaction of alkanolamide was carried out at optimum conditions (Reaction time of 2h, temperature at 40°C, amount of lipozyme of 0.1 g, and ratio of substrate, ketapang kernel oil/ethanolamine, 1:12,5). The percentage yields of alkanolamide obtained at these optimum ammonolysis reaction conditions was 60.07 %.

4. Conclusion

The optimum conditions of the enzymatic ammonolysis reaction of alkanolamide using Terminalia catappa L have been successfully investigated by a one-step reaction to produce higher yields. The GC-MS spectrum indicated that the type of alkanolamide formed is oleylethanolamide and palmitoylethanolamide

Acknowledgements

This project was financed by a grant from Directorate General of Research and Development Strengthening, The Ministry of Research Technology and Higher Education Republic of Indonesia, Scheme "Unggulan Perguruan Tinggi 2017".

References

[1] Bilyk A,. Piazza G.J,. Bistline R.G. Fearhailler, S.H, Haas M.J 1992 *J. Am. Oil chem. Soc.*, **69** **(5)**: 488
[2] Rahman MB, Yap CL, Dzulkefly K, Rahman RNZA, Salleh AB and Basri M 2003 *J. Oleo Sci.*, **52** (**2**) 65.
[3] Salleh AB, Basri M, Tan SW, Rahman MBA, Dzulkefly K, Rahman RNZ and Razak CNA 2001 *Malay. J. of Anal. Sci.*, **7** (**2**) : 281.
[4] Gunawan, E.R.. Basri M and Suhendra D 2011 *J. Naurt. Ind.*,**14 (1):** 37-41
[5] Patil H.V, Kulkarni R.D and Mishra S 2008 Synthesis and Characterization of Ethanolamides from Natural Renewable Resources and Utilization in Surfactants Synthesis, School of Chemical Technology, North Maharashtra University, Jalgaon
[6] Wang X. Cen Y. Jin Q. Wang X 2013 *J. Oleo. Sci.***62** (**6**): 427.
[7] Feairheller SH, Bistline RG, Bilyk A, Dudley RL, Kozempel MF and Haas MJ 1994 *J Am Oil Chem Soc* **71 (8)** :863.
[8] Suhendra D, Gunawan ER. Handayani SS, Ardianto T and Andriyany RS 2010 *Proceeding of Internasiona lSeminar On Economic, Culture and Environment* Mataram, Indonesia, 68.
[9] Gunawan ER. Suhendra D, Nurita AD, Komalasari D 2017 *Asian Journal of Chemistry* **29 (10)** 44
[10] Suhendra D, Wan Yunus WMZ, Haron MJ, Basri M and Silong S 2005 *J. of Oleo Sci.* **54** (**1**) 33.
[11] Wu H, Liang Y, Shi Jiafu J, Wang X, Yang D and Jiang Z 2013 *Mater. Sci. Eng.* C. **33** (**3**) 1438.
[12] Kitaguchi H, Fitzpartick DA, Huser JE and Klibanov AM1999 *Am. Chem. Soc.*, **111**(**8**): 3094.
[13] Krishna, S.H, Divakar S, Prapulla SG and Karanth NG 2001 *J. Biotech.*, **87 (3):**193.
[14] Garcia T, Sanchez N, Martinez M and Aracil J 1999 *Enzyme Microb. Technol.* **25** (**7**): 591.
[15] Adewuyi A, Oderinde A R, Rao B V S K and Prasad R B N 2012 *J. Surfact Deterg* **15** (**1**): 89.

Abilities of Co-Cultures of White-Rot Fungus *Ganoderma lingzhi* and Bacteria *Bacillus subtilis on* Biodegradation DDT

Erly Grizca Boelan[1] and Adi Setyo Purnomo*[2]

[1] Department of Chemistry Education, Faculty of Teacher Training and Education Science, Universitas Katolik Widya Mandira, Kupang, Indonesia.
2 Department of Chemistry, Faculty of Natrual Science, Institut Teknologi Sepuluh Nopember, Surabaya, Indonesia.

adi_setyo@chem.its.ac.id

Abstract. The insecticide 1,1,1-trichloro-2,2-bis(4-chlorophenyl)ethane (DDT) has been used extensively since the 1940s for controlling agricultural pest. Despite a band on DDT use in most industrialized since 1972, DDT and its related residues (DDTr) still persist in the environmental and pose animal as worst as human health risk. In this study, abilities of co-cultures of white-rot fungus *Ganoderma lingzhi* and *Bacillus subtilis* on biodegradation of DDT was investigated. *B. subtilis* at various volumes of 1, 3, 5, 7 and 10 mL (1 mL ≈ 6.7 x 10^8 CFU) were mixed into 10 mL of *G. lingzhi* culture for degrading DDT during a 7-days incubation period. The addition of 10 mL of *B. subtilis* showed the highest DDT degradation about 82.30% during 7-days incubation period. DDD (1,1-dichloro-2,2-bis(4-chlorophenyl)ethane) and DDMU (1-chloro-2,2-bis(4-chlorophenyl)ethylene) were detected as metabolite products. This study indicated that co-cultures of *G. lingzhi* and *B. subtilis* can be used for degradation of DDT.

1. Introduction

1,1,1-trichloro-2,2-bis(4-chlorophenyl)ethane (DDT) is an organochlorine pesticide that was extensively used for agricultural pest and control insect borne diseases such as malaria and typhus since World War II because low price (1,2). DDT and related compounds are highly hydrophobic and almost insoluble in water. These traits give rise to accumulation of DDT through the food chain. Although DDT had good capability as pesticides, in 1970 DDT was banned in most countries because their persistence (low biodegradability), bioaccumulation, carcinogenicity and negative impact on wildlife (2). Even though DDT has been banned, its residue still detectable in environmental samples and living organisms all over the world. DDT residues are lipophilic therefore high levels of DDT and its metabolites have been detected in human adipose tissues, blood plasma, liver, brain, placenta and even in breast milk along the food chain (2,3). DDT residues have been shown to persist in the environment predominantly in the form of DDT, 1,1-dichloro-2,2-bis-(4-chlorophenyl)ethylene (DDE) and 1,1-dichloro-2,2-bis(4-chlorophenyl)ethane. Because of negative impact of DDT, a method is needed to eliminate or reduction the residual of DDT and its metabolite products.

Biodegradation is one option treatment method to destroy or render harmless various contaminant using microorganisms (fungi and bacteria), because effective, economical and environment-friendly. There have been many studies about the ability bacteria and fungi to transform or degrade pollutants. White-rot fungi (WRF) have been studied extensively for their ability to transform or degrade a wide spectrum of hazardous chemical pollutants found in environment. *Ganoderma lingzhi* (before 2012 known as *Ganoderma lucidum*) is one of the most important and widely distributed WRF in the world and is associated with degradation of a wide variety of woods (4). Coelhoe *et al* (5,6) reported that *G. lingzhi* can degrade herbicide bentazon in solid and liquid state cultures approximately 90% and 55%

in 10 days incubation period respectively. Besides, *G. lingzhi* also have ability to degrade trichloroethylene and polycyclic aromatic hydrocarbon (PAHs) (7,8).

One kind of method to optimize the degradation of DDT by the fungus is with the addition of bacteria (9,10). Co-cultures of bacteria and fungus would be able to enhance the degradation of DDT with shor incubation period. Purnomo *et al* (9) reported DDT was degraded approximately 86% during 7 days incubation period by co-cultures of *Pleurotus ostreatus* and *Pseudomonas aeruginosa.* Sariwati *et al* (10) reported co-cultures of brown-rot fungus *Fomitopsis pinicola* and *B. subtilis* can degraded DDT about 86% during 7 days incubation. In this study, the optimization degradation of DDT by *G. lingzhi* with addition of bacteria *B. subtilis* was investigated.

2. Materials and Methods

2.1 Chemicals

DDT and Pyrene were purchased from Tokyo Chemical Industry Co. Methanol, sodium sulfate anhydrous and dimethyl sulfoxide (DMSO) were purchased from Merck Milipore (Darmstadt, Germany). *n*-hexane and acetone were purchased from Anhui Fulltime Specialized Solvent and Reagent Co., Ltd (Anhui, China).

2.2 Culture Conditions

White-rot fungus *G. lingzhi* BMC9057 were grown on 9 cm diameter potato dextrose agar (PDA; Merck Darmstadt, Germany) plates and incubated statically at 30 ^{o}C. The mycelia (1cm diameter) from agar plate were inoculated into 10 mL of a liquid medium potato dextrose broth (PDB; Merck, Darmstadt, Germany) in 100 mL Erlenmeyer flasks and then pre-incubated at 30 ^{o}C for 7 days (9-11).

Stock cultures of *B. subtilis* NRBC3009 were grown on 9 cm diameter nutrient agar (NA, Merck, Darmstadt, Germany) plates and incubated at 37 ^{o}C. The colony was inoculated into 60 mL of Nutrient Broth (NB, Merck, Darmstadt, Germany) medium in 100 mL Erlenmeyer flasks and then were pre-incubated at 37 ^{o}C for 20 h with a shaker (180 rpm) (9,10).

2.3 Biodegradation by G. lingzhi

Cultures *G. lingzhi* after pre-incubation 7 days were added 10 mL of PDB medium (final volume 20 mL) and 50 µL of 5 mM DDT in DMSO. The headspace of each flask was flushed with oxygen. To prevent the volatilization of substrate Erlenmeyer was sealed with a glass stopper and sealing tape and then incubated at 30oC for 7 days. As a control, after pre-incubation cultures were killed by autoclaving (121 ^{o}C, 15 min) (9,10).

2.4 Biodegradation DDT by B. subtilis

Cultures after pre-incubation were inoculated into PDM medium with various volumes 1, 3, 5, 7 and 10 mL (1 mL ≈ 6.7 x 10^8 bacteria cells) (final volume 20 mL) and then 50 µL of 5 mM DDT in DMSO (final concentration 12.5 µM) added to each bacterial-inoculated flask. The headspace of each flask was flushed with oxygen. To prevent the volatilization of substrate, Erlenmeyer was sealed with a glass stopper and sealing tape. The cultures were incubated statically for 7 days at 30 ^{o}C. As a control, after pre-incubation cultures were killed by autoclaving (121 ^{o}C, 15 min) (9).

2.5 Biodegradation DDT by Co-Culture of G. lingzhi and B. subtilis

Culture *B. subtilis* after pre-incubation with various volumes of 1, 3, 5, 7 and 10 mL was inoculated separately into 10 mL of *G. lingzhi* culture (final volume 20 mL). The headspace of each flask was flushed with oxygen. To prevent the volatilization of substrate, Erlenmeyer was sealed with a glass stopper and sealing tape. The cultures were incubated statically for 7 days at 30 °C. As a control, after pre-incubation cultures were terminated by autoclaving 121 °C for 15 min. The best degradation of co-cultures was selected for further experiments for identification of the metabolic products (9,10).

2.6 DDT Recovery
The culture flask was added 50 µL of 5 Mm pyrene in DMSO (final concentration 12.5 µM) as an internal standard, followed washed with 20 mL of methanol and 5 mL of acetone. The mixtures were homogenized and then centrifugated at 3000 rpm for 10 min. The supernatant was filtrated using Whatman filter paper 41 (GE Healthcare Life Science, UK). The filtrates were evaporated at 64 °C and extracted with 200 mL *n*-hexane and then the organic fraction was collected and over anhydrous sodium sulfate. The extracts were evaporated at 68 °C and concentrated to dryness under reduced pressure. The concentrate was diluted with methanol and the analysed by high-perfomanfe liquid chromatography (HPLC; Jasco, Japan) to quantify the amount of DDT and its metabolic products. The multiwavelength detector (Jasco, Japan) fitted with an Inertsil ODS-3 column (150 mm) with an inner diameter of 4.6 mm (GL Science, Japan). The samples were eluted by 82% methanol in 0.1% trifluoroaccetic acid aqueous solution at a flow rate of 1 mL min^{-1} (9–11).

2.7 Metabolic Detection
To identity metabolic products, samples were further diluted with *n*-hexane and the analysed by gas chromatography/mass spectrometry (GC/MS; HP, USA). The oven temperatures were programmed at 80 °C for 3 min, followed by a linear increase to 320 °C at 15 °C min^{-1} and held for 5 min the injector temperature was set at 250 °C. injection was splinted less approximately 1 µL (10).

2.8 Statiscal Analysis
Student's test was used to detect any significant differences between or within groups during DDT degradation. Differences between mean values at a confidence level of 5 % ($P < 0.05$) were considered to be statically significant. The results were calculated as average of triplicate determination.

3. Results and Discussion

3.1 Degradation of DDT by G. lingzhi
Fungus is a strong organism that has a longer life and can tolerate higher pollutant concentration more than bacteria. Many fungi especially white-rot fungi have been reported capable of degrading DDT. *Phanerocheate chrysosporium, Phlebia lindtneri* and *Phlebia brevispora*, can degrade DDT by approximately, 50% (30 days incubation period), 70% (21 days incubation period) and 30% (21 days incubation period), respectively (12,13). *G. lingzhi* is one of the most broadly distributed white-rot fungus. The ability of *G. lingzhi* BMC9057 to degrade DDT was determined in PDB medium. Selection of the PDB as a media inoculation was due to consists of potato extract and dextrose, which source of carbon and energy for fungus growth. DDT was eliminated by approximately 53% during the 7 days incubation period.
The ability of *G. lingzhi* to degrade pollutants was based on their ability to produced ligninolytic enzymes. These enzymes have been applied in various field including biodegradation of xenobiotic compound. Besides having the ligninolytic enzymes, *G. lingzhi* also has other enzymes involved in biodegradation, such as glyoxal oxidase, catalase, cellobiose dehydrogenase, endoglucanase,

cellobiohydrolase, alginate, lyase, chitinase, carbohydrate esterase, glycoside hydrolase and exo-1,3-beta-glucanase (12,14).

3.2 Degradation of DDT by Co-cultures G. lingzhi and B. subtilis

In this study *G. lingzhi* was used ad degrading fungus that was optimized by addition of biosurfactant-producing bacteria *B. subtilis*. This bacterium has been reported as biosurfactant as biosurfactant producing bacteria (15,16). For optimization process *B. subtilis* was added into *G. lingzhi* culture at 1, 3, 5, 7 and 10 mL (1 mL ≈ 6.7 x 10^8 bacteria cells). From HPLC analysis, the amount of degradation of DDT by *G. lingzhi* with addition of *B. subtilis* at 1, 3, 5, 7 and 10 mL volumes of bacteria were 14.35%, 37.61%, 71.79%, 32.43% and 82.30% respectively. Figure 1 showed the ability of *G. lingzhi* and co-cultures to degrade DDT. Based on this Figure, the addition of 5 and 10 mL of *B. subtilis to G. lingzhi* cultures enhanced the degradation of DDT about 71.79% and 82.30%. Because of DDT was hydrophobic compound, surfactin biosurfactant which produced from addition *B. subtilis*, can increase the solubility of DDT to optimize the DDT degradation process. These results indicated that addition of *B. subtilis* enhanced the biodegradation of DDT by *G. lingzhi*. The level of biodegradation of a pesticide depends on the rapid speed off microorganisms able to grow by utilizing pesticides as an energy source.

Degradation of DDT by co-cultures fungus-bacterium has been reported. Purnomo *et al* (9) reported that degradation of DDT by white-rot fungus *P. ostreatus* only about 19.34%, but co-cultures *P. ostreatus-B. subtilis* (5 mL) and *P. ostreatus* - *P.aeruginosa* (3 mL) enhanced degradation of DDT about 43.00% and 85.74 % respectively, during 7 days incubation. Sariwati *et al* (10) reported that co-cultures of bwron-rot fungsi *F. penicola* and *B. subtilis* 10 mL removed DTT about 85.87% during 7 days incubation.

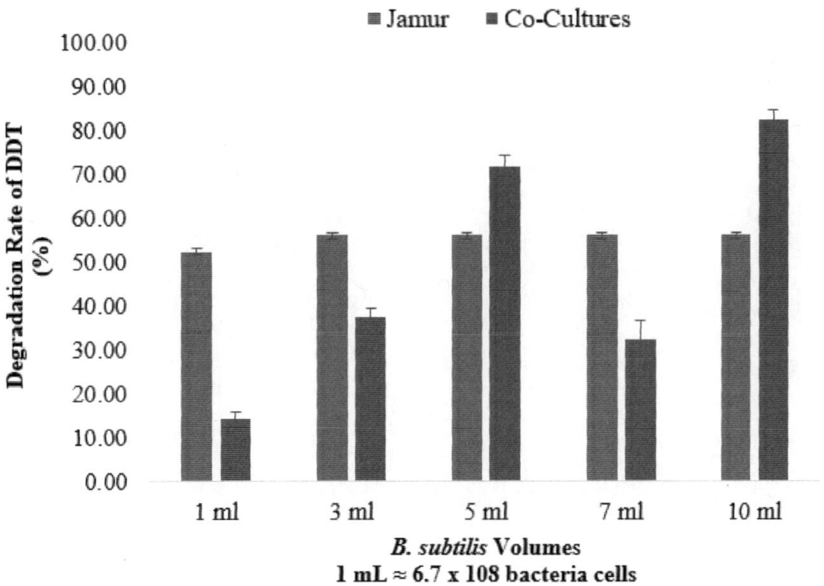

Figure 1. Percent Degradation of DDT by *G. lingzhi* and co-cultures *G. lingzhi* and *B. subtilis*

3.3 Identification Metabolite Product

To obtain trace metabolite of DDT by co-cultures *G. lingzhi* and *B. subtilis,* cultures extracts were analysed with GC/MS by comparing their retention times and mass spectra with those of available standard compounds. GC/MS analysis of samples co-cultures *G. lingzhi* and *B. subtilis* yielded two metabolites. Retention time (r_t) of them were 11.3 and 13.4 min (Figure 2). The metabolite at r_t = 11.3 min had a base peak *m/z* = 212 and three ionization peaks at *m/z* = 282, 247, 212 and 176 (Table 1), that correspond to DDMU (1-chloro-2,2-bis(4-chlorophenyl)ethylene). Since DDT have a base peak ion at *m/z* 235 (M+ -CCl_3 or -$CHCl_3$). The base peak ion at r_t =13.4 min corresponds to DDD, that metabolite had *m/z* = 318, 235 (base peak), 199, and 165 (Table 1). The standard internal peak was appeared in control and treatment samples at r_t = 11.6 which had m/z = 202, which identified as pyrene. In biodegradation of DDT by *G. lingzhi*, DDD and DDE were detected as metabolites products. In other research, white-rot fungus *P. lindtneri* and *P. brevispora* metabolized DDT to DDD, DDA (2,2-bis (4-chlorophenyl) acetic acid) and DBP (12). Purnomo *et al* (9) reported that degradation DDT by co-cultures *P. ostreatus* - *B.subtilis* yielded some metabolite product such as DDD, DDE and DDMU.

On the basis of the information obtained from the results, a degradation pathway co-cultures *G. lingzhi* and *B. subtilis* is proposed, DDT was transformed to DDD and DDMU. The mechanisms of microbial attack have already been described. Most study have been reported that DDT is reductively dechlorinated to DDD and then to DDMU. The results presented in this paper clearly demonstrate the capability of the co-cultures *G. lingzhi* and *B. subltilis* to degrade DDT in vivo. This research showed that co-cultures can be useful in the environmental pollution caused by DDT.

Table 1. Mass spectra and GC retention times of standard compounds and metabolite products

Substrate or metabolite	GC retention time (min)	Mass spectrum *m/z* (relative intensity)
DDMU	11.30	282, 247, 212, 176,
Pyrene	11.60	202,174, 150, 101,
DDD	13.40	318, 235, 199,165
DDT	14.80	354, 282, 235, 165

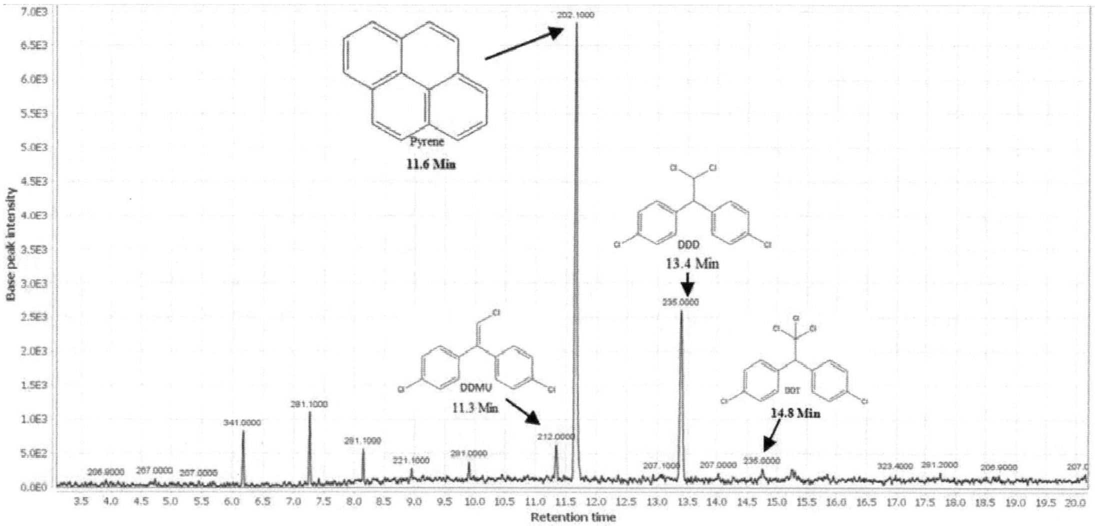

Figure 2. GC/MS chromatogram of DDT degradation by co-cultures of *G. lingzhi* and *B. subtilis*

4. Conclusion

Co-cultures of 10 mL of *B. subtilis* and *G. lingzhi* showed the highest DDT degradation about 82.30% on 7 days incubation period. The addition of *B. subtilis* 10 mL were higher than DDT degradation by *G. lingzhi* without addition of *B. subtilis*, which only 52.52%. DDD and DDMU were detected as metabolite products from DDT degradation by co-cultures of *G. lingzhi* and *B. subtilis*. This study indicated that addition of *B. subtilis* can be used for optimization of DDT degradation by *G. lingzhi.*

Acknowledgements

This work was supported by a grant from the Research Project for International Research Collaboration and Scientific Publication 2017, Number: 01714/IT2.11/PN.08/2017, from the Directorate of Research and Community Service, Directorate General of Strengthening Research and Development, Ministry Research, Technology and Higher Education, Indonesia.

References

[1] Booij P, Holoubek I, Klánová J, Kohoutek J, Dvorská A and Magulová K 2016 *Sci Total Environ.* **550** 231.

[2] Turusov V, Rakitsky V and Tomatis L. 2002 *Environ Health Perspect [Internet].* **110(2)** 125.

[3] Purnomo A.S, Mori T, Kamei I and Kondo R. 2011 *International Biodeterioration & Biodegradation* **.65(7)** 921.

[4] D'Souza TM, Merritt CS and Adinarayana C. 1999 *Lignin-Modifying Enzymes of the White Rot Basidiomycete Ganoderma lucidum Lignin-Modifying Enzymes of the White Rot Basidiomycete Ganoderma lucidum.* **65(12)** 5307.

[5] Da Silva Coelho J, De Souza CGM, De Oliveira AL, Bracht A, Costa MAF and Peralta RM. 2010 *Curr Microbiol.* **60(5)** 350.

[6] da Silva Coelho J, de Oliveira AL, Marques de Souza CG, Bracht A and Peralta RM. 2010 Int *Biodeterior Biodegrad* **64(2)** 156.

[7] Ting WTE, Yuan SY, Wu SD and Chang B V. 2011 *Int Biodeterior Biodegrad.* 65(1) 238.

[8] Punnapayak H, Prasongsuk S, Messner K, Danmek K and Lotrakul P. 2009 *African J Biotechnol.* **8(21)** 5897.

IOP Conf. Series: Journal of Physics: Conf. Series **1095** (2018) 012015 doi:10.1088/1742-6596/1095/1/012015

[9] Purnomo AS, Ashari K and Hermansyah F 2017 *J Microbiol Biotechnol* 27(7) 1306 .
[10] Sariwati A, Purnomo AS and Kamei I 2017 *Curr Microbiol.* **74(9**) 1068.
[11] Purnomo AS, Kamei I and Kondo R. 2008 *J Biosci Bioeng* **105(6)** 614.
[12] Xiao P, Mori T and Kamei I. 2011 *A novel metabolic pathway for biodegradation of DDT by the white rot fungi , Phlebia lindtneri and Phlebia brevispora.* **859** 67.
[13] Liu D, Gong J, Dai W, Kang X, Huang Z and Zhang HM 2012 *PLoS One*. 7(5).
[14] Bumpust J and Austt SD 2008 *Biodegradation of DDT [1 , 1 , 1-Trichloro-2 , 2-Bis (4-Chlorophenyl) Ethane] by the White Rot Fungus Phanerochaete chrysosporium.* Society. **53(9)** 2001.
[15] Zouari R, Ellouze-Chaabouni S and Ghribi-Aydi D. 2014 *Achiev Life Sci* **8(2)** 162.
[16] Pornsunthorntawee O, Arttaweeporn N, Paisanjit S, Somboonthanate P, Abe M and Rujiravanit R 2008 *Biochem Eng J.* **42(2)** 172.

*** Formatting Issue - Best Available Paper/Graphic ***

Bioethanol Production from Cassava (*Manihot esculenta*) Peel Using Yeast Isolated from Durian (*Durio zhibetinus*)

Hermansyah[1], Tounaly Xayasene[2], and Nguyen Huu Tho[2], Miksusanti[1], Fatma[1], Almunadi T.Panagan[1],

[1] Chemistry Dept, Faculty of Mathematics and Natural Sciences, Sriwijaya University, Jalan Raya Palembang-Prabumulih KM 32 Indralaya, Ogan Ilir, South Sumatera, Indonesia 30662. Phone (+62) -711-580269.
[2]Environmental Science and Management, Thai Nguyen University of Agriculture and Forestry, Vietnam.

E-mail : hermansyah@unsri.ac.id

Abstract. The process of ethanol production generally involves pretreatment, hydrolysis of lignocellulosic biomass to fermentable sugars followed by fermentation of such sugars to ethanol. Waste of the cassava peel (*Manihot esculenta*) was hydrolysed by using sulphuric acidThis research aimed to produce of bioethanol as an alternative source of fuel using cassava peels as raw materials.. Yeast isolated from Durian fruit (Durio zhibetinus) was used in the experiment for fermentation and the concentration of sulphuric acid of hydrolysis process was fermented by yeast for 1 ; 2 ; 3 ; 4 ; 5 ; 6 ; 7; and 8 days. 50 ml of Sodium Hydroxide NaOH was prepared to be added at this step to adjust the pH of the slurry until 5 and the temperature was kept at 25 °C. Nine samples were prepared at different three hydrolysis times at 121°C for 30 minutes, 45 minutes and 60 ninutes. For glucose consumption and ethanol product analysis, 2 ml of the sample were *taken* out at *every* 2 days *interval until* 8 days. During this fermentation process, sugar consumption was measured by DNS method, while quantification of ethanol was analyzed by Gas Chromatography. The result of this study obtained that the best time of hydrolysis process was 45 minute, where the result of concentration of glucose was 11.189 %. By virtue of that, fermentation process was influenced by shaking incubator at 6 days. the optimum concentration of sulphuric acid of the hydrolysis process was 30 minute, and duration time of fermentation process by shaking incubator was 8 days, while the concentration of bioethanol for the highest of hydrolysis and fermentation process was obtained 1.63 % ethanol.

Keywords : Bioethanol, Fermentation, Yeast, *Durio zhibetinus*

1. Introduction

The world's economy today highly depends on fossil energy sources such as coal, oil, natural gas which are used to produce fuels, electricity, chemicals, and other goods. The increase of both human population and industrial prosperity will simultaneously be increase of global energy consumption. The utilization of these conventional fossil energy sources in the long run cause depletion of energy source and is not sustainable. Thus, we should provide raw material for the industry and human needs in sustaible way, and this is the greatest challenges for us. Furthermore, fossil fuels is non-renewable energy sources, limited stock, and have a considerable negative environment impact.

Bioenergy from renewable resources becomes an alternative replacing or supplement for fossil fuels. The raw material for alternative renewable bioenergy source is available and abundant the raw materials. Almost all petroleum-based fuels can be replaced by renewable fuels produced from biomass such as bioethanol, biodiesel, biohydrogen, etc [1-3].

Bioethanol is one of prominent bioenergy which has some advantages such as becaused it has a high octane fuel, and it reduces polluting emission. It is a clean and renewable biofuel with major environment benefit, burning of the oxigenated the fuel mixture consisting of ethanol and gasoline results more completely burnt and reduces polluting emissions.

IOP Conf. Series: Journal of Physics: Conf. Series **1095** (2018) 012016 doi:10.1088/1742-6596/1095/1/012016

Today, almost all the current fuel ethanol in first generation is generated from edible sources contain sugars and starch such as corn, cassava, potatoes etc. In the second generation, lignocellulosic biomass has drawn much attention in recent times. However, efficient technology development to convert lignocellulosic biomass into fermentable sugars are the key areas of development in second generation bioethanol production [4], such as utilization of simultaneous saccharification and fermentation (SSF) with a batch and fed-batch methods [5].

Cassava peels is abundant waste food in the developing country, where the high number harvest of cassava so more and more high waste of cassava peels. It has been acknowledged worldwide that cassava peel is one of the best choice to replace edible sources for fuel ethanol production, without endangering food security.

In conversion of carbohydrate into bioethanol, microbial agent plays an essential role especially in fermentation step. A common yeast used in fermentation process producing ethanol used yeast *Saccharomyces cerevisiae* as microbial agent [6-7]. This research, yeast isolated from durian (*Durio zhibetinus*) fruit was applied in fermentation of cassava peels.

2. Methodology

2.1. Sample and Material

Cassava peel was obtained from a 26 *ilir* traditional market located in Palembang. Chemicals used in this study consisted of, YPD medium (1% yeast extract, 2% peptone, 2% glucose), fermentation medium (2% yeast extract, 2% ammonium sulfate, 2% magnesium sulfate, and 4% potassium dihydrogen phosphate), bacto agar, sulfuric acid, and sodium hydroxide, 2,5-dinitro salicylic acid (DNS), and sodium potassium tartrate tetrahydrate.

2.2. Preparation of Sample

The cassava peels was thoroughly washed with tap water and cut into smaller pieces. Then cassava peels was dried in oven at 80°C for 5 days. Once dried, the cassava peels was grinded using the grinding machine to get pul. The cassava peels sample was sealed in the seal bag or poly bag and stored in room conditions.

2.3. Pretreatment and Hydrolysis

Pretreatment and hydrolysis of cassava peel were conducted as protocol described in [8] with small modification. 30 grams of cassava peel flour was weighted into separate conical flasks and mixed with 150 ml of distilled water. The mixtures were covered with alumminium foil and then sterilized in an autoclave at 121°C for 1 hour and allowed to cool and remove water. After that diluted 100 ml of 1% Sulfuric acid into each sample then autoclave in three different times at 121°C for 30 min, 45 min and 60 min.

2.4. Inoculum preparation

Yeast isolated from Durian used for inoculation was grown in 15 ml reaction tube with 10 mL of YPD medium containing 1% yeast extract, 2% peptone, 2% glucose. After incubating in a room temperature with shaking 150 rpm for 1 day, the cell culture was aseptically transferred into fermentation medium to reach OD = 0.2 to start the fermentation.

2.5. Determination of glucose content using DNS method

Glucose content was determined using DNS method according to [9] with small modification. Both 0.3 mL of pre-heated substrate solution in 0.1 M Na-acetate and 0.3 mL of enzyme solution at 50°C for 5 min)were mixed, and incubating it at 50°C for 10 min. The mix solution was then added with 0.9 mL of 3,5 Dinitro salicylic acid (DNS) for 5 min, cooled at room temperature, and the absorbance was measured at 540 nm.

2.6. Fermentation

Fermentation was performed according to [10] with modification. The fermentation media was prepared as follows, 2 g yeast extract, 2 g ammonium sulfate, 2 g magnesium sulfate, and 4 g potassium dihydrogen phosphate dissolved completely in 500 ml water in a conical flask then autoclave at 121°C for 15 min. 25 ml of this mixture media was added to each of the samples. 50 ml of 0.1M sodium hydroxide was prepared to be added to adjust the pH of the slurry until 4.5 – 5 and the temperature was kept at 25 °C.

2.7. Determination of ethanol content using Gas Chromatography
Ethanol content produced from fermentation process was measured by Gas Chromatography (GC 2010 Shimadzu) complemented with flame ionization detector (FID), column size was 30 m and 0.25 mm for length and diameter, respectively. The condition of experiment was rate of carrier gas, N2 136.3 mL/min, 100 kPa, temperature of injector 150°C, column 60-240°C, detector 200°C.

3. Results and Discussion

3.1. Preparation of Cassava peel sample
Preparation of cassava peel covered washing, cutting, drying and grinding as shown in Figure 1. The cassava peels were washing with trap water to eleminate contamination. The cleaned samples were cut in order to easier in drying. The dried samples were grinded to obtaine the cassava peel powder. In powder form can increase degradation of lignin during pretreatment process, increase interaction between sample with sulfuric acid during hydrolisis process, and increase access enzymatic reaction during fermentation process.

Figure 1. Preparation of cassava peel sample consisted of washing, cutting, drying and grinding process.

3.2. Pretreatment and hydrolysis of cassava peel sample
Cassava peel is one of lignocellulosic material, mainly contains carbohydrate such as cellulose and hemicellulose, and non carbohydrate material lignin. Lignin should be reduced before fermentation of hydrolysate carried out. Pretreatment of cassava peel was conducted by physical method which is autoclaving 20% cassava peel – water mixture at 121°C for 1 hour. This physical method was choosed because it is easier and make pH alteration significantly. Unfortunately, quantitative analysis of lignin content was not determined.

Pretreated cassava peel was hydrolyzed using 1% sulfuric acid to produce a reducing sugar, such as glucose. In this experiment, hydrolysis was conducted in variation time 30, 45 and 60 minutes which produce different results (Figure 2). The highest reducing sugars was produced in 60 min hydrolysis time, 11.189% compared with 30 and 45 min hydrolysis time, and the latter producing the least amount. The result of the hydrolysis process of cassava peel was prepared as many as 12 samples to be fermented at room temperature and pH 5. Bioethanol from each sample in the fermentation process was collected at the 2nd, 4th, 6th, and 8th day each sample. Results showed that the optimum time of hydrolysis process was 60 min, which resulted 11.189%. Determination of glucose from hydrolysis process was done by DNS method. Principle of this method was reducing sugars have the property to reduce many of the reagents. Fermentation process by using durian isolate yeast which could convert glucose to bioethanol.

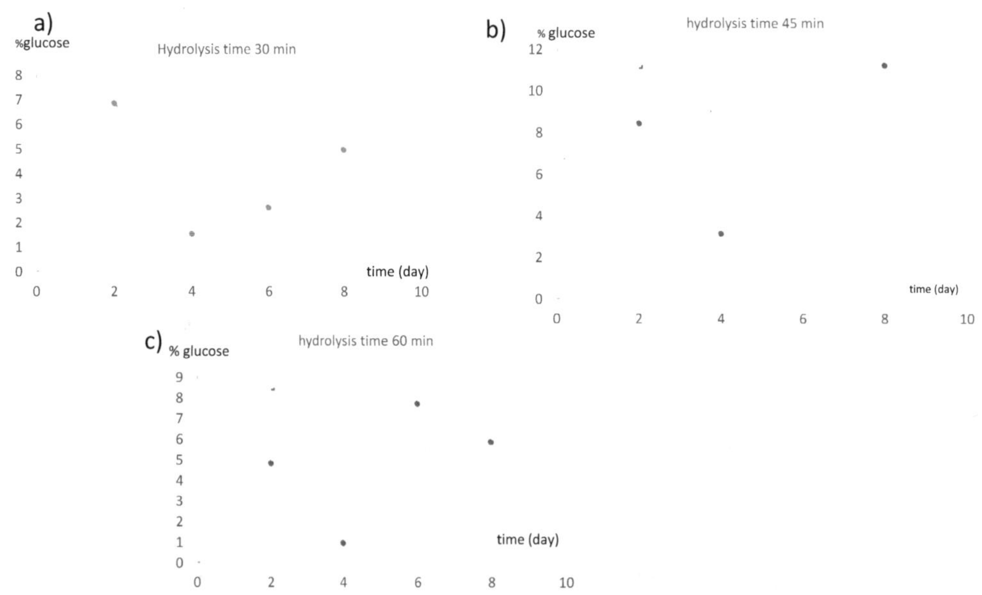

Figure 2. Glucose contents generated from hydrolysis of cassava peel with hydrolysis time variation 30 min, 45 min, and 60 min.

Comparison of the amount of reducing sugars expressed as hydrolyzed feedstock per 30 g raw feedstock, produced during the hydrolysis of the different times using 1% Sulfuric acid and 8 days shaking incubator process displayed the glucose content is start from high to low concentration, then continue increasing again in the last day of fermentation. So the result was not optimum concentration. This may be due to the disability of the pretreatment process did not convert cellulose and hemicellulose to glucose as well as it could be. Soaking in a aqueous ammonia treatment increased surface area and the pore size, and it lead to higher ethanol production [11].

The low amount of reducing sugars obtained from cassava peel during pretreatment may be attributed to the preparation procedure since they were not crushed but chopped, further confirming the importance of the preparative steps for pretreatment of the lignocellulosic feedstocks, along with the stringent preparative procedures required, illustrates the difficulty of using lignocellulosic feedstocks for ethanol production. The observed differences in the amount of reducing sugars produced from roots and other plant parts, further confirms this fact. Thus, the choice of feedstocks should encompass the time and effort required as well as the cost of preparation and hydrolysis, which has a bearing on the fermentation as well as the final product. The ability to produce sufficient amount of reducing sugars determines the importance of a particular feedstock for ethanol production. Feedstocks with ability to produce high amounts of glucose using simple hydrolysis procedures are important alternatives for biomass fuel production. All the factors mentioned above will in the end determine the final cost of the ethanol produced. The physiological state of the feedstock and the environmental conditions under which the feedstocks are grown, are also important factors to consider.

3.3. Fermentation of cassava peel using yeast isolated from durian

Yeast plays an essential role in fermentation process to convert carbohydrate into bioethanol. *S.cerevisiae* has well known as fermentation agent because of his ability to convert glucose into bioethanol. Yeast isolate used in this experiment was obtained from durian (*Durio zhibetinus*) fruit. Growth of yeast colonies when streaked onto YPDA agar media, it grew in small or large round, shiny, smooth and creamy morphology as shown in Figure 3. Utilization of yeast isolates from durian can explore indigenous yeast strain.

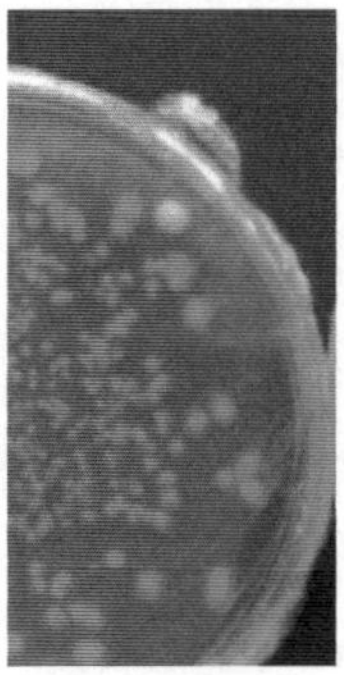

Figure 3. Colony of yeast isolated from Durian (*Durio zhibetinus*)

Fermentation of cassava peel hydrolysate was performed under room temperature condition with shaking 150 rpm. Bioethanol content generated by this fermentation was analyzed in the 2nd, 4th, 6th and 8th fermentation day time using Gas Chromatography. Based on the results obtained, 30 min hydrolysis with day 8 of fermentation showed the highest ethanol content in water which is 1.63 %, followed by 45 min hydrolysis in day 4 which is 1.42 %, then hydrolysis time at 60 min was 1.20% in day 8 . The lowest ethanol concentration in water with water was achieved at 30 min hydrolysis without shaking incubator time .

However this study shows lowest ethanol concentration in water at 30 minutes hydrolysis without shaking incubator time. This may be due to the disability of the yeast to fermentation at 0 day. There are other possibilities; the yeast that was used to conduct the experiment may be old. Old yeast will not carry out fermentation process efficiently compared to new yeast. This statement is not applicable for this study since at 30 minutes hydrolysis there is almost no ethanol production.

IOP Conf. Series: Journal of Physics: Conf. Series **1095** (2018) 012016 doi:10.1088/1742-6596/1095/1/012016
***** Formatting Issue - Best Available Paper/Graphic *****

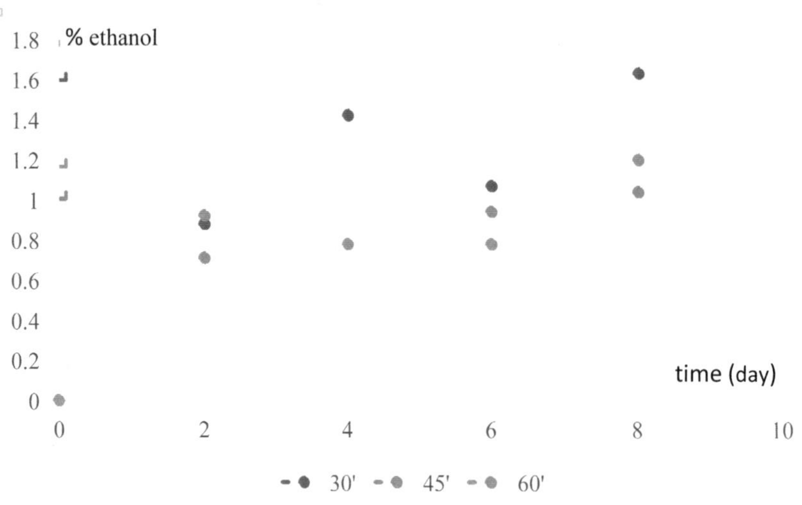

Figure 4. Bioethanol production from hydrolysate of cassava peel during fermentation, sample were analyzed in 2nd, 4th, 6th, and 8th day. Hydrolysis times were 30 min, 45 min, and 60 min.

Figure 4 shows three different hydrolysis times in affecting the bioethanol production. Overall, we can see that all of this have the same trend. They are on increasing trend. The production of bioethanol growed higher by days. However in the middle of timeline. In day 6 for 30 and 60 minute of the hydrolysis time, the production of ethanol seems to go down but go up again after that. But in the hydrolysis time 60 minute, in 4th day the production go down but go up again later. And all of productions in 3 conditions peak at the last days.

The high content of dry matter observed in the parts of cassava plants may be hydrolyzed into fermentable sugars. This has a bearing on the final yield of reducing sugars, since high contents of dry matter are desirable in ethanol production . The ethanol content between the different times of the hydrolysis and fermentation present study, 30 minutes hydrolysis in day 4th ,6th, 8th showing high levels of ethanol content compared to other 45 and 60 minutes hydrolysis, were notable since a significant relationship between ethanol properties and hydrolysis process in three different times existed. Long time hydrolysis content resulted in production of lower quantities of ethanol.

Fermentation efficiency depended on the ability of the yeast to utilize particular feedstocks based on their characteristics and compositional differences. The absence of differences in fermentation efficiency during the first 8th day was due to the yeast establishing itself in the fermenting solution, growing to a certain colony volume able to utilize the existing sugars. The variations in fermentation efficiency may be attributed to the type of sugars produced as well as substrate preferences by the fermenting organism, since different times of the hydrolysis produce different sugar types apart from. Although ethanol can be converted from glucose, however it reported that yeast cell growth and ethanol fermentation performance have no significantly difference from the cultures of glucose, corn stover hydrolysate liquid, and the pretreated corn stover solids as carbon sources, respectively [12]. Other study reported that increasing the yeast inoculum or cellulose concentration did not significant improve the ethanol yield or concentration [13].

The pretreatment process of feedstocks affect their hydrolysis, consequently affecting the type of reducing sugars produced and hence moderating the type of metabolism carried out by yeast under shaking in incubator. In particular, lignocellulosic materials in hydrolyzed peels may result in production of small molecular weight compounds such as furan derivatives, phenolic compounds, and amine-based compounds such as vanillin, all inhibiting fermentation. The low ethanol percentage obtained with progressive fermentation of sugar is most likely due to the fact which may inhibit its metabolism and hence reduce its efficiency.

4. Conclusion :

The use of cassava peel for bioethanol production as an alternative source of fuel provides a starting point for improvements in cultivation and adoption of cassava as well as improving food security. This study concluded that the optimum concentration of sulphuric acid of the hydrolysis process was 30 minute, and duration time of fermentation process by shaking incubator was 8 days, while the concentration of bioethanol for the optimum of hydrolysis and fermentation process was 1.63 %. These indicated that it needs to do more experiment to improve the higher yield.

Aknowledgment

This project was financed by a grant from Scheme "Hibah Kompetitif UNSRI 2017".

References

[1] Alvira P, Ballesteros M and Negro MJ 2009 *Bioresour. Technol.* **101(13)** 4851
[2] Antoni D, Zverlov V V and Schwarz WH 2007 *Appl. Microbiol. Biotechnol.* **77(1)** 23
[3] Gray KA and Zhao L 2006 *Curr Opin Chem Biol.* **10(2)** 141
[4] Muktham R, Bhargava SK, Bankupalli S and Ball AS 2016 *Journal of Sustainable Bioenergy System* **6** 72
[5] Li H, Kim N, Jiang M, Won J and Nam H 2009 *Bioresour. Technol.* **100(34)** 3245
[6] Dashtban M, Schraft H and Qin W 2009 *Int. J. Biol. Sci.* **5(6)** 578
[7] Hahn-hägerdal B, Karhumaa K, Fonseca C, Spencer-martins I and Gorwa-grauslund MF 2007 *Appl. Microbiol. Biotechnol.* **74(5)** 937
[8] Cardona CA, Quintero JA and Paz IC 2010 *Bioresour. Technol.* **101(13)** 4754
[9] Krivorotova T and Sereikaite J 2014 *Electronic Journal of Biotechnology* **17** 329
[10] Hermansyah, Novia, Sugiyama M and Harashima S 2015 *Microbiology Biotechnology Letters* **43(3)** 241
[11] Kim TH, Taylor F and Hicks KB 2008 *Bioresour. Technol.* **99(13)** 5694
[12] Qureshi AS, Zhang J and Bao J 2015 *Appl. Biochem. iBotechnol.* **175(6)** 3173
[13] Prasetyo J, Naruse K, Kato T, Boonchird C, Harashima S and Park EY2011 *Biotechnol. Biofuels* **4(1)** 35

Biodiesel effects on fuel filter; assessment of clogging characteristics

L N Komariah[1], F Hadiah[1], F Aprianjaya[1], and F Nevriadi[1]

[1]Department of Chemical Engineering Faculty of Engineering Universitas Sriwijaya, Jl. Raya Palembang-Prabumulih Indralaya South Sumatera, 30862 Indonesia

E-mail: leilynurul@unsri.ac.id

Abstract. Material compatibility is still one of the serious problems in the application of biodiesel on diesel engine. One of the main causes of the problem is related to the flow-ability and filterability of biodiesel. Due to its physical and flow properties, biodiesel potentially caused a blockage in the filter higher than petro-diesel. The profile and mechanism of filter clogging or blockage due to the use of to biodiesel and the blends are investigated in the present work. Typical fuel oil filter used is a cartridge with media made from porous synthetic materials. Biodiesel used in this work is palm oil-based which is applied purely and in the blends of petro-diesel with composition of 20:80 (B20). It was tested in on boiler fuel system. Fuel passed on the system through the filter with a contact time of 3500 hours. The filter media is also observed by immersing it in B20 and B100 in the same time period. The result showed that the filter contacted to biodiesel was clogged faster than diesel oil (B0). On endoscopy microscope image, the filter media surface was clearly degraded. It shown that the pores of filter media are mostly covered by oil sludge. The filter blocking tendency (FBT) value is used to predict filter clogging time limit. In this work, the use of B20 performing FBT in ranges from 1.51-1.84, whereas the use of B100 ranges of 1.58-2.42 in a contact period of 3500 hours.. The premature blockage was occurred in filter media contacted to B100, and a filter replacement is then required to conduct before 8 months. There is a significant relationship between the moisture content of biodiesel on the rate and characteristics of the filter clogging.

Keywords : Biodiesel, fuel filter, palm oil

1. Introduction

Biodiesel is considered as a renewable fuel which is derived from trans-esterification of vegetable oils. It is made from a diverse mix of feedstock including recycled cooking oil, soybean oil, and animal fats. Biodiesel is consisted of mono-alkyl esters of long chain fatty acid that conform to ASTM D6751 specifications for use in diesel engines. Biodiesel refers to the pure fuel before blending with diesel fuel. Biodiesel blends are denoted as "BXX" with "XX" representing the percentage of biodiesel contained in the blend (ie: B20 is 20% biodiesel, 80% petroleum diesel). Biodiesel can be used in existing diesel engines without modification as long as the fuel is technically suitable to engine performance specifications.

Currently in Indonesia, Biodiesel are widely used in automotive, industrial and commercial engines, it often in blends of up to 5 to 20 percent. As mandated by Indonesian government, it is targeted to use 30 percent of biodiesel in petro-diesel blends in 2020.

Fuel-grade biodiesel must be produced to strict industry specifications (ASTM D6751/EN 14214) in order to ensure proper performance in internal combustion engines. The stability and cleanliness of fuel is a key element in the context of its storage and the impact of the proper engine operation, in which it is burned, and it depends on the content and construction of compounds included in its composition.

Biodiesel may be believed to be a clean fuel due to its contours to emissions reduction and its biodegradability. On the other hand, biodiesel is a fuel that is prone to contamination. There is still considerable suspicion from users that biodiesel is safe and compatible with all diesel engine and its elements. There are also concerns that the use of biodiesel can shorten the durability of life span of the engine and its components, including filter.

Fuel filters serve a vital function in recent modern, tight-tolerance engine fuel systems. The fuel filter is a filter in the fuel line system that screens out any contamination from the fuel, normally made into cartridges containing a filter media. After use in a period of engine operation, the fuel filters need to be maintained. Technically, it is commonly proceed with disconnecting the filter from the fuel line and replacing it with a new one, although some specially designed filters can be cleaned and reused. Filter is commonly replaced regularly or anytime possibly when it is become clogged with contaminants and cause a restriction in the fuel flow. Filter clogging will cause serious engine operation problems. Clogging of filter phenomena is the primary failure mode leading to the replacement or cleansing of filter. Although the filter media can be replaced easily and the purchasing cost is low but replacement of the component cause an increase in the engine downtime frequency. It will cause in greater operating losses and reduced the engine durability.

Due to its characteristics and handling, the use of biodiesel in diesel engine may cause a very short fuel filter life. In this work, the fuel contamination contributed by biodiesel which affecting its filterability will be studied. It is important to know the clogging profile and mechanism on the a fuel filter due to the use of biodiesel in engine. Otherwise, it also crucial to find out the rate of clogging on filter media to estimate the appropriate time limit of filter use or maintenance requirement in the utilization of biodiesel as fuel.

1.1 Fuel filter system

The main role of oil filter is to cleanse oil from destructive contaminants within an engine. Most engine oil filters were of *cartridge* (or *replaceable element*) construction, in which a permanent housing contains a replaceable filter element or cartridge. Depending on engine model, the final fuel filter, mounted on the engine just before the injection pump, it commonly has a single-pass retention (pore size) rating of 2 to 10 microns. This is the size that most engine manufacturers figure will protect injector pumps and nozzle tips from damage [1]. The common media filter materials are cotton yarn, low twisted cotton yarn, accordion shape pleated paper, cellulose disks, bags woven with wool yarn and dense knit filter bags are used. The canister-type filters are the most common use in diesel engine [2]. Cellulose media are advantageous because it can absorb some water contamination.

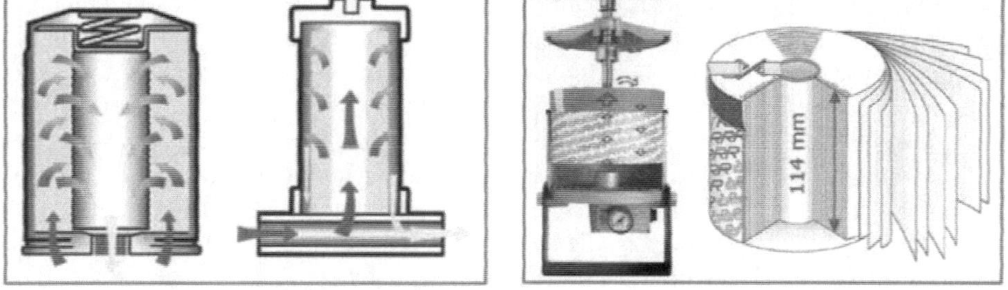

Figure 1. Filter types and media configuration

The filter media operate under several types of filtration mechanisms, including direct interception and depth entrapment, adsorption, inertial impaction, Brownian movement and gravitation effects [2]. Oil filters can be characterized by the method in which the contaminants are filtered or the method in which the oil flows through the housing. One technique used to control contamination in filters is through surface-type media. Fuel filter configuration in its system was responsible for the advanced performance of oil filtration technology according to engine operation requirement [3]. It supposed to consider the fuel characteristics as well.

1.2 Filter clogging mechanism and fuel contamination

Filter clogging/plugging/blocking is the failure mode seen most frequently in oil filters in varies of engine operation. Filter blocking problems must be seriously anticipated with appropriate techniques. Modern engine are incredibly sensitive to fuel contaminants. Engine filter can block and fuel injectors can suffer damage, resulting in total engine failure, it may also increase oil consumption during operation.

As known, fuel is expected leaving the refinery typically in clean and meet the specifications, but contaminations is always unavoidable. The principal source of contamination showed in particulates and water is in the transportation and storage. Contaminants are dissolved in fuel and most other captured by the filter.

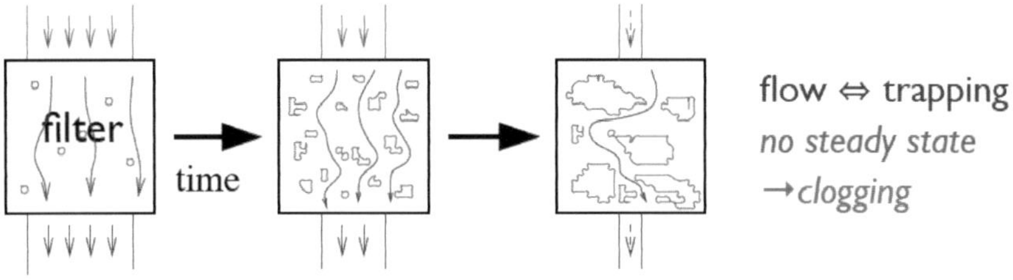

Figure 2. Typical filtration and clogging mechanism

The most critical factor in proper oil filtration characteristics is the structural integrity, contamination (dirt-holding) capacity, pressure loss and particle capture efficiency. The structural integrity relates to a filter's ability to prevent the passage of oil through an unfiltered flow path. Contamination capacity refers to the amount of contaminants that can be loaded onto the filter before the filters efficiency is limited. Pressure loss involves the overall differential pressure lost from the filter's placement on the system. The pressure loss will be influenced by the porosity of the filter and surface area.

There are several types of fuel contaminants found in a fuel filter, it may perform as water contamination, microbial growth, and solid particulate. Sometimes the contaminants are directly hard to detect without discharging the filter media from its house. Therefore the presence of filter contaminants can be observed through several engine performance parameters such as mechanical efficiency, engine power, fuel consumption, heavy smoke from exhaust, and clogging filters.

1.3 Biodiesel and the filter clogging

Contaminants in diesel engine systems can be found as in organic and inorganic. Organic contamination, otherwise known as sludge, is made up of all the by-products of combustion and comprises approximately 75% of the total contaminants found in diesel engine oil as reported by several filter manufacturer such as cummins filtration, Parker Racor, etc. Inorganic contaminant, referred to as dust, is made up of dirt, gasket material, core sand and wear metal. Those contaminants will plug a filter, limiting its ability to filter out other harmful contaminants.

As known, any kind of oil fuel may cause contamination that clog filters during operation, including petrodiesel fuel. The use of petrodiesel blends or neat biodiesel are suspected to exacerbate the occurrence of blockage in the filter a reported many researchers [5-8]. Introducing higher blends will make the occurrence of diesel bugs more prolific.

Biodiesel fuel properties are quite comparable to petro-diesel, but some of it is potentially trigger the formation of more contaminants. There are two major problems related to use of biodiesel as fuel, they are its oxidation stability and cold flow performance. Oxidation instability can led to the formation of oxidation products like aldehydes, alcohols, shorter chain carboxylic acids, in solubles, gum and sediment in the biodiesel [7;8]. Oxidation products originated with biodiesel affect storage life and contribute to deposit formation in tanks, fuel systems and filters. During a period of storage time, oxidation may occurred due to biodiesel having the chain of unsaturated fatty acids and the presence of double bonds in the molecule, which can cause the reacticity of the compound to increase oxygen, especially when placed in a container that allows contact with air or water [7]. Basic problem with biodiesel is that it oxidizes while in contact with environment with respect to time which further lead to increase in fuel viscosity and other physical properties as studied by [9, 10]. In most cases, the oxidation of biodiesel is known to cause deposits and corrosive materials [11]. Tang *et al.* [12] states that the formation of precipitates in biodiesel blends may have serious implications for diesel engine fuel delivery systems.

Several tests were conducted to show that filters are considered to be faster clogging due to contact with biodiesel rather than petrodiesel fuel as reported by [9; 13]. Based on severeal researches, engine requires quicker fuel filter change while running on biodiesel due to clogging by sediments and wear debris and pitting corrosion [9]. Filter analysis showed that blocking were caused by some insoluble components coming from FAME (Saturated monoglycerides) that precipitated at low temperatures [14].

Contaminants indicated to be problematic for fuel filters were detected. High concentrations of the monoglycerides of saturated C16 and C18 fatty acids, monopalmitin and monostearin were found on plugged diesel fuel dispenser filters [15] stated that biodiesel related contaminants such as saturated monoglycerides, sterol glucosides, and carboxylate salts due to its chemical properties [8;15].

Biodiesel has some solvent properties and will act as a solvent in the fuel. Blends greater than B20 may have enough of a solvent effect to break down the varnish deposits on the walls of the existing fuel storage tanks or fuel systems. The break-down of these varnish deposits will contaminate the fuel with particulate, which can cause fuel filters to plug rapidly [3]. Biodiesel may also degrade many non-metallic materials such as natural rubber, and other kind of polymers. Hydrocarbon fuels typically form a layer of deposits on the inside of tanks, hoses, etc. Biodiesel blends loosen these deposits, causing them to block fuel filters [16].

Diesel fuel should meet ASTM D 975 specification diesel fuel related to flowability in engine system. Petroleum diesel fuels are plagued by the growth and agglomeration of paraffin wax crystals when ambient temperatures fall below the fuel's cloud point (CP). These solid crystals may cause start-up problems such as filter clogging when ambient temperatures drop [17].

Cloud point and Cold flow issues with diesel containing FAME (biodiesel) and FAME material has been studied extensively. The fatty acid composition of biodiesel is the main factor that affects its cloud point, pour point and cold filter plugging point. The properties of biodiesel at low temperatures are poorer than those of diesel as conducted by [17; 18]. The poor cold flow properties result in crystallization of fuel particles. It may result in obstruction in fuel flow which causes a failure of engine operation by choking the fuel lines and filters. The cold flow problem arises due to higher number of saturated fatty acid present in the biodiesel as reported by some authors [18; 19]. The cloud point, is the temperature at which a liquid fatty material becomes cloudy due to the formation of crystals and solidification of saturates [18].

Biodiesel is hygroscopic, due to it has a polar chemical structure of carboxylate ester groups and presents the characteristics of absorbing more water than diesel. The water content of fuel was gradually increased with the increase of its blends [10]. The addition of biodiesel to diesel increases the retention capacity of the blend [20]. Humidity is one of important thing that affect in water content on fuel. Water that is chemically dissolved or absorbed into fuel and distributed molecule by molecule [21]. The *maximum* amount of allowed *water content in biodiesel* as specified in ASTM standard *D6751* is 500 *ppm.* The water present in B100 is considered to be free water.

As higher Biodiesel content in fuel blends, the risk of micro-organism growth increases due to less chemical stability and higher water solubility in biodiesel. The life activity of microorganisms in fuel is connected with the presence of water, dissolved oxygen concentration, the temperature range optimal for development, the neutral or alkalescent water environment reaction, the presence of appropriate compounds in fuel, the use of some types of enhancing additives, especially those that contain nitrogen [22].

Biodiesel is an organic fuel so it provides an ideal environment for microscopic fungi, yeast and bacteria to feed and grow. This environment provides dissolved water for germination, carbon for food, oxygen and sulphur for respiration, trace elements for growth and propagation.

In the presence of water in Biodiesel, it may promote the development of bacterial flora during storage as reported by [8;23]. Additional water accumulates in tanks as atmospheric moisture condenses. Moisture accumulates in diesel tanks as condensate droplets on exposed tank surfaces, as dissolved water in the fuel and as water bottoms beneath the fuel Microbes depend on this water for growth. Additionally, microbes depend on the organic and inorganic molecules in diesel fuel for nutrition. Dodos *et al.* [24] were investigated the effect of microbiological contamination on several parameters of biodiesel fuel and its blends during storage for 16 weeks. They found that FAME concentration was increased up to 20% v/v contributed to microbial stability reduction. The biodiesel blends produced considerably higher acid value. The minimization of sulfur content seems to allow enhanced microbial activity. The biodegradation of fuel, in support of microbial growth, is a direct impact of contamination. Color, heat of combustion, pour point, cloud point, detergent and anti-corrosive properties change as microbes selectively attack fuel components and potentially found in fuel filter.

2. Material and Methods

The fuel system used in this work is direct combustion engine for fire tube type boiler package. The set of apparatus is shown in Figure 3.

Figure 3. Experimental set up

The biodiesel used in this work is palm oil based which has met EN14214. The fuel is stored at average of 6-7 months since it was manufactured in a stainless steel tank. The fuel sample then analyzed to see the difference on several physical properties. The blending of biodiesel with diesel oil is done by splash blending technique. The fuel tested in this study was on a mixing of 20% biodiesel (B20), and B100. Filter media used is a type of canister with media paper (cellulose). A digital endoscope microscope camera is used to analyze the findings on the surface of the filter media with magnification up to 500x. The flow system was set in static (immersion) and dynamic at a fuel rate of 10 ml/min. Samples are evaluated every 30 days. After 3500 days of contact, surface morphology of filter media was observed.

The water content of each sample BXX and B0 was analyzed using the Karl Fischer Coulometric (KF) titration Metrohm KF coulometer method. Acid Value is measured using PP Colorimetric

Titration according to EN 14104. Pore size distribution then evaluated principly in close accordance with ASTM E 1294-89 and ASTM F316-03 standards.

A change in filtration performance after storage or pre-treatment, and operation in engine system must be evaluated to indicate potential filter blocking problems or small changes in fuel condition. It will be represented by the filter blocking tendency (FBT). Petro-diesel, biodiesel and its blends were evaluated using bench-top filtration testing equipment to determine its filter blocking tendency (FBT) values. Fuel Blocking Tendency (FBT) is a calculated dimensionless value that defines the tendency of particulates in a fuel to plug or block a filter. Standard Test Method for Determining Filter Blocking Tendency-Active Standard ASTM D2068 and IP 387. For this test, 300 ml of sample is pumped at a constant flow rate (20 mL/min) through a specified filter medium. Both pressure difference across the filter and the volume of fuel passing the filter are monitored until the pressure reaches 105 kPa or the volume of fuel passing the filter medium reaches 300 mL. When 300 ml of fluid is pumped, the end pressure is used to calculate the FBT number. However, when the pressure reaches 105kPa before the 300ml is passed, the volume of fluid pumped at this point is used to calculate the FBT number.

3. Results and discussion

The quality of fuel in this study is related to the level of cleanliness of the fuel according to its contamination occured. Before detecting the contaminants, the characteristics of biodiesel and the blends are evaluated, based on representative physical properties. In TABLE 1, it shown the change of fuel properties between initial condition and before being used in the engine system. The initial condition is represented from analysis of fuel after the production process. After several months of storage, Biodiesel properties has changed in density, viscosity, acid value, and water content. The significant changes were found in B100. A tolerable changes found in petrodiesel fuel (B0). Water content and acid numbers of biodiesel indicated a value that potentially exceeds the maximum limit of the specified standard.

Table 1. Properties of biodiesel, diesel fuel and the blends (B20)

Properties	Units	Initial Condition			Before Use Condition		
		petro-diesel (B0)	B20	B100	petro-diesel (B0)	B20	B100
Density at 15°C	kg/m^3	822	843	873.1	831	852	895
Viscosity at 40°C	mm^2/s	2.546	3.205	4.83	2.586	3.308	5.477
Acid Value	mg IOD/g	0	0.59	0.48	0	0.61	0.569
Water content	mg/kg	197	265	430	227	334	627

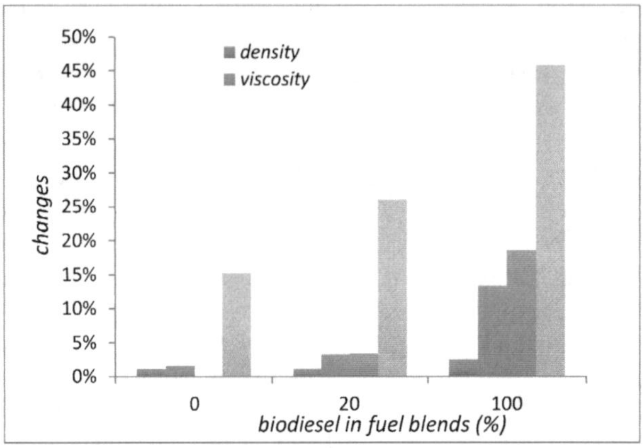

Figure 4. Properties changes on biodiesel and the blends

Environmental factors are very likely to aggravate the storage conditions of B100, particularly in the high increase in water content. As seen on Figure 4, the highest properties change was on water content. Biodiesel may absorbed more humidity since fatty acid methyl esters (FAMEs) are hygroscopic compounds. It is convincing stated that biodiesel getting more hydrophilic without a proper handling and storage. It is surely due to biodiesel has a polar chemical structure of carboxylate ester groups, it presents the characteristics of absorbing more water and more in triggered by environmental conditions. The humidity in the storage area were measured averagely in ranges from 80-90%. The larger changes in physical properties occurring in the biodiesel are expected to be equal to the increased risk of contaminant formation in the oil, thus allowing more blockages to be retained on the surface of the filter media.

Due to high amount of saturated fatty acid in the fuel, it potentially leads to nucleation and growth of solid crystals. Prolonged exposure of the fuel particularly in low temperature causes crystals to grow and form interlocking networks. Unfortunately it was not captured in the tank visually due to some technical difficulties. The solid crystals in oil fuel was not detected easily in the flow due to the sample was discharged from the tank with at ambient air temperature of 26-30°C. However, in any appropriate condition solid materials will cease the fuel flow and lead to starvation of fuel in the engine.

The incidence fuel filter clogging was captured from the media. A diagnose filter blocking problems caused from biodiesel visually presented in Figure 6.

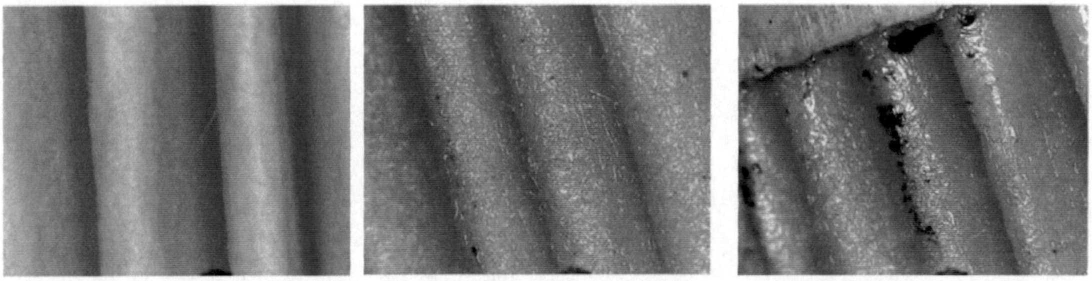

(a) fuel filter canister (initial) (b) after contact B20, 1500 hours (c) after contact B20 2500 hours

Figure 5. Visual impression of clogged filter for engine fuelled B20

Majority clogging in filter that contact with B20 appeared as typical slimy deposits. Within a longer contact period, sticky asphaltine and dirt are appeared in the filter media. Quartz, rust and silica were also found and considered as dirt (particulate), but not on the entire surface of the filter. This is very likely triggered by the higher biodiesel blending ratio with petrodiesel. A mixture of of biodiesel causes water uptake to increase but according to its proportions, the growth of slimy deposits formed in most part of media surface.

On the filter media contacted to B100, it shows more clogging and thicker deposits. The filter surface and the edge of the catridge are almost entirely covered by oil sludge that crystallized. Some particles are trapped inside and form a permanent gell as shown in FIG 6. Such particulates may arise from the biofilms formed in the tank that are thrown off and carried by the fuel line. Asphaltines are generally less than 2 micron in size, they can agglomerate into larger particles which can easily block engine filters or damage injectors. These particles collect at the bottom of a fuel tank and can form an oily sludge that is often confused with microbial contamination.

The oxidation stability of biodiesel is detected problematic in this work, due to it was measured >6 hours. It is inferior to that of petro-chemical diesel. As the fuel comes into contact with oxygen, chemical reactions break down the diesel into peroxide, organic acids and gummy sediment. These soft, sticky substances can stick to fuel filters and engine components and cause acid erosion as found in filter media.

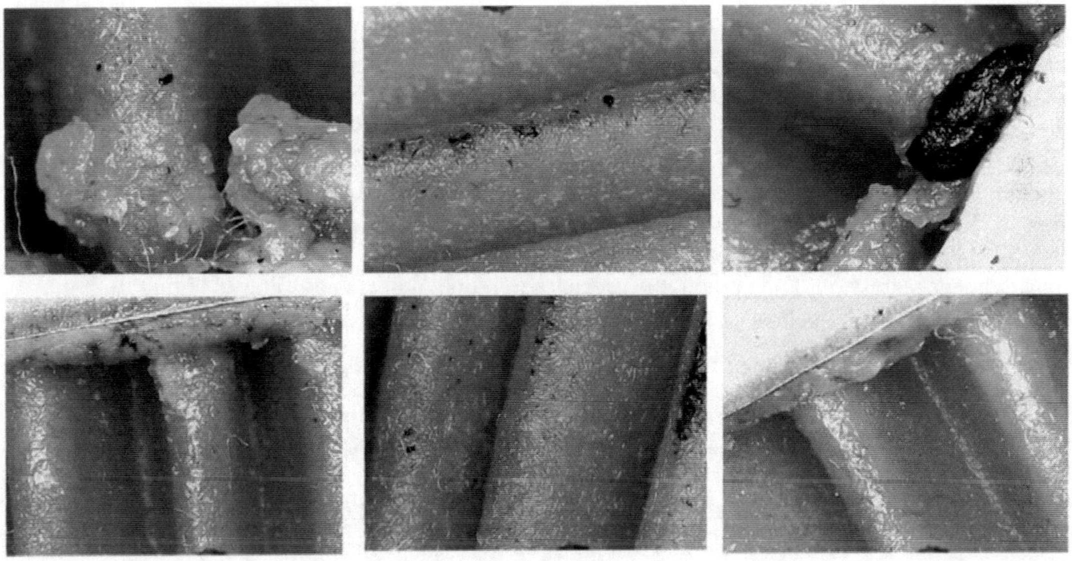

Figure 6. Visual impression of clogged filter contacted to B100 for 3500 hours

The presence of slime deposits due to the use of B100 strongly indicates an increase in the contaminants that strongly possible shown in in the form of sterol glucosides. Sterol glucosides (SGs) occur naturally in vegetable oils and fats in the acylated form. During the biodiesel production process, they are converted to nonacylated SGs. The presence of SGs in biodiesel may contribute to flowability problem due to the high melting point of SGs 240 °C (=464 °F). Otherwise, SGs performed insoluble in biodiesel or diesel fuel. It can essentially be considered as dispersed fine solid particles in biodiesel. These dispersed SG particles may also promote the crystallization of other compounds [25]. The presence of SGs at double-digit parts per million (ppm) levels may cause the formation of a cloud-like haze in biodiesel, even at room temperatures. Under the microscope, the cloud-like substance appears as agglomerates of various sizes composed of discrete particles of 10 to 15 microns [15].

The presence of sterol glucosides in biodiesel has promoted filter blockage. Testing indicates that as the level of sterol glucosides increases, the frequency of Filter Blocking Test (FBT) values increase.

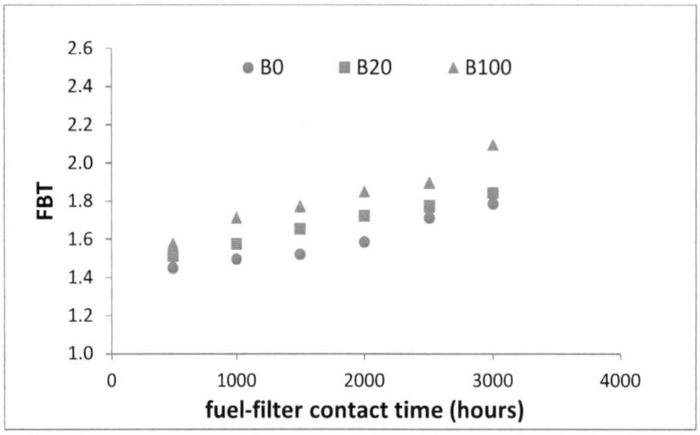

Figure 7. Fuel blocking tendency for various biodiesel blends

The regeneration filter is related to the limit of FBT. The FBT value limit specified as an indication of clogged filters for regeneration has not been widely regulated, due to the variety of filters type and media specification. However some manufacturers and users set some limits related to the sensitivity of the machine and the risks that are important to avoid. According to Tamson Instrument, high blocking tendency can be detected when FBT in range of 3.16 to 7.55. Otherwise, according to a British Standards Institution seen by Platts, all diesel engine user need to respect an FBT limit of 2.52. In this work, the use of B20 ranges in FBT ranges from 1.51 - 1.84, whereas the use of B100 shows FBT 1.58-2.42 in a contact period of 3500 hours.

Using linear regression of the graph, the regeneration time indicated by filter clogging limit may be predicted with the mathematical equations as stated in Table 2.

Table 2. Time limit projection based on filter blocking tendency

Fuel used	Linear Equation	Max Time limit [hours]
B0	y = 1E-04x + 1.3953	11,500
B20	y = 0.0001x + 1.4483	10,500
B100	y = 0.0002x + 1.4947	5000

From the prediction equation models, it requires replacement of fuel filter after 5000 hours contact time with neat biodiesel. It is represented a dramatical clogging on filter under 8 months after B100 being used in the system. In typical of cellulose filter media, the life span for B0 and B20 are respectively 16 months and 14 months. More operation time is still possible as long as petrodiesel fuel qualities kept with proper storage and transport handling before entering the engine system.

Depending on the type of micro-organisms, acids or bacteria sludge is also formed in the filter media. In order to descibe in details microbial growth occurring on the filter media, the type, characteristics and kinetics of growth of each kind of microbes require further testing in the laboratory. Including determining the number of microbial colonies grown per mL of oil sample and oil sludge (Colony Forming Unit) corresponded to the dirt (contamination) holding capacity on a fuel filter .

4. Conclusion

The magnitude of changes in the properties occurring in the biodiesel and its blends are aligned with the increased risk of contaminant formation in the oil. It may increase the possibility of a clogging or blockage on the surface of the filter media. The higher blending ratio of biodiesel used in the engine operation, more potential blockage of the filter occured. Filter regeneration rate becomes faster. Fuel Blocking Tendency (FBT) can be the representative value to predict the maximum time limit of filter use before it is blockage or clogged. The contribution of sterol glucosides to filter blocking is considered in addressing filter problems with biodiesel and biodiesel blends. From the macro image, it showed that the surface of the filter media contacted with biodiesel is more dominantly covered by oil sludge in the form of slime. It indicates that the content of fatty acids in biodiesel is high and forms a contaminant compound that is easy to become gell. This is due to low oxidation stability so it is easily crystallized and become solid.

Acknowledgement
This work was conducted on New and Renewable Energy Sriwijaya University at Indralaya. Authors are grateful to PT Sumiasih Oleochemicals Bekasi West Java Indonesia for providing some materials and the necessary information related.

References
[1] Seagrant.uaf.edu. 1998 *Fuel Filters: Choose Well and Prosper From Pacific Fishing*
[2] Fitch B, 2013 Noria Corporation. Anatomy of Fuel Filter. Machinery Lubrication Magazine
[3] Filter Manufacturers Council 2011 *Technical Service Bulletin* **06-1** 1
[4] Redner S, Datta S. 2000 *Phys. Rev. Lett.* **84** 6018
[5] Dunn R, Knothe VG, and Krahl 2005 *The Biodiesel Handbook AOCS Press*
[6] NREL (National Renewable Energy Laboratory) U.S. Department of Energy 2009 *Biodiesel Handling and Use Guide*
[7] Kapilan N, Babu TP, Reddy RP 2009 *International Journal of ChemTech Research* **1(2)** 278
[8] Schmidt, Margret. DGMK 2009 German Interindustry Study: Filter Blocking, Biofiels, Microbs. Shell Global Solutions. Presentation File
[9] Jain S, Sharma MP 2010 *Renew Sustain Energy Rev* **14** 763
[10] Komariah LN, Marwani, Aprisah S, Rosa YSL 2017 *Matec Web of Conferences* 101. 02012
[11] Terry B. 2005. *Impact of Biodiesel on Fuel System Component Durability.* The Associated Octel Company Limited, Coordinating Research Council. The Associated Octel Company Limited.
[12] Tang H and Salley SO 2008 *Fuel* **87** 3006
[13] Jain SK, Kumar S, Chaube A 2011 *International Journal of Chemical Engineering and Applications* **2(2)**
[14] PM Tamson Instruments. User-Manual Filter Blocking Tendency. https://tamson-instruments.com
[15] Fersner A, Galante-Fox J 2014 *SAE Int. J. Fuels Lubr.* **7 (3)** 783
[16] Wardle DA 2003 *Renew Sust Energy Rev* **7** 1
[17] Dwivedi G, 2013 *et al., International Journal of Renewable Energy Research* **3(4)**
[18] Bhale BV and Nishikant V 2009 *Renewable Energy* **34**, 794
[19] Carraretto C, Macor A, Mirandol A, et al. 2004 *Energy* **29** 2195
[20] Fregolente PBL, Maciel W, and Oliveira LS 2015 *Braz, J. Chem. Eng.* **32**
[21] Williams A, McCormick RL, Hayes R, Ireland J 2006 *NREL.*
[22] Glab J 2016 *Journal of Konbin* **39**
[23] Christensen E, McCormick RL 2014 *Fuel Processing Technology* **128** 339
[24] Dodos GS, Konstantakos T, Linginos S, Zannikos F 2011 *Effecst of Microbiological Contamination in the quality of Biodiesel Fuels,* ICEST 12th Proceedings (2011).
[25] Biodiesel magazine 2017 The Role of Sterol Glucosides on Filter Plugging. Evonik Industries

IOP Conf. Series: Journal of Physics: Conf. Series **1095** (2018) 012018 doi:10.1088/1742-6596/1095/1/012018

Batch Study, Kinetic and Equilibrium Isotherms Studies of Dye Adsorption of Jumputan Wastewater onto Betel Nuts Adsorbent

L Cundari[1], K F Sari[1], L Anggraini[1]
[1]Chemical Engineering Department, Faculty of Engineering, Universitas Sriwijaya, Palembang 30139, Indonesia

E-mail: liacundari@ft.unsri.ac.id

Abstract. Adsorption is an alternative process which can reduce the contaminant of the Jumputan wastewater. Betel nuts activated carbon is one of bio-sorbent that can be utilized for dye adsorption process. The activated carbon is made with carbonization temperature of 500 °C and the HCl concentration of 0.5 M. The adsorption process applies in a batch system with the agitation speed of 150 rpm. The aim of this research is to get kinetic and equilibrium isotherm of dye adsorption in the Jumputan wastewater. This research is done by varied the stirring time (5, 10, 15, 20, and 25 minutes), the mass of the adsorbent (5, 10, 15, 20, and 25 grams). The dye concentration is analyzed by using a portable spectrophotometer. The results show that the adsorption follows pseudo-second-order kinetic with R^2 of 0.929 and k value of 0.00008 min^{-1}. Equilibrium isotherm describes by Langmuir model with R^2 of 0.999 and maximum adsorption capacity of 12.99 mg/g.

1. Introduction

Industrial is a large and complex wastewater producer. It produces a variety of wastes from its production process. Textile industry wastewater can easily be known from its color. Pollution contamination varies both the type and the amount. Napthol is a synthetic dye commonly used in batik traditional Palembang such as Songket, Jumputan, and Blongsong, because of its durability and easy to apply. Napthol contains high chemical organic compounds that cause environmental pollution. It is necessary to eliminate the dyestuff before discharged the waste into the water bodies by using biological, chemical, or process with an adsorbent [1].

One of the adsorbents that can be used to adsorpt contaminant in Jumputan wastewater is betel nuts activated carbon. Cundari, et al. [2-4] used betel nuts activated carbon as an adsorbent to proceed Jumputan wastewater. The optimum condition to produce the adsorbent was 500 °C and 0.5 M HCl for carbonization temperature and concentration of the activator respectively [2]. Batch application of the adsorbent to Jumputan wastewater reduced COD 98.61%, BOD 98.5%, TSS 87.5%, and Crom 96.30%[2]. Variation of bed height and sampling time was done to decrease BOD and COD level on continuous Jumputan wastewater treatment. The optimum condition was 25 cm of bed height and 150 minutes with 99.77% BOD level reduced [3]. COD level was also analyzed with variation of bed

[1] Lia Cundari.

height and sampling time in a Continuous Fixed-Bed Adsorber [4]. The reduction of COD level was 99.69% [4]. The influence of particle size of the adsorbent and sampling time carried out to determine degradation of COD, BOD, and TSS level on continuous adsorption of Jumputan wastewater. The result of this research shows the best adsorbent is 0,5 mm with decreased of COD level to 21.13% at 90 minutes of sampling time, BOD to 20.01%, and TSS to 90.28% at 30 minutes of sampling time [5].

Kinetics adsorption is a very important thing in the process of design, operation, and control that received less attention than the adsorption isotherm [6]. The kinetics adsorption describes the rate of adsorption between adsorbent and adsorbate. Characteristic of adsorption can be seen from its adsorption rate constant (k) and the reaction order resulting from an adsorption kinetics model. The adsorption rate testing can be performed by guessing the reaction order [7].

Some researchers conducted kinetic and isotherm adsorption. Equilibrium modeling and kinetic studies on the adsorption of basic dye by a low-cost adsorbent: coconut (Cocos nucifera) bunch waste was done by varied methylene blue concentration and agitation time [8]. The results obtained were then modeled using three isotherm models: Langmuir, Freundlich, and Temkin. The results showed that the adsorption fitted to the Langmuir isotherm model, with adsorption capacity of 70.92 mg/g and the adsorption kinetics obtained pseudo-second order.

Sururi, et al [9] worked on kinetic study of NOM adsorption process on surface water with zeolite and activated carbon. The adsorption referred to second-order model with q_e value 0.30 mg/g for activated carbon and 0.336 mg/g for zeolite. The value of q_e was closer to q_e experiment value of 0.34 mg/g. Zeolite adsorbed better than the activated carbon.

Other research about equilibrium and kinetics of adsorption of selenium using rice husk ash (RHA) in batch system conducted by [10]. The results of this study showed the optimum weight of rice husk ash 6 g/l with initial concentration of Selenium (IV) 100 mg/l. The adsorption of Se (IV) to rice husk ash followed second-order kinetics and fitted to Freundlich isotherm model.

Kinetic study of adsorption of chromium metallic ion solution (Cr) using banana charcoal (Musa paradisiacal) was done by [11]. They varied concentration, time, and pH. The results showed that the surface area of adsorption of banana charcoal was 3.4559 m^2/g. The adsorption capacity of banana charcoal to Cr^{6+} was 0.8019 mg/g obtained at equilibrium time of 40 minutes with an initial concentration of Cr^{6+} about 75 ppm. The maximum adsorption capacity obtained 0.9088 mg/g at pH 4.00. The adsorption kinetics pattern followed the pseudo second order kinetics with the value of the adsorption rate constant was 0.0008 $min^{-1}ppm^{-1}$.

Betty Hidayati, et al [12] conducted kinetics study of adsorption of Cu^{2+} by using activated natural zeolite. The results showed the kinetics adsorption followed first-order kinetics model with k = 0.0119 min^{-1}. The k value was 0.0111 min^{-1}, 0.0123 min^{-1}, and 0.0145 min^{-1} for temperature of 30°C, 40°C, and 50°C respectively. The value of activation energy for the adsorption process was equal to 10.841 kJ/mol.

Taha M. Elmorsi, et al [13] determined Kinetic and equilibrium isotherms studies of adsorption of Pb (II) water onto natural adsorbent. The adsorption fitted to Langmuir isotherm model with capacity as 83 mg/g and the adsorption kinetics followed the second order kinetics.

The adsorption kinetics is the rate of adsorption of a fluid by the adsorbent in a time period. The adsorption kinetics of a substance can be determined by measuring the change in the concentration of the adsorbed agent, and analyzing the k value (slope), and plotting the data on the graph. Many kinetic models describe the order reactions of the adsorption system based on concentration and capacity of the adsorbent. This kinetic equation has two commonly used types of reaction kinetics, that is one and two reaction order kinetics [14].

a. Pseudo Zero Order

A reaction is said to have a zero order if the magnitude of the reaction rate is not affected by any change in the concentration of the reactant. This means that any increase in reagent concentration will not affect the magnitude of the reaction rate. The linear equation of the zero reaction order is expressed by the following equation.

$$C_A = C_{Ao} - kt \quad (1)$$

If equation (1) above is plotted in the graph of y versus x, then it looks like the figure 1(a).

b. Pseudo First Order

The first-order reaction is a reaction whose velocity depends only on one of the substances which react or is proportional to one of the forces of the reactant. The linear equation of one reaction order is given by the following equation [14]:

$$\ln C_A = -kt + \ln C_{AO} \quad (2)$$

If equation (2) above is plotted in the graph of y versus x, then looks like the figure 1(b).

c. Pseudo Second Order

The second-order reaction is a reaction whose speed is directly proportional to the product of the concentration of the two reactants or directly proportional to the square of the concentration of one of its reactants. If the adsorption mechanism is a second order reaction in which the adsorption rate is directly proportional to the two concentrations of followers. The rate of the kinetics of a two-order reaction can be expressed by the following equations [14]:

$$\frac{1}{C_A} - \frac{1}{C_{AO}} = kt \quad (3)$$

If equation (3) above is plotted in the graph of y versus x, then it looks like the figure 1(c). Where C_A = concentration A at time t = t, C_{Ao} = concentration A at t = 0, k = kinetics constant (min^{-1}), and t = time (minutes).

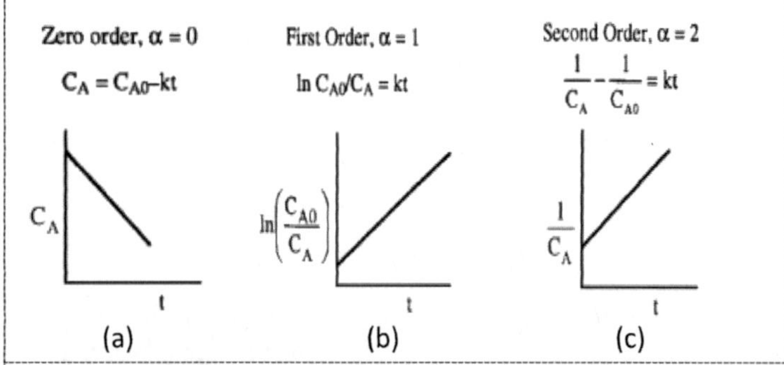

Figure 1. Order Kinetic Graph

The aim of adsorption isotherm is to connect the concentration adsorbate in large quantities and the amount adsorbed on the interface [15]. There are several isotherm equations available for analyzing experimental adsorption equilibrium parameters, the most common are Langmuir and Freundlich models.

a. Isotherm Langmuir Model

The Langmuir isotherm model is based on the assumption that there are a limited number of active sites distributed homogeneously through the surface of the adsorbent. This active site has the same affinity for the adsorption of single-layer molecules and there is no interaction between the adsorbed molecules [16]. The linear form of the Langmuir equation can be expressed as follows:

$$\frac{C_e}{q_e} = \frac{C_e}{q_m} + \frac{1}{q_m b} \quad (4)$$

Where C_e = concentration of solute at equilibrium (mg/L), q_e = number of absorbed solutes per unit mass of adsorbent (mg/g), q_m = maximum adsorption capacity (mg/g), b = a constant related to the affinity of the binding sites (L/mg).

b. Freundlich Isotherm Model

The Freundlich isotherm model applies to adsorption on heterogeneous surfaces by an interaction between the adsorbed molecules, and not limited to the formation of a single layer (monolayer). This model assumes that the concentration of adsorbate increases, the concentration of adsorbate on the surface of the adsorbent also increases, the adsorption energy exponentially decreases at the completion of adsorption [17]. The equation for the Freundlich model is as follows:

$$\text{Log } q_e = \text{Log } K_f + \frac{1}{n} \text{ Log } C_e \quad (5)$$

Where K_f = Freundlich constant and $\frac{1}{n}$ = Heterogenity factors related to capacity and intensity of adsorption.

2. Research methodology

The apparatus are furnace, electric oven, analytical balance, mortar, beaker glass, measuring cup, funnel, desiccator, and pH meter.The materials are betel nut, activator HCl 0.5 M, Jumputan wastewater, and aquadest.

Variable of this research are stirring time (0, 5, 10, 15, 20, and 25 minutes), weight of adsorbent (5, 10, 15, 20, and 25 gr). Production of betel nuts activated carbon based on [2]. Partikel size of adsorbent is 1 mm (18 mesh). Amount of adsorbent is mixed to 200 mL Jumputan wastewater with agitation speed 150 rpm. The batch system occurs to adsorpt dye contaminant on Jumputan wastewater. All samples analyzed with a portable spectrophotometer, with procedures according to [18].

3. Result and discussion

Based on Table 1, the rate of zero-order adsorption kinetics can be seen in figure 2. The graph shows the concentration value (C_A) of the dye in the Jumputan wastewater versus the stirring time is obtained zero-order kinetics rate based on equation (1). The optimum value of R^2 approaches 0.813 with 5 gr of adsorbent.

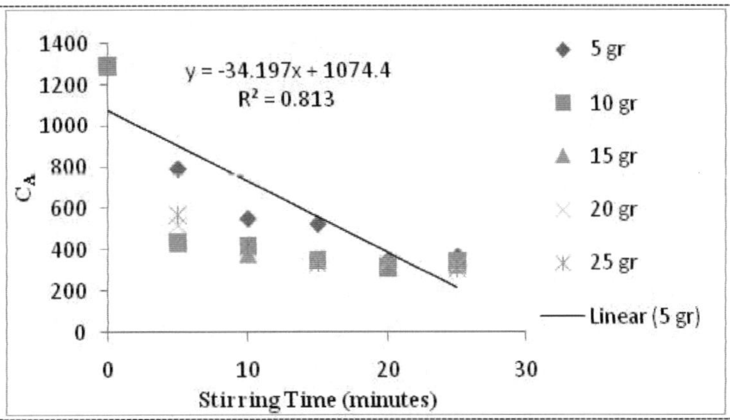

Figure 2. Pseudo-zero-order kinetics for the adsorption of Jumputan's dye onto betel nuts activated carbon

Figure 3 shows the rate of first-order adsorption kinetics on betel nuts activated carbon. The graph shows the value of ln C_{Ao} / C_A versus stirring time. It obtains first order kinetics rate based on equation (2). At the first-order, the value of R^2 approaches 0.907 with 5 gr adsorbent addition.

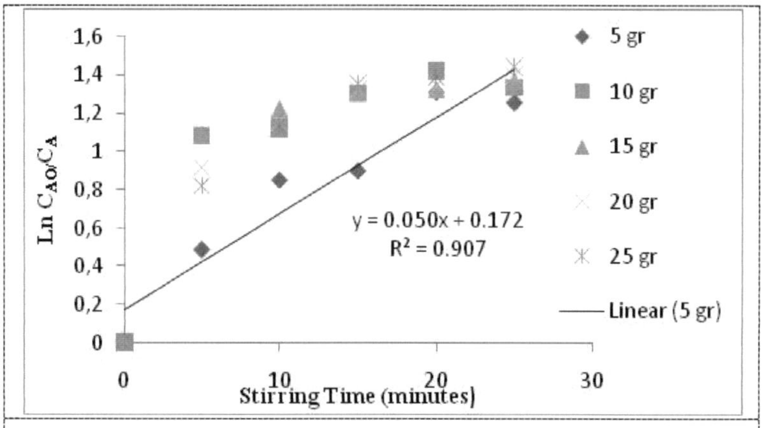

Figure 3. Pseudo-first-order kinetics for the adsorption of Jumputan's dye onto betel nuts activated carbon

Table 1. Kinetics Adsorption of Betel Nuts Activated Carbon

Adsorbent Weight (gr)	Stirring Time (minutes)	C_{AO} (ppm)	$C_{A\,t=t}$ (ppm)	$\frac{1}{C_A}$	$Ln\frac{C_{AO}}{C_A}$
Initial Concentration		1293	1293	0	0.001
5	5	1293	793	0.489	0.001
	10	1293	552.5	0.850	0.002
	15	1293	526.5	0.898	0.002
	20	1293	348.5	1.311	0.003
	25	1293	368	1.257	0.003
10	5	1293	438	1.083	0.002
	10	1293	421	1.112	0.002
	15	1293	350	1.307	0.003
	20	1293	311	1.425	0.003
	25	1293	339	1.339	0.003
15	5	1293	434.5	1.091	0.002
	10	1293	379.5	1.226	0.003
	15	1293	349.5	1.308	0.003
	20	1293	343.5	1.326	0.003
	25	1293	327.5	1.373	0003
20	5	1293	517.5	0.916	0.002
	10	1293	410	1.149	0.002
	15	1293	345	1.321	0.003
	20	1293	337.5	1.343	0.003
	25	1293	319.5	1.398	0.003
25	5	1293	569.5	0.820	0.002
	10	1293	415	1.136	0.002
	15	1293	334	1.354	0.003

IOP Conf. Series: Journal of Physics: Conf. Series **1095** (2018) 012018 doi:10.1088/1742-6596/1095/1/012018

20	1293	321.5	1.392	0.003
25	1293	305	1.444	0.003

Figure 4 shows the value of $1/C_A$ versus stirring time. The rate of second-order kinetics is determined based on equation (3). In the second order, the value of R^2 obtains 0.929 on the addition of 5 gr adsorbent.

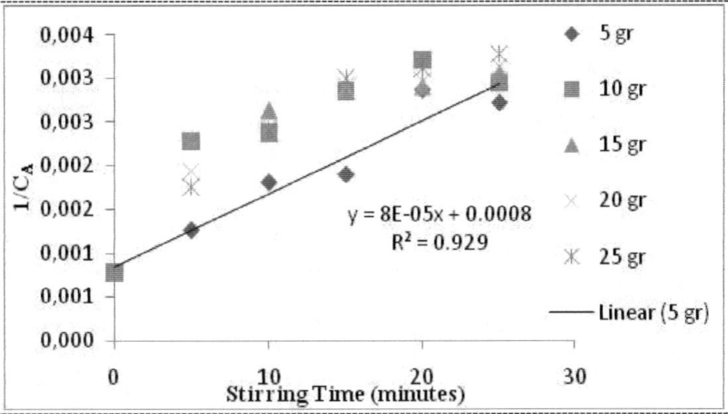

Figure 4. Pseudo-second-order kinetics for the adsorption of Jumputan's dye onto betel nuts activated carbon

From figure 2, 3, and 4, kinetic adsorption rate in this research follows on the second-order with maximum R^2. The adsorption of Jumputan's dye onto betel nuts activated carbon is greatly represented by the pseudo-second-order kinetics. The applicability of second-order to the adsorption data of Jumputan's dye onto betel nuts activated carbon indicates that the concentration of both adsorbent and dye are involved in the rate determining step and the adsorption process may be chemisorption. Similar trends were shown for the adsorption of methylene blue onto coconut's activated carbon [8]. So, the value of adsorption constant (k) of this research is 0.00008 min^{-1}.

Table 2. Comparison of Adsorption Rate Kinetics

Kinetic Order	R^2	k
Pseudo-Zero-Order	0.813	- 34.19
Pseudo- First -Order	0.907	0.05
Pseudo- Second -Order	0.929	0.00008

To know the maximum capacity of Jumputan's dye adsorption which can be absorbed by betel nuts activated carbon occurs by using the equation of Langmuir and Freundlich isotherms. The Langmuir isotherm equation is performed to determine the adsorption capacity using an adsorption isotherm curve prepared by plotting the dye concentration in equilibrium (C_e) versus the amount of the precipitated dye (C_e/q_e). Langmuir's isotherm equation curve is shown in figure 5.

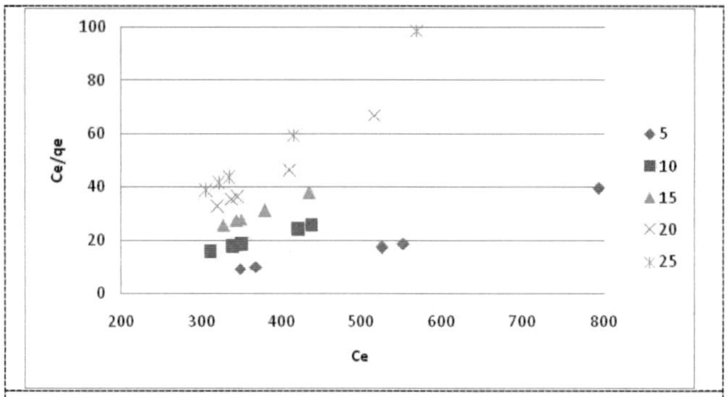

Figure 5. Langmuir Isotherm Adsorption

Table 3. Isotherm Adsorption

Adsorbent Weight (gr)	Stirring Time (minutes)	C_o (ppm)	C_e	C_t	q_e	C_e/q_e	Log C_e	Log q_e
5	5	1293	793.0	500.0	31.7	39.7	2.90	2.79
	10	1293	552.5	740.5	22.1	18.7	2.74	2.76
	15	1293	526.5	766.5	21.1	17.2	2.72	2.69
	20	1293	348.5	944.5	13.9	9.2	2.54	2.54
	25	1293	368.0	925.0	14.7	9.9	2.57	2.49
10	5	1293	438.0	855.0	8.8	25.6	2.64	2.67
	10	1293	421.0	872.5	8.4	24.1	2.62	2.60
	15	1293	350.0	943.0	7.0	18.6	2.54	2.56
	20	1293	311.0	982.0	6.2	15.8	2.49	2.54
	25	1293	339.0	954.0	6.8	17.8	2.53	2.53
15	5	1293	434.5	858.5	5.8	38.0	2.64	2.58
	10	1293	379.5	913.5	5.1	31.2	2.58	2.54
	15	1293	349.5	943.5	4.7	27.8	2.54	2.52
	20	1293	343.5	949.5	4.6	27.1	2.54	2.49
	25	1293	327.5	965.5	4.4	25.4	2.52	2.47
20	5	1293	517.5	775.5	5.2	66.7	2.71	2.50
	10	1293	410.0	883.0	4.1	46.4	2.61	2.52
	15	1293	345.0	948.0	3.5	36.4	2.54	2.48
	20	1293	337.5	955.5	3.4	35.3	2.53	2.47
	25	1293	319.5	973.5	3.2	32.8	2.50	2.45
25	5	1293	569.5	723.5	4.6	98.4	2.76	2.51
	10	1293	415.0	878.0	3.3	59.1	2.62	2.48
	15	1293	334.0	959.0	2.7	43.5	2.52	2.47
	20	1293	321.5	971.5	2.6	41.4	2.51	2.45
	25	1293	305.0	988.0	2.4	38.6	2.48	2.44

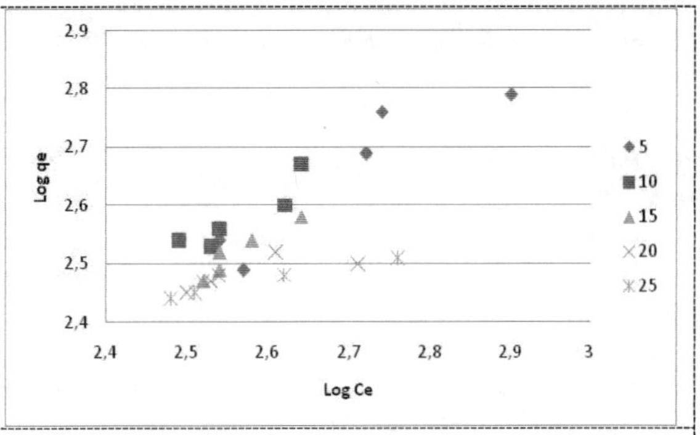

Figure 6. Freundlich Isotherm Adsorption

Figure 6 is a Freundlich isotherm model for the adsorption of Jumputan wastewater using betel nuts activated carbon. The graph representing the linear plot of the Log C_e versus Log q_e based on equation (5). The value of K_F and 1/n (Table 4) were determined from the intercept and the slope respectively. Although, the value of R^2 (0.901) of Freundlich is slightly lower than the value of R^2 (0.999) of Langmuir isotherm. The value of 1/n (indicative of favorability when $0.1 < 1/n < 1$) is 0.855, which is close to the unity and indicates the favorability of the adsorption process. Therefore, Freundlich model is still a good model to describe the adsorption data. The results of Langmuir and Freundlich implies that the adsorption of Jumputan's dye onto betel nuts activated carbon show a complex mechanism involving both monolayer and heterogeneous surface condition. Previous reports indicated that adsorption onto different bio-sorbents such as coconut bunch [8], banana charcoal [11], and natural adsorbent [13] show similar results.

Table 4. Langmuir and Freundlich Constant

Weight Adsorbent (gr)	Langmuir			Freundlich			
	R^2	**q_m**	**b**	**R^2**	**1/n**	**n**	**K_f**
5	0.965	14.93	0.0041	0.86	0.85	1.18	2.30
10	0.999	12.99	0.0093	0.818	0.812	1.23	3.13
15	0.999	8.47	0.0088	0.901	0.855	1.17	2.12
20	0.996	5.85	0.0075	0.55	0.238	4.20	73.96
25	0.994	4.42	0.0071	0.914	0.228	4.39	75.86

The value of q_m, b and R^2 are presented in Table 4. The high value of R^2 as 0.999 indicated minimal deviation from the fitted equation showing that the adsorption data would follow Langmuir equation. The data in Table 4 indicated that the maximum adsorption capacity of betel nuts activated carbon for Jumputan's dye was calculated as 12.99 mg/g. It can be mentioned that the surface of betel nuts activated carbon is homogeneous and the adsorption of Jumputan's dye formed a monolayer on its outer surface. B.H. Hameed, et al [8] found the adsorption of methylene blue onto coconut bunch activated carbon followed Langmuir model and formed a monolayer.

4. Conclusion

In this work, the ability of betel nuts activated carbon to remove dye from Jumputan wastewater is investigated. The adsorption follows pseudo-second-order kinetic with R^2 of 0.929 and k value of 0.00008 min^{-1}. Equilibrium isotherm describes by Langmuir model with R^2 of 0.999 and maximum adsorption capacity of 12.99 mg/g. The betel nuts activated carbon is a promising material to reduce dye from Jumputan wastewater.

Acknowledgments

The authors acknowledge the research grant provided by the Universitas Sriwijaya under SATEKS scheme (No. 1013/UN9.3.1/PP/2017)

References

[1] Ramdja A F 2008 *J. Teknik Kimia*, **15**(4) 1.
[2] Cundari L 2015 *Added Value of Energy Resources (AVoER VII)*. Palembang: Universitas Sriwijaya.
[3] Cundari L 2016 *Applicable Innovation of Engineering and Science Research (AVOER VIII)*, 190.
[4] Cundari L 2017 *MATEC Web of Conferences*, **101**, 0200.
[5] Cundari L 2016 *J. Teknik Kimia*, **22**, 4, 19.
[6] Morgan D.A1985 *J. Am. Oil Chem. Soc.*, **62**, 292.
[7] Muslich 2010 Kinetic adsorption of *β-Karoten from rigid palm olein by using Bentonite*. The technology of Agriculture Industrial Department (Bogor – IPB.
[8] Hameed BH 2007 *J. of Hazardous Materials*, **158**, 65.
[9] Sururi M and Rangga 2009 *J. Sains dan Teknologi Lingkungan*, **1**, 2, 107.
[10] Gulipalli and Sekhararao CH 2011 *J. of Engineering Science and Technology*, **6**, 5, 586.
[11] Widihati and Ida Ayu G 2012 *J. Kimia*, **6**, 1, 8.
[12] Hidayati and BettyKinetic 2012 adsorption study of Cu^{2+} *by using natural zeolite,* Chemical Engineering Department, Faculty of Engineering, Universitas Riau.
[13] Elmorsi and Taha M 2014 *J. of Environmental Protection*, **5**, 1667.
[14] Emrah B, Mahmut O and Ayhan S 2008 *Microporous and Mesoporous Material, Elsevier*, **115**, 234.
[15] Eastoe J and Dalton J.S 2000 *J. Colloid Interface Sci.*, **85**,103.
[16] Langmuir I 1918 *J. American Chem. Soc.*, **40**, 1361 .
[17] Freundlich H 1906 *Z.Phys. Chem.*, **57**, 384.
[18] Rusdianasari 2012 *Chemical analytic and Instrumentation modul*, 30.

Growth Retardation of *Saccharomyces cerevisiae* by Noni Fruit (*Morinda citrifolia*) Extract Occurred in G1 to S Transition of the Cell Cycle

Hermansyah[a)*)], Susilawati[b], and Dyah Subositi[c]

a Chemistry Dept, Faculty of Mathematics and Natural Sciences, Sriwijaya University, Jalan Raya Palembang-Prabumulih KM 32 Indralaya, Ogan Ilir, South Sumatera, Indonesia 30662. Phone (+62) -711-580269
b Faculty of Medicine, Sriwijaya Univerity, Jalan Dr.Moh. Ali Komp. RSMH Palembang, South Sumatera, Indonesia, 30126. Phone . (+62). 711-352342
c Research and Development of Medicinal Plants and Traditional Medicinal Center, Karanganyar-Tawamangu, Jawa Tengah, Indonesia.

*email. hermansyah@unsri.ac.id

Abstract : We previously reported that methanolic extract of noni fruit (*Morinda citrifolia*) can inhibit growth of *Saccharomyces cerevisiae*. This finding suggested that noni fruit has anti-proliferative effect. Previous data based upon FACS (Fluorescence Activated Cell Sorting) showed that growth of *S.cerevisiae* inhibited in G1 to S transition phase in the presence of 1% methanolic extract of noni fruit. Further analysis to explore genes involved in this slow transition step, we conducted real time PCR. Expression some genes changed in the presence of 1% methanolic extract of noni fruit. Transcriptional level of gene *CDC28* (cyclin dependent kinase-28), *CLN2*, *CLN3*, dan *SWI6* changed 2.25, 2.79, 4.87 and 7.21 folds, respectively. This data suggested that over transcription of those gene inhibit cell cycle progression. *CDC28* is master regulator of mitotic and meiotic cell cycles, *CLN2* and *CLN3* are G1 cyclin which play an important role in regulation of cell cycle, and SWI6 is transcription cofactor where forms complexes with *SWI4* and *MBP1* in regulation of transcription at G1 to S phase transition.

Keywords : *Morinda citrifolia, Saccharomyces cerevisiae*, G1 phase, Antiproliferative

1. Introduction

Noni plant (*Morindra ciytrifolia L*) is a small tropical tree that well known as volk medicine for some diseases. Almost all parts of this plant such as fruit, leaf, bark, root, flower, and seed are widely used in traditional medicine including diabetes, hypertension, arthritis, infections, gastric ulcers, depression, senility, menstrual difficulties, headaches, sprains, muscles aches, colds, cancer, poor digestion, atherosclerosis, blood vessels problems, drug addiction, antibacterial, antiviral, anti-tumor, antihelmin, analgesic, hypotensive, antiinflammatory, immune enhancing properties [1-2]. Noni plant grows widely in Polynesia. Unfortunately, many of these property claims are still needed scientific experiment. We reported antiproliferative effect of methanolic extract of noni fruit using *Saccharomyces cerevisiae* cell organism model [3].

Antiproliferative property of noni fruit is an essential property to inhibit growth of cells. Many antiproliferative mechanism involve genetics and epigenetics alteration, change in cell cycle progression mechanism, DNA replication, response of genomic integrity stress. Therefore *Saccharomyces cerevsiae* has a potent as organism model in screening medicine candidate for anti cancer and elucidating genes function cellular mechanism for antiproliferative property [4].

Cell cycle is the series of events to produce two daughter cells divided in four phases in the following order G1, S, G2, and M phases. DNA replication occurs in S phase, while replicated DNA molecule are segregated to daughter cell in M phase. During G1 phase, the cell grows and prepares itself for S phase, while G2 phase is the gap between the end of replication and the beginning of mitosis. These cell cycles progression have to run properly and control tightly [5-6]. The regulatory subunits of the cyclin-dependent kinases (CDKs) or cyclins, that regulate progression through the cell ceycle, and cyclin D1 overexpression is found in more than 50% of human breast cancers and causes mammary cancer in transgenic mice [7].

This study aimed to explore genes involved in anti-prolifertaive effect of noni fruit to *S.cerevisiae* as organism model, *S.cerevisiae* wild type strain was used as an organism model. Our recently research based upon FACS data resulted that in the presence of 1% extract methanol of noni fruit inhibited growth of *S.cerevisiae* wildtype strain FY833 slightly inhibited in G1 phase. mRNA level some genes were determined using quantitative PCR. From our study indicated genes *CDC28*, *CLN2*, *CLN3*, and *SWI6* involved in arrested in G1 phase.

2. Materials and Methods

2.1 Materials

Isolates, strains, and media *S.cerevisiae* used in this experiment was wild type strain FY833 with genotype *Mata his3Δ200 ura3–52 leu2Δ1 lys2Δ202 trp1Δ6*. YPDA broth medium comprised of 10 g/L yeast extract, 20 g/L peptone, 20 g/L glucose, and 400 mg/L adenine, while 20 g/L bacto agar was added to prepare YPDA agar. Extract methanol of noni fruit was mixtured into autoclaved YPDA media.

2.2. Experimental procedure

2.2.1 Culture

Routine cultivation of yeast strain was conducted by streaking yeast on YPDA agar media plate incubated at 30°C for 1-2 day.

2.2.2 Spot culture

A 5 mL cells grown on YPDA broth culture until reach OD_{660} 1.0, cells were then collected using centrifuge 5000 rpm, and resuspended cell with a sterile water adjusted until getting cells with concentration ten fold dilution series started from highest concentration 10^8 cells/mL, and 5 μL cell suspensions were spotted on YPAD agar media plate with and without 1% methanolic extract of noni fruit, Incubated cells at 30°C for 1-2 day.

2.2.3 Synchronization of cellcycle

Synchronization of cell cycle were performed according to manufacture's protocol with slight modification. FY833 strain cells were grown in YPAD to mid exponential phase, and washed with a 30 ml sterile water twice by centrifugation 2000 rpm 5 minutes to remove a protease (Bar1p) produced *MAT***a** cells that destroyes the α-factor, then were arrested in G1 phase using 6 μg/ml α-factor (Sigma) incubated with shaking three hours, after which cells were washed five times with fresh YPAD and transferred into fresh medium in the presence and absence of noni fruit methanol extract.

2.2.4 FACS (Fluorescent-Activated Cell Sorting) analysis.

Cells were grown in YPAD medium overnight and transferred into YPAD fresh medium to reach exponential phase culture (OD_{660} = 1.0 or 10^7 cells/ml) at 30°C. It then was prepared to performed

FACS analysis according to [8] with slight modification. After washed three times with 30 ml sterile water, the cells were released by resuspending them into the same volume of fresh YPAD broth media with and without 1% (w/v) extract methanol of noni fruit. Sample culture was taken and transferred into 2 ml eppendorf tube every 15 or 20 minutes. Cells were collected by centrifugation for 1 min at 12,500 rpm and resuspend in 300 μl of 0.2 M Tris-HCl pH 7.5, and then 700 μl of cold ethanol (- 20°C) were added gradually with vortexing, and stored it at -20°C until using fixed cells (> 12 h). Cells samples were washed twice with 0.2 M Tris-HCl pH 7.5, sonicated with minimum output (set the level "1") for 5 second (2 second x 2 times with interval. Cells were harvested by centrifugation for 5 min at 15,000 rpm and resuspended in 100μl of RNAse solution (1 mg/ml RnaseA in 0.2 M Tris-HCl pH 7.5) at 30°C for 3-4h. Afterwards samples were collected by centrifugation for 5 min at 15,000 rpm and resuspended in 100 μl of PI solution (Trisodium citrate dehydrate 20 mg, NaCl 11.6mg, NP-40 20 μl, PI 1 mg, H_2O 20 ml) on ice for 15 min, added 400 μl of 0.2 M Tris-HCl pH 7.5 and chill on ice until measuring DNA contents. Suspension cells were filtered using 5 ml polystyrene round bottom tube within cell stainer cap. Finally, treated samples were analyzed using FACS (Fluorescent-Activated Cell Sorting).

2.2.5 Determination of mRNA level using quantitative PCR (RT-PCR

RNA was isolated by the hot phenol method [9] with some modification. Briefly, cells grown in independent cultures to mid-log phase at 30°C in YPAD in the presence or absence of 1% (w/v) noni fruit methanol extract were harvested by centrugation 2000 rpm 5 minute. Total RNA was isolated by sequential treatment of TES/ distilled water saturated phenol, Trizol-LS (BRL), chloroform, and isopropanol prior to precipitation with sodium acetate buffer (pH 5.2) and 99.9% Ethanol mixture. The resulting RNA was dissolved in pre-chilled distilled sterile water.

First-strand cDNA was synthesized using a High Capacity cDNA Archive kit (Applied Biosystems) and used as the template for quantitative RT PCR with a 7300 Real Time PCR system with SYBR green PCR Master Mix (Applied Biosystems). The RT-PCR primers used in this study were designated using Primer Express (Applied Biosystems). The primers sequence are shown in the Table 1:

Table 1. Primer pairs used in this study

No	Primer	Nucletide Sequence	Sequence Amount
1	KfRT-*ACT1*	5'- TGGTATGTGTAAAGCCGGTTTTG -3'	(23)
	KrRT-*ACT1*	5'- CATGATACCTTGGTGTCTTGGTCTA -3'	(25)
2	KfRT-*CDC28*	5'-TGGGAACGCCGAATGAAG-3'	(18)
	KrRT-*CDC28*	5'-GAGGAAAGCTTGGCTTGAAATC-3'	(20)
3	KfRT-*CLN1*	5'-AATGGTCCTGTAAGAGAAAGTCACAA-3'	(26)
	KrRT-*CLN1*	5'-TTTGCGGTGAACTACTGATATGTTC-3'	(25)
4	KfRT-*CLN2*	5'-TTACGGGACCAAGCCAAATT-3'	(20)
	KrRT-*CLN2*	5'-TTACAACCGCCCCAAGTTTT-3'	(20)
5	KfRT-*CLN3*	5'-CGGCAACATTCGACTCCTA-3'	(19)
	KrRT-*CLN3*	5'-GGACAACGCCAGGCGATA-3'	(18)
6	KfRT-*SWI6*	5'- TGAAGGCGATATTTCCAAACTATTAG-3'	(26)
	KrRT-*SWI6*	5'- CTGCGGGAAGTTGATTTCGT-3'	(20)

IOP Conf. Series: Journal of Physics: Conf. Series **1095** (2018) 012019 doi:10.1088/1742-6596/1095/1/012019

7	KfRT-*WHI5*	5'- GGCCCATCCGCGAAAT-3'	(16)
	KrRT-*WHI5*	5'- TGCAGCTTGACTAACGCGTAA-3'	(21)

3. Results and Discussions

3.1 Phenotypic Assay

Anti-proliferative effect induces cell cycle arrest. Recently, we reported that the growth of *S.cerevisiae* wild type strain FY833 was strongly inhibited in the presence of 3% noni fruit methanol extract, and cell cycle completely inhibited at G1 to S transition progression [3]. This indicated that cell arrested either before or after the establishment of DNA synthesis.

In this article, we report that in the presence of 1% noni fruit methanol extract on YPAD media, the growth of *S.cerevisiae* cell slightly slow down. The growth was observed slower comparing with cells cultivated in the absence of extract and incubated at 30°C for 1 day as shown in Figure 1.

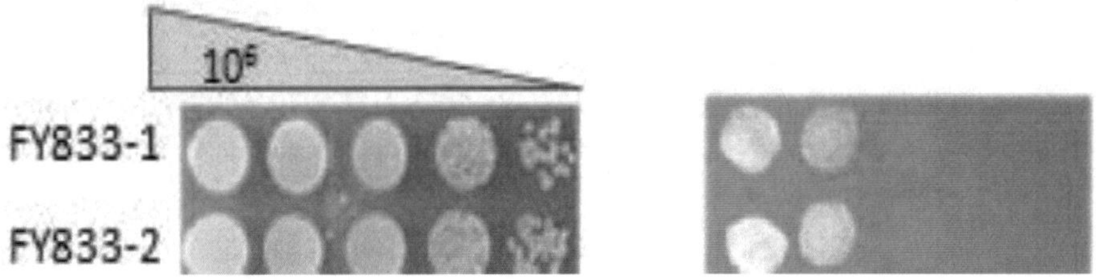

Figure 1. Spot analysis growth of yeast *S.cerevisiae* strain FY833 in YPAD agar media in the presence of 1% methanolic extract of noni fruit.

3.2 *Yeast cells synchronization with α-factor pheromone*

This experiment aimed to obtain cells with unbudding cells form because cell cycle delayed in G1 to S phase. After 6μg/mL α-factor pheromone was added into **MAT**a cells cultured in the YPAD medium. Based upon morphological observation stated that 80% cells in the unbudding form and 20% cells in the budding indicated that cells were inhibited in G1 to S progression (Figure 2a). Oppositely, in the absence of α-factor pheromone, 40% cells in the unbudding form and 60% cells in the budding indicated that mostly cells were in G2 to M phase (Figure 2a).

The α-factor pheromone effect synchronized cells into one form at G1 phase and arrested of cell cycle progression. Morphological observation showed that α-factor pheromone was inducible responses in unbudding, especially the shmoo form. Response of **MATa** yeast cells to the α-factor pheromone also induce the arrest of cell cycle progression. The pheromone response is a suitable production phase, and that it may be useful for informing synthetic biology design [10].

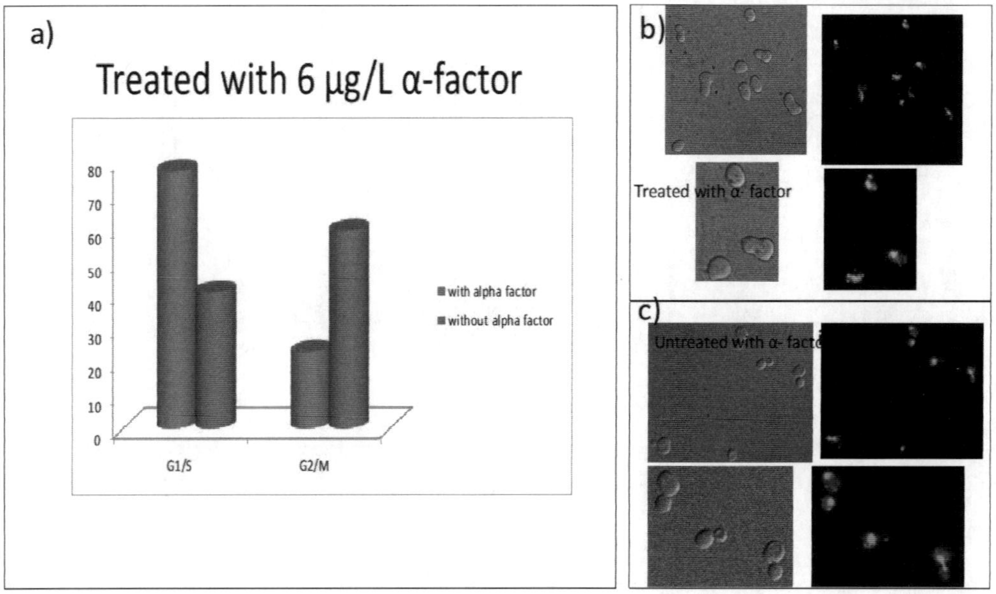

Figure 2. Yeast cells synchronization with α-factor pheromone. Cells treated with α-factor arrested at G1 to S transition where those cells were higher at G1/S phase than at G2/M2 phase (Figure a). Cells treated with α-factor generated unbudding with shmoo cells form .

3.3 FACS analysis of cells gown in YPDA medium with methanolic extract of noni fruit

Further experiment to verify whether growth inhibition or sensitive phenotype of yeast cell when grown in YPDA containing 1% methanolic extract of noni fruit, we conducted Fluorescence Activated Cell Sorting (FACS) analysis.

Cells synchronized with α factor, then cells were released in YPDA medium with this noni fruit extract. FACS analysis data resulted that in the presence of 1% methanolic extract noni fruit, cells were difficulty to form budding or enter into G2 to M phases of cell cycle (Figure 3a and Figure 3b). Oppositely, when α factor-synchronized cells ware release in YPDA media without noni fruit extract, cells progress to form budding or enter into G2 to M phase properly (Figure 3c and Figure 3d). These resulted indicated that methanolic extract of noni fruit inhibited G1 to S transition progression.

Noni ethylacetate extract caused an increase in the proportion of cell in sub G1 phase, which reflects the induction of apoptosis [11].

IOP Conf. Series: Journal of Physics: Conf. Series **1095** (2018) 012019 doi:10.1088/1742-6596/1095/1/012019

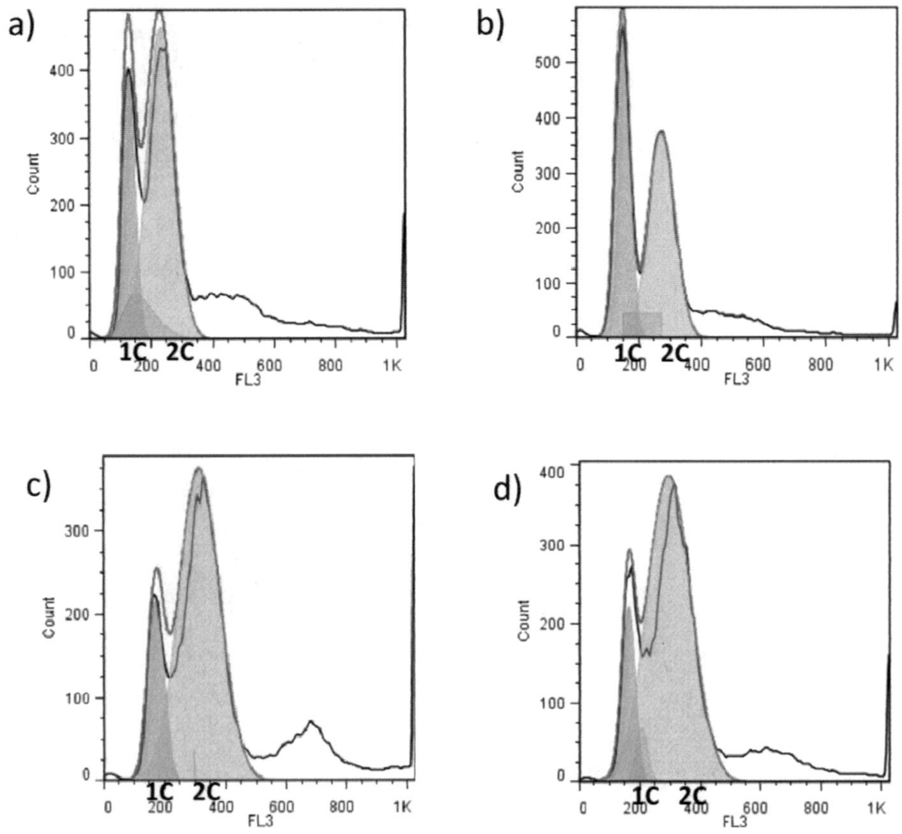

Figure 3. FACS analysis data. The α-factor-treated cells were analysed using FACS, cells after synchronization (a) and(c); and synchronized cells were released in YPAD media containing methanolic extract (b) compared with without noni extract (d).

3.4 *Alteration of some genes involved in cell cycle, CDC28, CLN2, CLN3, WHI5, and SWI6*

To better understand anti-proliferative activity from noni fruit extract, the cultured cells with lower concentration (1% w/v) was then analyzed the expression level of genes cyclin dependent kinase Cdc28 which is crucial genes for cell cycle progression and related genes cyclins (*CLN1*, *CLN2*, *CLN3*), *WHI5* and *SWI6*.

Based upon this RT-PCR data, expression level of *CDC28*, *CLN2*, *CLN3*, and *SWI6* increase 2.2, 2.8, 4.9, and7.2, respectively while expression level of *WHI5* was similiar between presence and absence of extract methanol of noni fruit (Figure 4). These resuts indicated that genes *CDC28*, *CLN2*, *CLN3*, and *SWI6* involved in arrested in G1 phase.

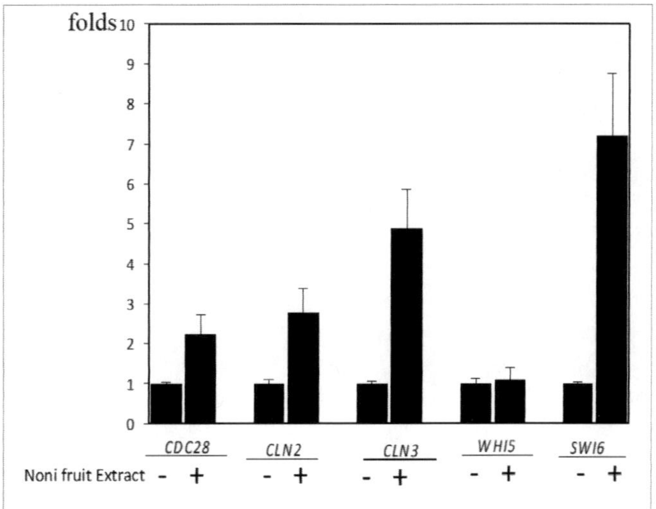

Figure 4. RT-PCR analysis genes involved in G1 to S transition cell cycle

Cyclin genes play an essential role in cell cycle progression. As described in www.yeastgenome.org stated that *CDC28* is Cyclin-dependent protein kinase; interacts with regulatory subunit Cyb1; determination of cell morphology during the cell cycle; phosphorylated mostly by Swe1 and phosphorylation is regulated by Hsl1; 5'-UTR intron; Spider biofilm repressed. *CLN2* is G1 cyclin involved in regulation of the cell cycle; activates Cdc28p kinase to promote the G1 to S phase transition; late G1 specific expression depends on transcription factor complexes, MBF (Swi6p- Mbp1p) and SBF (Swi6p-Swi4p). *CLN3* is G1 cyclin involved in cell cycle progression. *WHI5* is repressor of G1 transcription that binds to SCB binding factor (SBF) at SCB target promoters in early G1; phosphorylation of Whi5p by the CDK, Cln3p/Cdc28p. *SWI6* is Transcription cofactor that activates transcription by RNA polymerase II during heat stress and during the G1/S mitotic transition; also regulates meiotic recombination; component of both the MBF and SBF complexes. Other study using human liposarcoma SW872 cells were characterized by strong antiproliferative effect and modulating oxidative stress [12].

4. Conclusion

In this research described that expression some genes changed in the presence of 1% methanolic extract of noni fruit. Transcriptional level of gene *CDC28* (cyclin dependent kinase-28), *CLN2*, *CLN3*, dan *SWI6* changed 2.25, 2.79, 4.87 and 7.21 folds, respectively. This data suggested that over transcription of those gene might inhibit cell cycle progression.

Acknowledgment

The authors would like to thank Research and Development of Medicinal Plants and Traditional Medicinal Center B2P2TOOT for supporting through RISTOJA research grant. Special thank to Asc.Prof Minetaka Sugiyama for facilitating RT-PCR analysis.

References

[1] Wang MY, Su C 2000 *Annal New York Academy of Sciences*, 61
[2] Samoylenko V, Zhao J, Dunbar DC, Khan IA, Rushing JW MI Damnacanthal. J 2008 *Agric Food Chem* **54(17)** 6398
[3] Hermansyah, Susilawati 2017 *J Mol* **12(1)** 37
[4] Qaddouri B, Guaadaoui A, Bellirou A, Hamal A, Melhaoui A, Brown GW Bellaoui M 2011 *Evidence-Based Complementary and Alternative Medicine* **2011** 1
[5] Vermeulen,K.,Van Bockstaele,D.R.,Berneman ZN, Vermeulen K, Van Bockstaele DR, Berneman ZN 2003 *Cell Prolif.* **36(3)** 131
[6] Amigoni L, Colombo S, Belotti F, Alberghina L, Martegani E 2015 *Cell Cycle* **14(15)** 2429

[7] Casimiro MC, Crosariol M, Loro E, Li Z, Pestell RG 2013 *Genes and Cancer* 3(11-12) 649
[8] Schönthal AH, Pozarowski P, Darzynkiewicz Z. Analysis of Cell Cycle by Flow Cytometry 2004 *Checkp Control Cancer* **281** 301
[9] Ausubel FM, Brent R, Kingston RE, Moore DD, Seidman JG, Smith J 2003 *Current Protocols in Molecular Biology* **1** 146
[10] Williams TC, Peng B, Vickers CE, Nielsen LK. 2016 Metab Eng Commun 6(3) 142
[11] Sharma K, Pachauri S, Khandelwal K, Ahmad H, Arya A, Biala P, 2015 *Drug Res* **66(3)** 141
[12] Ruhomally Z, Somanah J, Bahorun T, Neergheen-Bhujun VS 2016 *J Tradit Chinese Med Sci* **6(3)** 299

Moisture Sorption Isotherm Characteristics of Fermented Cassava Flour By Red Yeast Rice

M N Cahyanti[1], M N Alfiah[1], S Hartini[1]

[1]Chemistry Department, Science and Mathematics Faculty, Universitas Kristen Satya WacanaJl. Diponegoro 52-60 Salatiga, Indonesia

E-mail : margareta2486@gmail.com

Abstract. This study aims to study the characteristics of moisture sorption isotherm of fermented cassava flour by red yeast rice at various fitting models and to determine the best fit model. Samples were analyzed at 303K, 308K, and 313K and equilibrium relative humidity (ERH) from 10%-99% using the gravimetric method. The experimental data were fitted into five equations, e.g., Guggenheim-Anderson-de Boer (GAB), Henderson, Halsey, Oswin, and Chen-Clayton model. The best-fitting model was analyzed using Mean Relative Determination (MRD) analysis. The results showed that the GAB model was the best fitting model for 303K and 308K with MRD coefficient 1.98% and 3.11%. Henderson model was the best fit model for 313K with MRD coefficient 3.32%.

Keywords: moisture sorption isotherm, cassava flour, red yeast rice

1. Introduction

The moisture sorption isotherm curve is a curve that connects the equilibrium moisture content to the relative humidity of the environment at a given temperature. This curve is modeled with various mathematical models to determine the best model to further learns. The modeling curve can be used to study the moisture adsorption characteristics of the product. The modeling curve is also used to determine the shelf life of a product.

Previous research states that the GAB is the best model to explain the moisture adsorption by fermented cassava flour [1]. This research has a limitation because it uses only three models namely GAB, BET, and Caurie. Other research on sorption isotherms of moisture in cassava and its derivative products use another modeling.

Other studies [2] examined the isotherm of water sorption on instant cassava using BET, Oswin, Halsey, Henderson, Caurie, Chen-Clayton, and GAB models. The results show that Henderson model is the best model to explain the water absorption by instant cassava. Koua et al. use GAB, BET, Halsey, Oswin, Smith, Iglesias-Chirife, and Khun model to study moisture adsorption on cassava. Results demonstrated that the GAB model adequately predicted equilibrium moisture content of cassava for the range of temperatures and water activities studied [3]. Adebowale et al. use Bradley, Exponential, GAB, Iglesias-Chirife, and Oswin to research moisture adsorption on three different varieties of cassava. However, the GAB model gave the best fitfor sorption isotherm of tapioca from either the Odongbo or Okoiyawo varieties, while the exponentialmodel was the best for that from TMS30572 variety [4].

The purpose of this research is to determine the characteristics of water sorption isotherm from fermented cassava flour using various modeling.The experimental data were fitted into five equations, e.g., Guggenheim-Anderson-de Boer (GAB), Henderson, Halsey, Oswin, and Chen-Clayton model.

2. Materials and Methods

The sample used in this research is fermented cassava with Angkak (Monascus purpureus). The chemicals used are seven types of salt used to regulate relative humidity (RH), i.e., NaOH, $MgCl_2$, K_2CO_3, $Mg(NO_3)_2$, KI, NaCl, KCl. Other chemicals used are distilled water. The tool used for this research is porcelain cup, chamber to determine sorption isotherm, moisture analyzer to measure moisture content, analytical balance, incubator, thermometer, hygrometer, drying cabinet and sieve for making flour.

2.1 Preparation of Fermented Cassava Flour [5]

Fresh cassava is cleaned from the skin then washed with running water. Cassava is steamed for approximately 60 minutes, then drained and cooled until cassava reaches room temperature. Cassava is cut and weighed as much as 1 kg then inoculated with Angkak with a concentration of 12%, then fermented for three days at room temperature. Fermented cassava is heated at a temperature of 55°C to dry. After that, the fermentation is milled and sieved with mesh sieve 61.

2.2 Preparation of Chamber [6]

The saturated salt solution is prepared using NaOH, $MgCl_2$, K_2CO_3, $Mg(NO_3)_2$, KI, NaCl, KCl. The salt is weighed with a certain mass and then put into a beaker glass that contains warm water temperatures of approximately 50°C and stirred until dissolved. If the salt is entirely soluble, the salt is added bit by bit until the salt does not dissolve. 50 ml of saturated salt solution is inserted in a chamber. The solution is allowed to stand for about seven days to obtain a constant RH.

2.3 Measurement of Moisture Content of Equilibrium [7]

5.00 grams of each sample is stored in a chamber with a specific RH. The chamber is stored at a temperature of 30°C, 35°C, and 40°C. Weight sampling is done daily for seven days until steady state is reached. If the sample has molded before seven days then weighing can be stopped. Please follow these instructions as carefully as possible, so all articles within a conference have the same style to the title page. This paragraph follows a section title so it should not be indented.

2.4 Mathematical Modelling

Mathematical models used to study isotherm moisture sorption include the Guggenheim-Anderson-de Boer (GAB) model, Henderson, Halsey, Oswin, and Chen-Clayton. Each line equation for each model can be seen in Table 1.

2.5 Model Accuracy Test

The statistical validity of each model is determined using the MRD (Mean Relative Deviation) statistic parameter with the following equation:

$$\mathrm{MRD} = \frac{100}{n}\sum_{i=1}^{n}\frac{|\mathrm{Mi}-\mathrm{Mpi}|}{\mathrm{Mi}}$$

MRD: Mean Relative Determination (%)
n: Numberof observations
Mi: Experimental values of the equilibrium moisture content
Mpi: Predicted values of the equilibrium moisture content

Table 1. Models Used and Their Equations

Model	Equation	
GAB	$M = \frac{M_m CKa_w}{(1 - Ka_w)(1 - Ka_w + CKa_w)}$	M_m : Monolayer moisture content M : Equilibrium moisture content a_w : Water activity K, C : Contants
Henderson	$\ln M = \frac{1}{B}\ln(-\ln(1 - a_w)) + \frac{1}{B}\ln\frac{1}{AT}$	M : Equilibrium moisture content a_w : Water activity A, B : Contants T : Absolute temperature
Halsey	$\ln M = \ln A_1 - A_2 \ln(-\ln a_w)$	M : Equilibrium moisture content a_w : Water activity A_1, A_2 : Contants
Oswin	$\ln M = \ln P_1 + P_2 \ln\left(\frac{a_w}{1 - a_w}\right)$	M : Equilibrium moisture content a_w : Water activity P_1, P_2 : Contants
Chen-Clayton	$\ln\left(\ln\left(\frac{1}{a_w}\right)\right) = \ln P_1 - P_2 M$	M : Equilibrium moisture content a_w : Water activity P_1, P_2 : Contants

3. Results and Discussion

The relative humidity (RH), water activity (a_w) and equilibrium moisture content (M) on the fermented cassava flour are shown in Table 2. Moisture sorption isotherm curves connecting the water activity with equilibrium moisture content are shown in Figure 1.

Table 2. Relative Humidity (RH), Water Activity (a_w) danEquilibrium Moisture Content (M) at 30°C, 35°C,and 40°C

Salt	30°C			35°C			40°C		
	RH (%)	a_w	M (%db)	RH (%)	a_w	M_e (%db)	RH (%)	a_w	M (%db)
NaOH	11	0.11	2.55	10	0.10	2.37	10	0.10	2.14
$MgCl_2$	38	0.38	6.57	38	0.38	6.09	31	0.31	4.88
K_2CO_3	49	0.49	8.88	49	0.49	8.43	38	0.38	6.62
$Mg(NO_3)_2$	66	0.66	13.08	65	0.65	10.88	48	0.48	8.57
KI	79	0.79	18.54	79	0.79	16.85	64	0.64	12.23
NaCl	88	0.88	24.35	88	0.88	22.52	74	0.74	15.28
KCl	99	0.99	39.59	99	0.99	35.49	81	0.81	17.22

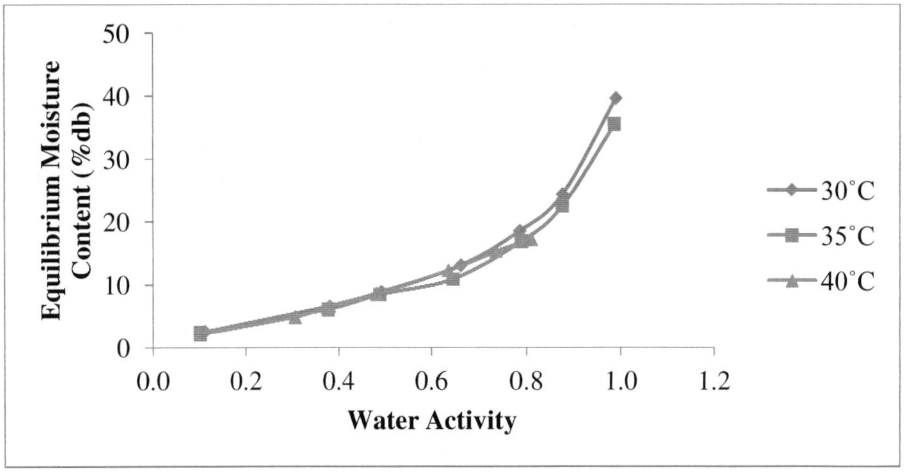

Figure1. Moisture Sorption Isotherm Curve of Fermented Cassava

Table 2. shows the effect of RH and temperature on the moisture content of fermented cassava starch equilibrium. At high RH the equilibrium moisture content tends to be high as well. This phenomenon is due to the higher RH, the more water vapor in the environment around the fermented cassava flour that interacts with the powder. The higher the storage temperature, the lower the moisture content. This phenomenon is because the rising temperatures cause the water molecules to be active and increase their energy levels, so they become less stable and tend to release the bonds with flour [8]. This result is following previous research that examined the equilibrium moisture content of cassava composite flour and red bead tree seed [9]. The moisture sorption isotherm curve is S-shaped so that the process of moisture absorption on the fermented cassava flour is included in type II.

The moisture sorption isotherm curve is modeled with various mathematical models. The moisture sorption isotherm curve of fermented cassava flour of GAB, Henderson, Halsey, Oswin and Chen-Clayton models can be seen in Figures 2, 3, 4, 5, and 6.

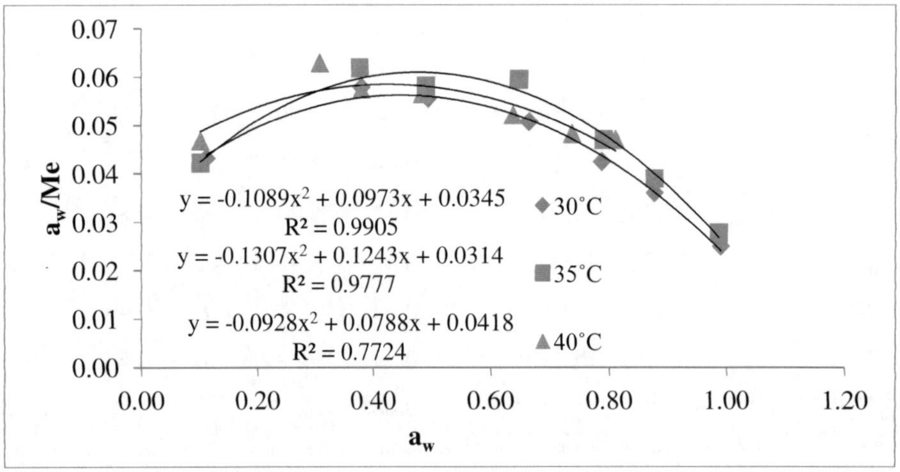

Figure 2. Moisture Sorption Isotherm Curve of Fermented Cassava Flour for GAB Model

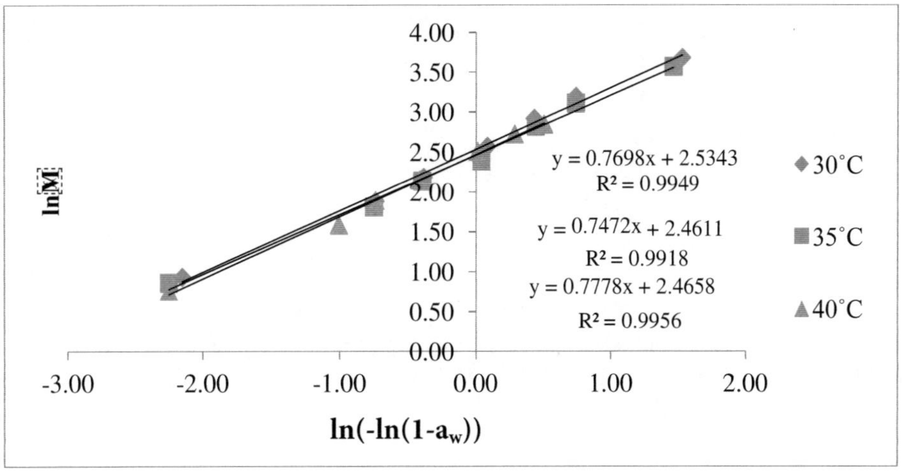

Figure 3. Moisture Sorption Isotherm Curve of Fermented Cassava Flour forHenderson Model

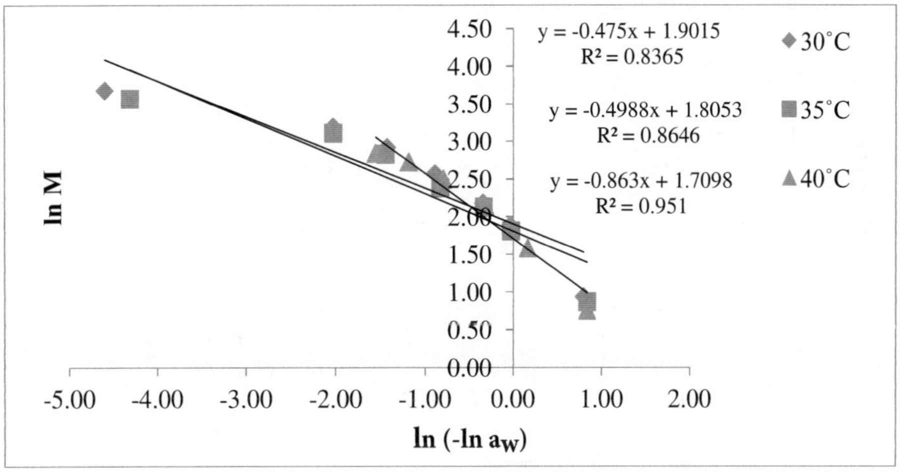

Figure 4. Moisture Sorption Isotherm Curve of Fermented Cassava Flour forHalsey Model

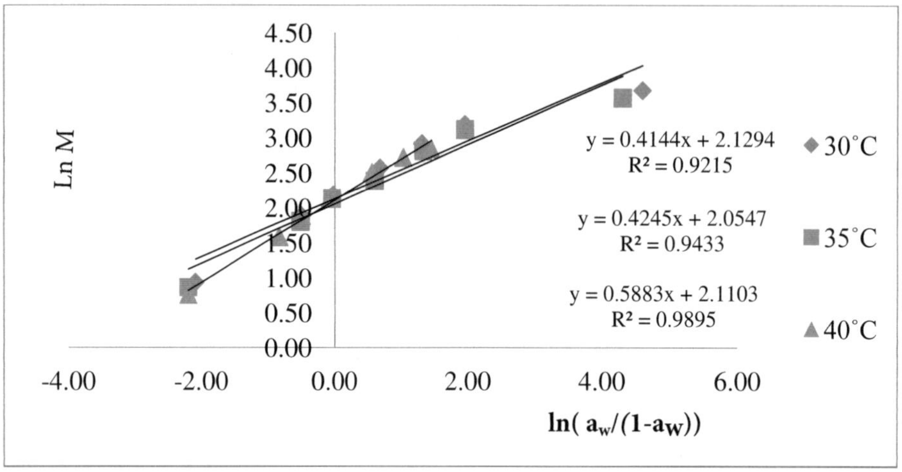

Figure 5. Moisture Sorption Isotherm Curve of Fermented Cassava Flour forOswin Model

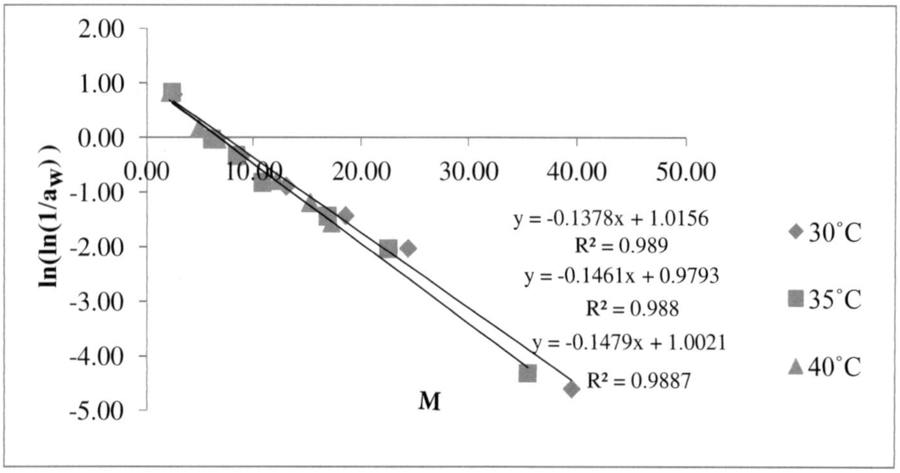

Figure 6. Moisture Sorption Isotherm Curve of Fermented Cassava Flour forChen-Clayton Model

The equation of the lines of each model for each temperature is used to determine the predicted values of the equilibrium moisture content used for the calculation of MRD to assess the accuracy of the model. Table 3 shows the MRD values for each model for each temperature.

Table 3.The MRD Values for Each Model for Each Temperature

Models	MRD (%)		
	30°C	35°C	40°C
GAB	1.98	3.11	4.36
Henderson	5.61	6.67	3.32
Halsey	32.00	28.39	12.97
Oswin	21.61	17.82	6.29
Chen-Clayton	12.26	14.83	10.97

Based on Table 3. GAB model has the smallest MRD values at 30°C and 35°C and Henderson at 40°C. The GAB model of GAB at 40 ° C is still below 5% indicating that the GAB model can be used to explain the phenomena of adsorption of water in fermented cassava flour. The results of this study are in line with previous studies on the isotherm of water sorption of various types of flour. The research of sesame flour at three temperatures using five models (modified Oswin, modified Halsey, modified Chung-Pfost, modified Henderson and GAB) states that the GAB model is the best model for describing water absorption in sesame flour [10]. Research on moisture sorption isotherm fufu and tapioca at temperatures using eight models (Bradley, BET, Oswin, Smith, Halsey, Henderson, GAB, and Chung-Pfost) state that GAB is the best model to describe water adsorption of fufu and tapioca flour [11].

4. Conclusion

The best mathematical model describes the phenomenon of water adsorption of cassava flour fermented at temperatures 30°C and 35°C is GAB and Henderson models at 40 ° C

References

[1] Alfiah MN, Hartini S and Cahyanti MN 2017 *Alchemy* **29** 40.
[2] Wijaya IMAS, Fitriani PPE, Gunam IBW and Wrasiati LP 2016. *International Conference on Agricultural and Food Engineering* (Kuala Lumpur – UPM).
[3] Koua BK, Koffi PME, Gbaha P and Toure S 2014 *J Food Sci Technol* **1711** 23.
[4] Adebowale AR, Sanni L, Awonorin S, Daniel I and Kuye A 2007 *International Journal of Food Science and Technology* **448** 52.
[5] Lakahina O, Saputri YLID andHartanto BD 2015 Mocaf Merah- *Pangan Kaya Antioksidan Berbasis Kearifan Lokal. Salatiga: Unpublished.*
[6] Hayati R, Abdullah A, Ayob MK and Soekarto ST.2004 *Jurnal Teknologi dan Industri Pangan.* **207** 13.
[7] Aini N, Prihananto V and Wijonarko G 2014 *Agritech* **50** 5.
[8] Chowdhury MMI, Huda MD, Hosain MA and Hassan MS 2006 *Journal of Food Engineering* **462** 7.
[9] Cahyanti MN, and Pattiserlihun A 2017 *Seminar Nasional Kimia dan Pendidikan Kimia IX* (Solo: UNS).
[10] Menkov ND and Durakova AG 2007 *Food Technol Biotechnol.* **96** 100.
[11] Sanni LO, Atere C and Kuye A 1997 *Journal of Food Engineering* **203** 12

Artificial Neural Network (ANN) for Optimization of Palm Oil Mill Effluent (POME) Treatment using Reverse Osmosis Membrane

Muhammad Said[1], Muneer Ba-Abbad[2], Siti Rozaimah Sheik Abdullah[3], Abdul Wahab Mohammad[4]

[1]Department of Chemistry, Faculty of Mathematics and Science, University of Sriwijaya, Indralaya 30662, Sumatera Selatan, Indonesia
[2]Department of chemical Engineering, Faculty of Engineering and Petroleum, Hadhramout University of Science &Technology, Mukalla, Hadhramout, Yemen
[3]Centre for Sustainable Process Technology (CESPRO), Faculty of Engineering and Built Environment, Universiti Kebangsaan Malaysia, 43600 Bangi, Selangor, Malaysia
[4]Department of Chemical and Process Engineering, Faculty of Engineering and Built Environment, University kebangsaan Malaysia, 43600, Bangi Selangor

E-mail: saidusman2011@gmail.com

Abstract. The treatment of palm oil mill effluent (POME) by Reverse Osmosis (RO) membrane was achieved experimentally. The biological oxygen demand (BOD), chemical oxygen demand (COD) and Color as final response of POME treatment process were selected. However, main influence factors on POME treatment process as concentration, transmembrane pressure and pH were tested on responses. The experimental results of responses were compared to the prediction by applying of Artificial Neural Network (ANN) as a simulation technique to create model of the process. Higher validation of ANN model was found for COD, BOD and Color which the prediction values very close compared to experimental result. The COD removal was investigated as a major affect factor in experiments. The best removal of COD was obtained at lower of POME concentration, pH, transmembrane pressure and time of contact. Therefore, these results showed that appreciate model created by ANN for POME treatment process which contributed easily to apply in industrial filed as future application.

Keyword : POME, RO, ANN, Pollutant removal

1. Introduction

Malaysia is one of the largest palm oil producers in the world with a total crude palm oil (CPO) production of 17.56 million tons per year. The CPO production always increases every year which means the increase in wastewater, called palm oil mill effluent (POME). POME is a hazardous liquid as wastewater that comes out of the discharge pipe at a temperature of 80-90oC which contains high biological oxygen demand (BOD) of 25 g/L, chemical oxygen demand (COD) of 53.6 g/L, Total suspended solids (TSS) of 19 g/L and low pH of 3.5-4.0 [1].

In the last decade, the application of membrane technology in POME treatment has been growing. Initiated by Said et al. [2], which investigate the pretreatment methods followed by membrane ultrafiltration and the performance shows more than 90% removal of total nitrogen, suspended solids, turbidity and color. Ahmad et al. [3] used combination treatment of coagulation and adsorption equipped by membrane UF and RO. From the investigation, the odor and color as well as the solid suspension (SS) and turbidity under different conditions almost removed. Yejian and Li [4] has been applied the biological system combine with membrane UF and RO which at the end of process, clear water was produced and could be used as boiler feed water. Other researcher, recently not only succeeded to removed the pollutant but also investigation of cleaning of membrane using chemical [5].

Since the success of a membrane filtration process depends on the relationship between many factors, the modeling has many problems difficultly solved by simple mathematical equations.

Therefore, there are alternative ways to solve the problem using the existing data with assistant a sophisticated tool. The conventional mathematical models which applied to estimate the parameters of cross flow filtration have limitations and disadvantages. For example, linear models have not provided a satisfactory description of the actual processes when occurring as a nonlinear process. Main disadvantage of the conventional model is difficult to eliminate confounding factors or noise that usually appeared in the filtration process which have negative affected on final result [6]. Therefore, artificial neural network (ANN) have been applied to solve this problem as new tool that can be constructed the experimental results to predict the performance of the membrane system.

Generally, ANN is technique inspired by biological neuron processing consist of three layers. The first layer is called the input layer which is the variables conditions of the experiment, in the middle there is a hidden layer where acts as feature predictors with more than one hidden layer and last layer is the output layer which is the target of an overall process. In turn, hidden nodes responses are weighted by the connections between the hidden and output layer which forwarded to nodes of the output layer [7].

Extensive research has been conducted to determine the capability of ANN to explain the membrane system phenomena. For example, ANN has been applied by many researchers to explain the membrane fouling [8-10], and also to predict the parameter to produce high water quality [11-14].

In this study, a feed-forward network was proposed. The input layer consists of four variables as concentration, transmembrane pressure, pH as well as time and the output layer contains four variables also as COD, BOD, TSS and Color. The objective of this study is to develop a neural network model to estimate the performance in POME parameters removal as COD, BOD, TSS and Color with compared to actual value of experiment.

2. Materials And Methods

2.1 Materials

For this study, a POME was collected from one of palm oil mill in Carey Island in Malaysia, and preserved in PVC containers at a temperature lower than 4°C. The sample was analyzed using DR/2010 portable data logging spectrophotometer (HACH, USA), covering the biological oxygen demand (BOD), chemical oxygen demand (COD), Total suspended solids (TSS), and Color. All the chemicals, required for the characterization of COD and BOD were supplied by HACH Chemical. Firstly, POME was treated with adsorption and ultrafiltration to remove large particles. Furthermore, POME was treated by RO membranes as final stage. The characteristic of POME was listed in Table 1.

Table 1 Characteristic of raw sample of POME

No.	Parameters	Content value
1.	BOD (mg/L)	174.67
2.	COD (mg/L)	244.67
3.	TSS (mg/L)	7
4.	Color (PtCo)	1263.33
5.	Turbidity (NTU)	43.57

The cross flow filtration set was equipped with adsorption, UF and RO membrane. The Montmorillonite was chosen as the adsorbent materiel. The hollow fiber made from Polysulphone for ultrafiltration while RO membrane used tubular type (RE-2012 LPF) from Woongjin, Korea. The pH of POME sample was adjusted by additional of NaOH and HCl solutions. All chemicals were supplied by Friendemann Schmidt Chemicals.

2.2 Methods

2.2.1 Experimental procedure

A 10- liter of POME sample was feed to adsorption column and fall drop by drop using the dosing pump with concentration variation of 10 to 90%. The product was collected at the bottom of the column and analyzed the content. The product was sent to feed sample tank of membrane module with analyzed. The first stage, POME was treated by UF membrane and continued with RO membrane. Permeate from UF was also analyzed the content and considers a simultaneously as the feed sample for RO membrane process. For the adsorption and UF membrane treatment stage, the sample was collected and analyzed for interval time of 1 hr. Furthermore, the samples pass through the RO membrane. Permeate was collected and retentate was recycled to feed sample tank. This process continues for 6 hrs and collected sample analyzed for every 2 hrs. A transmembrane pressure was varied between 0.5 to 2.5 kPa while the pH from 3 to 11 adjusted.

The product accumulates in basins that equipped with balance weight and connected to computer and recorded automatically every 1 min for 6 hrs. The results of all parameters were compared to the predict value from ANN calculation. The difference between actual and predict value as error of the process was recorded.

2.2.2 Artificial neural Networks (ANN)

An ANN consists of a set of connected cells known as the neurons. The number of neurons in the input and output layers is given respectively, by the number of input and output variables of the process. An additional bias input was added to the weighted sum for increasing or decreasing the net input to the activation function. The term of networks architecture is tandem with learning algorithm that using to train the networks. Generally, ANN architecture is consists of two activation functions, i.e. feed forward and feedback. One of important aspect in ANN is learning process as initial step. The learning process uses to predict the output which it having the minimum difference compared to the actual value. This disparity between actual and predict values of output is error of output over the target. Learning process consists of three learning algorithm, Back propagation, Quasi-Newton and Lavenberg-Marquardt. Majority of networks architectures use feed-forward networks where it processes the input data using error back-propagation algorithm.

In this study, the back propagation algorithm was trained using MATLAB 7.11.0.584 Neural Network Toolbox (R2010) from the Math Works, Inc. Four conditions of operating condition were applied in the RO membrane stage: concentration of POME (10%, 50% and 90% by volume), pH of solution (3.0, 7.0, and 11.0), transmembrane pressure (0.50 kPa, 1.50 kPa, 2.50 kPa) and time to collect sample (2.0 hr, 4.0 hr, 6.0 hr). The concentration and pH of POME were applied since the pretreatment step (on adsorption and UF membrane) and the others special for RO membrane. All experiments were conducted at different set of conditions and repeated for minimum two times. Thus, 81 runs of experimental conditions were made to conduct the optimization of RO membrane process.

3. Results and Discussions

3.1 Proposed of ANNs architecture

The important step in ANN is to determine the number layer that has specific number of neuron for each layer, which used to train the network optimally [15]. A two-layer feed-forward network with sigmoid hidden neurons and linear output neurons uses to fit multi-dimensional mapping problems well and given consistent data and enough neurons in hidden layer. Figure 1 shows the correlation of hidden neurons number and MSE. The optimization of neural network was determined based on the mean squared error (MSE) and the coefficient of determination (R^2). The number of hidden neuron that gives the smallest MSE is the best number of neurons. Hence, 6 neurons were chosen as the hidden neuron to optimize the neural network.

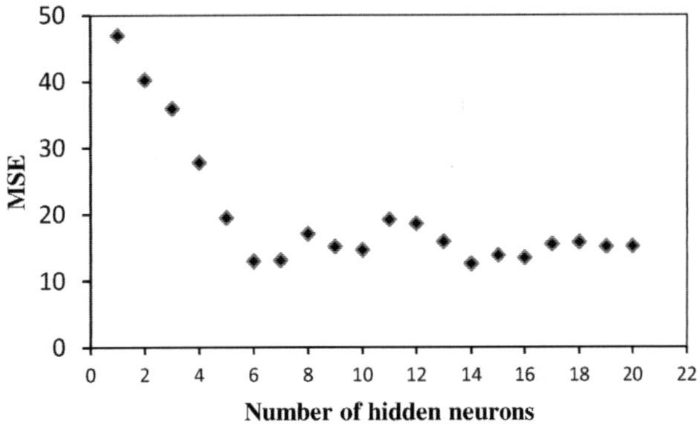

Figure 1. Correlation of neurons and MSE Number

In this study, the input data were applied a concentration, pH of POME, transmembrane pressure and time to collect sample. The input data were used to predict the BOD, COD and Color removal after treatment with RO membrane. The neural network was performing using back propagation algorithm. The architecture of a single hidden layer and an output layer was shown in figure 2.

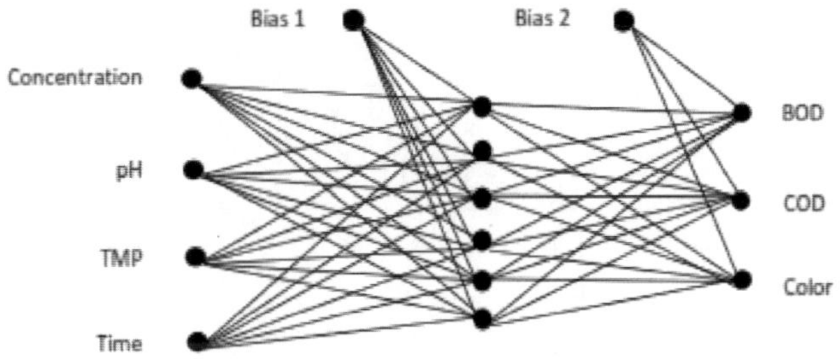

Figure 2. The architecture of ANNs

3.2 Neural network training

The relationship between the results from experimental data and ANN predictions of COD, BOD and Color removal are presented in figure 3. Plot distribution shows that predictive capability between the values of experiment and predicted quite satisfactory with slope line closed to 1. It is seen that the predicted value by neural network almost similar to the experimental data. The R^2 found to be more than 0.9 for all parameter of COD, BOD and Color removal which indicated the higher accuracy of the ANNs.

Some literature shows the similar trend of accuracy as Elmolla [16] obtained 0.997 for removal of COD, Hamzaoui [12] reached 0.9913 for removal of COD, Ghandehari [17] produced 0.9965 for removal of BSA and Bhatti [18] was reached to 0.982 for removal copper.

The development of ANN model included two steps as a training and validation. For the treatment with RO membrane, a total number of 324 experimental data were created. About 243 data apply for training and 81 data for rest for validation of proposed ANN model.

IOP Conf. Series: Journal of Physics: Conf. Series **1095** (2018) 012021 doi:10.1088/1742-6596/1095/1/012021

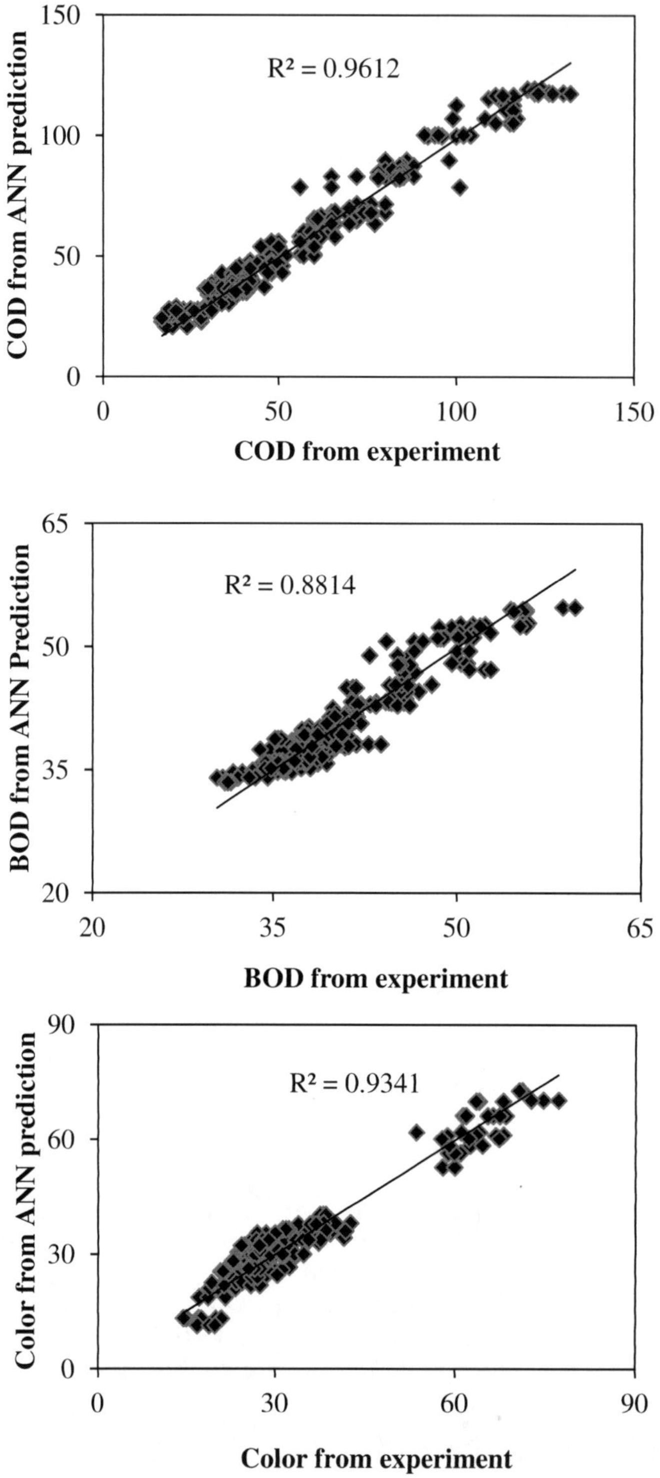

Figure 3. Correlation between experimental data and predicted value of (a) BOD, (b) COD, (c) Color

3.3 Neural network model

Neural network with six neurons in each hidden layer and four inputs are used to propose the simulate model of COD, BOD and color. Therefore, the ANN model is applied according to Equation (1):

$$\mathbf{Y} = \sum_{n=1}^{n} \left\{ \boldsymbol{W}\mathbf{o}_{(1,n)} \left[\frac{2}{1+\exp(-2(\sum_{m=0}^{n} (\mathbf{Wi}_{(n,m)}\, \mathbf{In}_{(m)} + \boldsymbol{b1}_{(n)})} - \mathbf{1} \right] \right\} + \boldsymbol{b}\mathbf{2}_{(l)} \quad (1)$$

where Y is the parameter forecast of COD, BOD and color, n is the number of hidden neurons, m is the number of inputs, l is the linear output, W_i and b_l is the weight and the bias in the input layer to the hidden layer, W_o and b_2 are the weights and biases in the hidden layer to the output layer. Hyperbolic tangent "TANSIG" (which is a function of the transition) is selected for the mapping of input to hidden layers while pure linear transfer function "secretary" for mapping hidden layer to the output layer. Weight bias circuit and connected to the removal of COD, BOD and color are described in Table 2 and 3.

Table 2. Weights and biases for the ANN model to describe the COD, BOD and color removal

Neuron	**W_i**				**b_1**
	Concentration (%)	**pH**	**Trans Membrane Pressure (bar)**	**Time (hr)**	
1	-1.6418	1.0695	-3.8884	-0.044911	3.5633
2	-4.0443	5.3732	0.25598	2.0763	0.84903
3	1.2197	1.1371	-0.31881	-0.0027809	-1.4653
4	3.5744	6.5205	2.0026	0.094434	6.8339
5	1.531	-0.95662	0.11425	-0.01517	-1.0172
6	-4.5128	5.621	0.13695	0.33648	0.64359

Table 3. Output of COD, BOD and color removal

Output	**$Wo_{(1,1)}$**	**$Wo_{(1,2)}$**	**$Wo_{(1,3)}$**	**$Wo_{(1,4)}$**	**$Wo_{(1,5)}$**	**$Wo_{(1,6)}$**	**b_2**
COD	-0.080188	-0.16399	0.029091	0.029091	1.0826	0.51319	0.27136
BOD	-0.27865	-0.30863	-0.66178	-0.32838	1.3025	0.70999	0.15934
Color	0.39307	-0.20635	-2.16	-0.55507	3.0361	0.96809	-0.3892

3.4 Prediction of COD, BOD and color removal

The ANN model is validated by independent data and the experimental data with simulated results are shown in figure 4. It is seen that the COD and Color values are increased while BOD relative constant. This result was not surprising because at the beginning of the sequence experiment, the lowered concentration of POME as 10 % was used and then further increased to 50 and 90%. An increasing of concentration means increased the organic and impurities particle in POME. Despite of RO membrane showed a good ability to filter particles, but the RO membranes were very sensitive to particle size. If the particles contains in the solution, a great shape and rough rapidly caused membrane fouling and it reduced the ability of the membrane.

In addition, the fluctuation of curve was influenced by differences of pH values from 3 to 11. When pH below than 5.05, the POME was more positively charged and the particles were attracted to the membrane and enter the pores and trapped. The trapped particle shows difficulty to pass through the membrane, where the particles contain a variety of organic molecules was observed. This mechanism caused to value of parameters in the permeate decrease. In general, RO membranes succeed to remove pollutants in POME which is characterized by impairment of BOD, COD and color value. These finding showed similar results compared other previous studies [11, 16, 19-21]

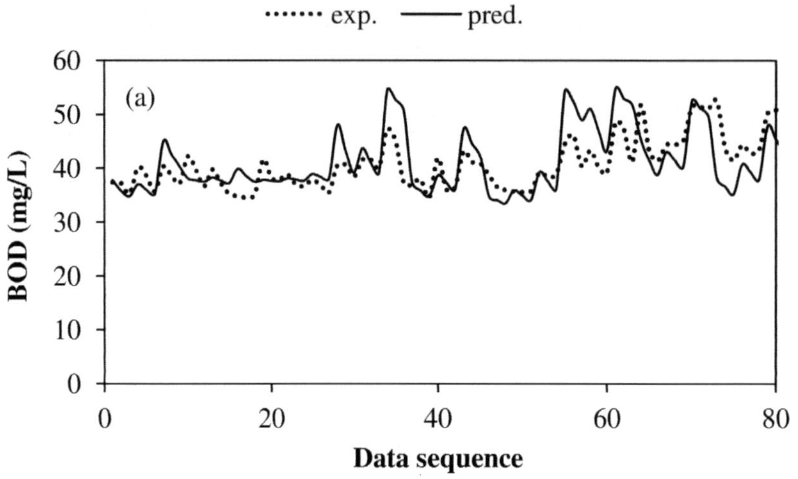

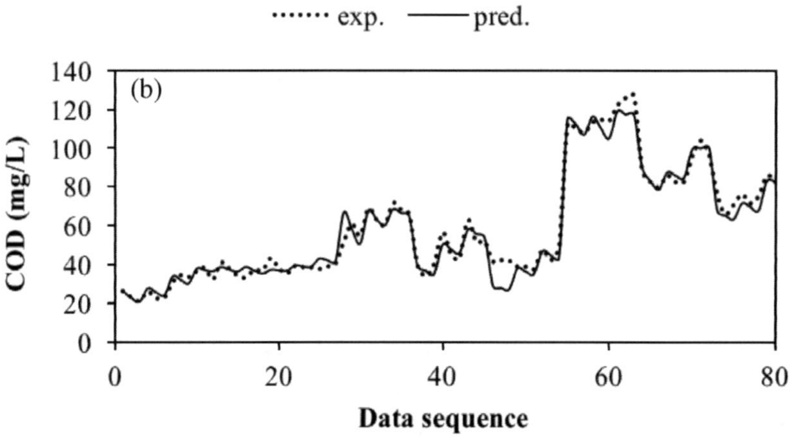

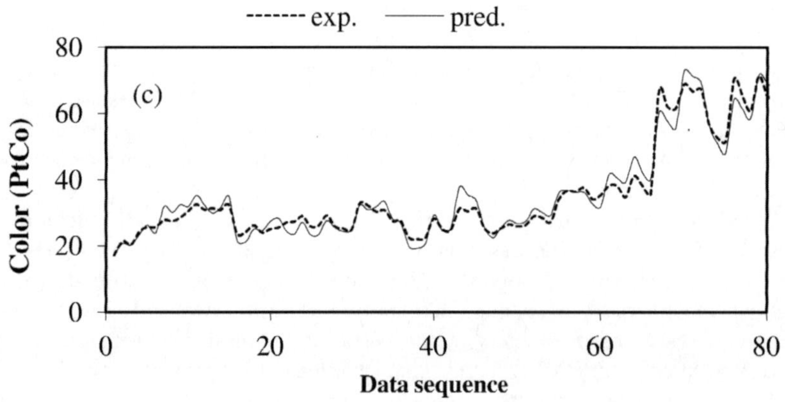

Figure 4. The prediction values by ANN model for (a) BOD, (b) COD, (c) Color

IOP Conf. Series: Journal of Physics: Conf. Series **1095** (2018) 012021 doi:10.1088/1742-6596/1095/1/012021

3.5 The 3D plot of COD removal

In this study, only a 3D plot for COD removal was investigated due to COD a major affect factor in experiments. Plot 3D for COD removal shows the effect of interaction with two variables of final results as given in figure 5.

The successfulness of COD removal was characterized by a decrease in COD value at the outlet of the reverse osmosis membrane. The negative effects on COD removal was observed when the pH at greater than 5.0. The same sequence happening, when the concentration of POME and transmembrane pressure were increased. These results attributed due to a smaller concentration of POME which least impurity content in POME solution beginning indicator of the final COD. In addition, lower pH value leads to changes in impurities surface which was resulted the more negative charge of both impurity and membrane surfaces. The similarity of these charges made a more hydrophilic nature of the membrane which the dirt not stick to surface of membrane and permeate flow rate was increased. According to these results, to improve COD removal by diluting the concentration of POME and lowering the pH was observed.

Furthermore, the implementation of low operating pressure, ranging from 0.5-1.0 kPa and long contact time were found to be an important factors removal COD of POME. However, in general by applying a low transmembrane pressure and long contact time was made the filtration process of impurities by the membrane become maximal. However, filtration POME for a long time also causes the short lifetime of the membrane which important to clean the membrane within the process necessary.

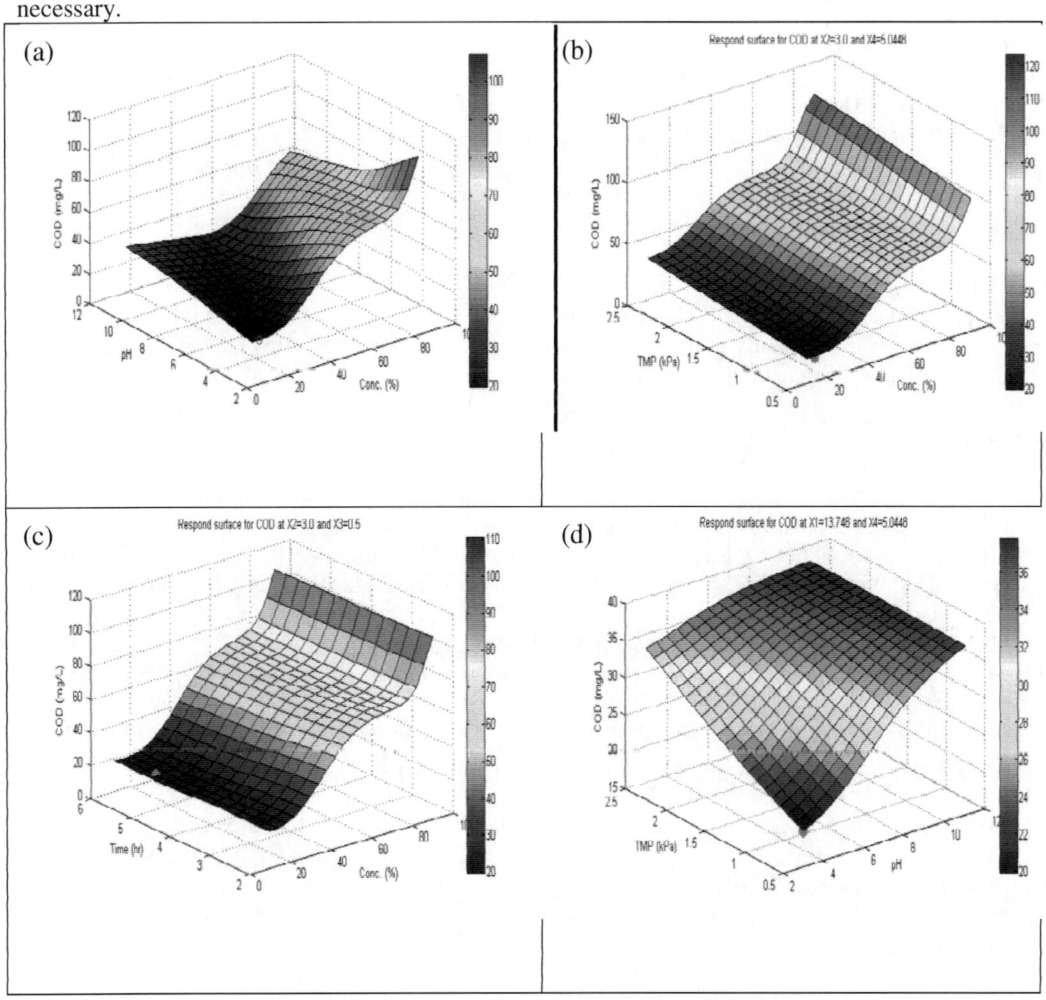

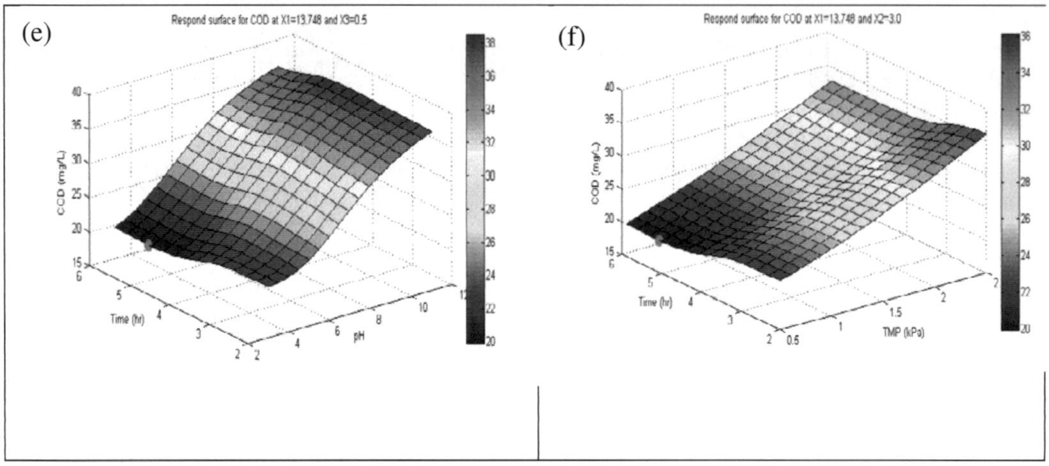

Figure 5. Plot 3D of the optimum for COD removal: a) effect of concentration and pH, b) effect of concentration and transmembrane pressure, c) effect of concentration and time, d) effect of pH and transmembrane pressure, e) effect of pH and time, f) effect of transmembrane pressure and time

4. Conclusion

In this study, the palm oil mill effluent (POME) treatment by Reverse Osmosis (RO) membrane was tested successfully. The factors of concentration, transmembrane pressure and pH on the final results of BOD, COD and Color were investigated. All predicted results by artificial neural network (ANN) were close to experimental results which confirm high accurate model produced to simulate the filtration process. These results indicate that artificial neural network (ANN) appreciated method to develop of filtration process using the membrane for waste water treatment in future.

Acknowledgment

The authors would like to thank the local palm oil mill in Malaysia for the supplying the POME samples to conduct in this study and also to Universiti Kebangsaan Malaysia (UKM) for supporting through the Beasiswa Zamalah Universiti Penyelidikan

References

[1] Lam M K, Lee K T 2004 *Biotechnol. Adv.* **29** 124
[2] Said M, Mohammad A W, Nor M T M, Hasan, H A 2015 *Sains Malaysiana* **44(3)** 421
[3] Ahmad A L, Ismail S, Bhatia S 2003 *Desalination* **157** 87
[4] Yejian Z, Li Y, Xiangli Q, Lina C, Xiangjun N, Zhijian M, Zhenjia Z 2008 *J. Env.Sci.* **20** 558.
[5] Said M, Mohammad A W, Nor M T M, Abdullah S R S, Hasan H A 2014 *Membrane Water Treatmen* **5(3)** 207
[6] Silva I N D, Flauzino R A, 2008 *Applied Soft Computing* **8** 590
[7] Haykin S 1999 *Neural network- A comprehensive foundation* (New Jersey: Prentice Hall International, Inc)
[8] Vincent N D, Cabassud C, Cabassud M, Bourlier L D, Laine J M 2000 *Desalination* **131** 353
[9] Delgrange N, Cabassud C, Cabassud M, Bourli L D, Laine J M 1998 *Desalination* **118** 213
[10] Chellam S 2005 *J.Membrane Sci.* **258** 35
[11] Akratos C S, Papaspyros J N E, Tsihrintzis V A 2008 *Chem. Eng. J.* **143(1–3)** 96
[12] Hamzaoui Y E, Hernández J A, Silva-Martínez S, Bassam A, Álvarez A, Lizama-Bahena C 2011 *Desalination* **277(1–3)** 325
[13] Bello M M, Nourouzi M M, Abdullah L C, Choong T S, Koay Y S, Keshani S 2013 *J. Hazard.Mat.* **262** 106
[14] Aber S, Amani-Ghadim A R, Mirzajani V 2009 *J. Hazard. Mat.* **171(1–3)** 484
[15] Sahoo G B, Ray C, 2006 *J. Membrane Sci.* **283** 147

IOP Conf. Series: Journal of Physics: Conf. Series **1095** (2018) 012021 doi:10.1088/1742-6596/1095/1/012021

[16] Elmolla E S, Chaudhuri M, Eltoukhy M M 2010 *J. Hazard. Mat.* **179** 127
[17] Ghandehari S, Rahmati M M M, Asghari M 2011 *Desalination* **277** 348
[18] Bhatti M S, Kapoor D, Kalia R K, Reddy A S, Thukral A K 2011 *Desalination* **274** 74
[19] Mjalli F S, Asheh S A, Alfadala H E 2007 *J. Envi. Manag.* **83** 329
[20] Pendashteh A R, Razi A F, Chaibakhsh N, Abdullah L C, Madaeni S S, Abidin Z Z 2011 *J. Hazard. Mat.* **192** 568
[21] Zhaobo C, Aijuan Z, Nanqi R, Yu T, Dongxue H 2009 *J. Envi.Sci.* **21** 1503

Dual-column Switching Ion Chromatography for the Simultaneous Determination of Inorganic Cations and Anions (particularly Thiocyanate) in Human Urine and Saliva Samples to Identify Smokers Types

Muhammad Amin[1,*], Budhi Oktavia[2], Anang Sedyohutomo[3], Lee Wah Lim[4], and Toyohide Takeuchi[4]

[1]Department of Chemistry Education, Faculty of Teacher Training and Education, Universitas Khairun, Ternate 97723, North Maluku, Indonesia
[2]Department of Chemistry, Faculty of Mathematic and Natural Sciences, Universitas Negeri Padang, Padang 25131, West Sumatera, Indonesia
[3]Department of Chemistry, Hamamatsu University School of Medicine, 1-20-1 Handayama, Higashi-ku, Hamamatsu, Shizuoka Prefecture 431-3122, Japan
[4]Department of Chemistry and Biomolecular Science, Faculty of Engineering, Gifu University, 1-1 Yanagido, Gifu 501-1193, Japan

Email: muh_amin@unkhair.ac.id

Abstract. An ion chromatography (IC) have been developed for the simultaneous determination of inorganic cations (Na^+, NH_4^+, K^+, Mg^{2+}, and Ca^{2+}) and inorganic anions (PO_4^{3-}, Cl^-, NO_3^-, and SCN^-) using a single pump, a single eluent and a single detector. The method is based on column switching means allows anions and/or cations could be determined in a single analysis system. While one ion-exchange column is being operated, the other ion-exchange column is being conditioned, i.e., the columns are always ready for analysis at any time. When the combination of 10 mM tartaric acid was used as the eluent, and operated at eluent flow rate of 0.5 mL/min, five cations and four anions could be separated on the cation-exchange column and the anion-exchange column, respectively. The separation of cations was completed within 15 min, whereas the separation of anions was completed within 40 min. The detection limits was calculated at S/N=3 were 3.10−12.14 ppb (μg/L) for the cations and 16.00−88.10 ppb for the anions. The relative standard deviations of the all ions were less than 4.81%, 6.30%, and 6.22% for peak height, peak area, and retention time, respectively. The developed technique was successfully applied to the simultaneous determination of inorganic cations and anions in human saliva and urine samples.

Keywords : Ion Chromatography, simultaneous, inorganic cations and anions

1. Introduction

Since it was first introduced by Small et al. in 1975, IC has become a reliable analytical technique for determining cations and anions various samples until ppb level of concentration or more. IC has been widely applied in agriculture, food, environmental, industrial, medicine [1–4], however, in the field of health is still not much application, and that is why it needs to be given more attention.

Dual-column switching ion chromatography (DSIC) means that the separation of cations and/or the separation of anions can be done at different times in a single of analysis systems. The flow can be switched from the cation-exchange to the anion-exchange column by simply rotating the direction of the switching valve.

Mostly, cations and anions are separated separately and use different eluents. So, the analysis time will be longer to get completed data of the cations and anions.

A number of articles related to the separation of cations and anions in a single analysis system and using different packing material columns, it can be divided into 2 types of methods, namely (1) the cation-exchange column and the anion-exchange column are arranged in series and/or parallel [5], and (2) the both ion-exchange columns are also arranged in series and connected via the switching valve [6–12]. For both methods, a suitable eluent should be chosen for the simultaneous separation of cations and anions. But, it should be noted that without a switching valve, these (cations and anions) peaks sometimes overlap and the method is not applicable widely. Otherwise, the method using a switching valve, peak overlap can be avoided.

Urine and saliva may contain many ions, elements, compounds that can provide information of health and clinical conditions of a person, as well as to identify smoker type purposes. So, it is important to know qualitatively and quantitatively of the ions, elements and compounds contained in these samples. And new approaches in development of IC technique for the determination of these ions in this field always get priority and attention [13–19].

In this study, discuss 1) the improvement of IC based on column switching to determine cations and anions with a single eluent in single analysis system, 2) the improved IC then applied to identify the type of smokers (heavy smoker, moderate smoker, light smoker, passive smoker, and non-smoker) by determining the concentration of inorganic cations and anions in their urine and saliva samples. However, prior to that, SCN^- ion is the main target of ion should be determined, while the other inorganic cations (Na^+, NH_4^+, K^+, Mg^{2+}, and Ca^{2+}) and inorganic anions (PO_4^{3-}, Cl^-, and NO_3^-) were also examined.

2. Materials and Methods

2.1 IC instrument

The experimental apparatus was assembled from a PU-2080i plus HPLC pump (Jasco, Tokyo, Japan), a model 5095 loop injector (Rheodyne, Cotati, CA, USA) with an injection-loop volume of 20 µl, a CM-8020 conductivity detector (Tosoh, Tokyo, Japan), a Shodex IC YK-421 column (150×4.6 mm I.D.), a TSK_{gel} IC-Anion-SW column (50×4.6 mm I.D.), a Model 7610-600 10-port switching valve (Rheodyne) and a Computer Aided Chromatography data processor (Nippon Filcon, Tokyo, Japan). The two separation columns were connected in series via the 10-port switching valve. The flow system allows us to select the position of the switching valve for peak separation. The diagram of the apparatus employed in this study is illustrated in Fig. 1.

2.2 Chemicals

Analytical reagent grade chemicals were purchased from Nacalai Tesque (Kyoto, Japan), unless otherwise noted. Standard solutions were prepared by using deionized water. The deionized water throughout this study was prepared in the laboratory using a GS-590 water distillation system (Advantec, Tokyo, Japan). Stock standard solutions for anions and cations containing a mixture of interest ions were prepared. The details of analysis condition, as described in Table 1. The standard solution of cations was prepared by dissolving 1.0 mM of each NaCl, NH_4Cl, KCl, $MgCl_2$, and $CaCl_2$, whereas the standard solution of anions was prepared by dissolving 1.0 mM of each $NaHPO_4$, NaCl, $NaNO_3$, and NaSCN. All standard solutions were stored in polyethylene containers and kept under refrigeration at 4°C. Tartaric acid was obtained from Wako (Osaka, Japan), and used for the eluent. The eluent was prepared before use.

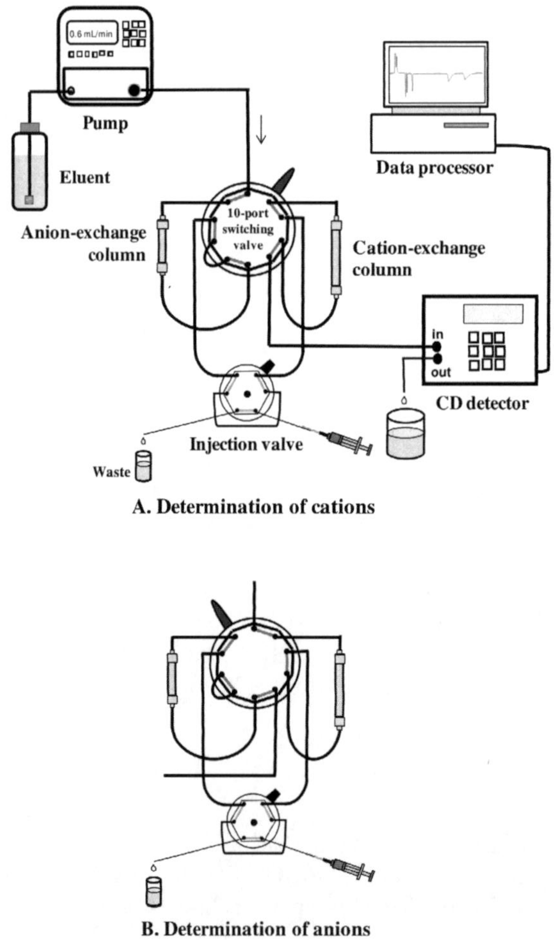

Figure 1. Flow scheme of the instruments and the positions of the switching valve and the both columns.

Table 1. Analysis Conditions

Parameter	Analysis conditions
– Eluent	10 mM Tartaric acid
– Analytical Columns	Shodex IC YK-421 (for cation separation) TSK_{gel} IC-Anion-SW (for anion separation)
– Detector type	Conductivity
– Flow rate of eluent	0.5 mL/min
– Total pressure	6.5 MPa
– Injector volume	20 μL

The pH of tartaric acid eluent was measured with a Toa Electronics Ltd. (Tokyo, Japan) IM-20E ion meter with a glass electrode.

2.3 Dual-column switching work mechanism
Figure 1 shows the flow scheme of the system, the position of ion-exchange columns, injection valve, and 10-port switching valve. The anion-exchange and cation-exchange columns were connected serially via 10-port switching valve to allow separation being carried out in either column. The injection valve is placed between the ion-exchange columns. The separation system performed according to the procedure below:

Position A: for separation of cations refers to the switching valve for the separation of cations and the eluent passes through both anion-exchange and cation exchange columns until a straight baseline are obtained. A 20μL volume of the sample solution was then injected to the IC system. The analytes will reach the cation-exchange column and then the cations will be separated. While the separation of cations on the cation-exchange column is carried out, the anion-exchange column is being conditioned with the eluent. Therefore, the anion-exchange column is ready for use at anytime.

Position B: for separation of anions refers to the switching valve for the separation of anions. A 20μL volume of the sample solution was then injected to the system. The analytes will reach the anion-exchange column and then the anions will be separated. While the separation of anions on the anion-exchange column is carried out, the cation exchange column is being conditioned with the eluent. Therefore, the cation-exchange column is ready for use at anytime.

Eluent. The eluent used for the determination of anions and cations was 10 mM tartaric acid. The pump was operated at a flow rate of 0.5 ml/min to keep the inlet pressure lower than 7.0 MPa.

2.4 Preparation of urine and saliva samples
Urine sample was just diluted with deionized water for 5 times and then filtered with a 0.45-μm membrane filter, while the saliva sample was also diluted 5 times with deionized water, and were centrifuged at 2000 rpm for 5 min, followed by filtration with a 0.45-μm membrane filter prior injection. All samples were then stored in a refrigerator at 4°C.

3. Results and discussion

3.1 Effect of tartaric acid concentrations
In this study, tartaric acid was used as the eluent to determine both cations and anions. By considering of monovalent ions (Na^+, $NH4^+$, K^+, Cl^-, NO_3^-, and SCN^-), divalent ions (Mg^{2+} and Ca^{2+}) as well as

trivalent ion (PO_4^{3-}), the difference of eluent concentration should be given more attention. In IC, especially non-suppressed systems, weak acids such as tartaric acid are often used as eluents to separate the inorganic cations and anions. The eluent used, must have a strong affinity for ion exchange resins, so that the separations can be more effective. Tartaric acid has a carboxyl group which can help form a complex compound between ions and eluent.

The effect of the difference of eluent concentration on the retention time of inorganic cation and anions were observed in this study. With the five different eluent concentrations were used in range of 5-15 mM, the retention time for these ions increased as the eluent concentration decreased, as in Figure 2. The retention behavior is influenced by hydrophobic absorption of the ion-exchange (polymer and silica) resins and the pH of eluent.

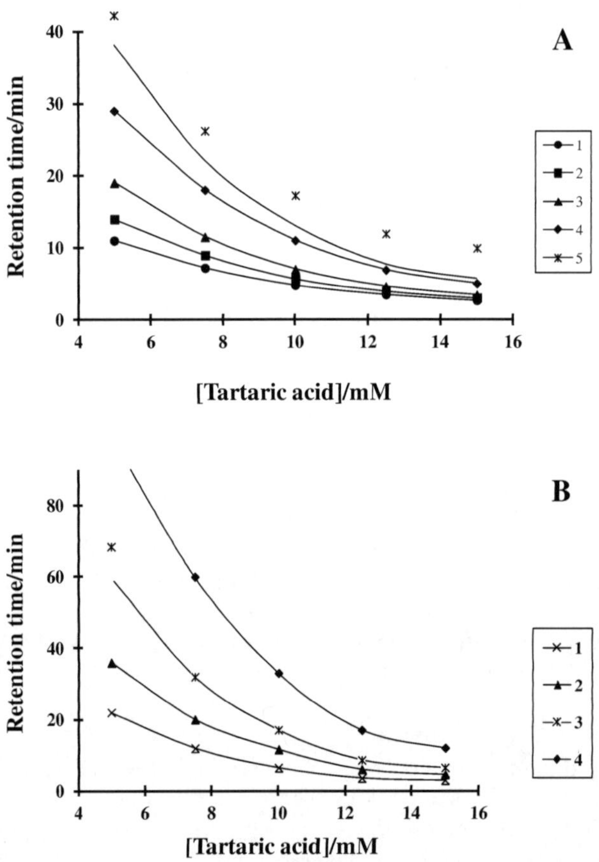

Figure 2. Effect of tartaric acid concentration in the eluent on retention time of cations (A) and anions (B). Eluent: 10 mM tartaric acid. Columns: Shodex IC YK-421 (150 x 4.6 mm I.D.) and TSK_{gel} IC-Anion-SW (50 x 4.6 mm I.D.). Eluent flow-rate: 0.5 mL/min. Column temperature: 35°C. Injection volume: 20 µL. Lines for cations: 1=Na^+; 2=NH_4^+; 3=K^+; 4=Mg^{2+}; 5=Ca^{2+}. Lines for anions: 1=PO_4^{3-}; 2=NO_3^-, 3=Cl^-; 4= SCN^-.

3.2 Determination of inorganic cations using standard samples

Figure 3 shows the determination of cations using standard sample. Based on cation-exchange mechanism, five cations (Na^+, NH_4^+, K^+, Mg^{2+}, and Ca^{2+}) could be eluted on the stationary phase within 15 min. The five cations determined in those order by different of their ion size and charge and their interaction with stationary phase.

As in the chromatogram, all cations appeared as negative peaks. By this result was due to the difference in the conductivity between the analyte ions contained in the samples and the driving eluent ion contained in the tartaric acid.

3.3 Determination of inorganic anions using standard samples

Figure 4 shows the determination of anions using standard sample and 10 mM tartaric acid was used as the eluent. Based on anion-exchange mechanism, four anions (PO_4^{3-}, Cl^-, NO_3^-, and SCN^-) commonly contained in urine and saliva samples were well eluted in within 40 min. The order of these anions was occurred by different of their ion size and charge and interaction with stationary phase.

All anion peaks were appeared as positive peaks. By this result was due to the difference in the conductivity between the analyte ions and the driving eluent ion. Compare to cation separation, anion separation took longer time. And it is expected that due to the different type of resins from the both columns.

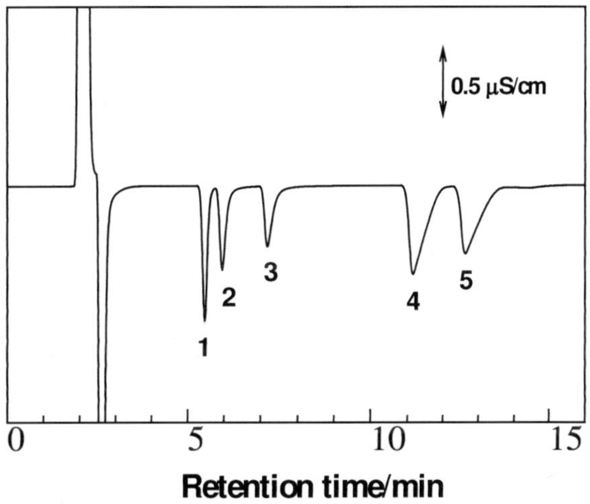

Figure 3. The representative chromatogram of inorganic cations using standard sample. Eluent: 10 mM tartaric acid. Column: Shodex IC YK-421 (150 mm x 4.6 mm I.D.). Eluent flow-rate: 0.5 mL/min. Column temperature: 35°C. Injection volume: 20 μL. Analytes (concentration, 1.0 mM each): 1=Na^+, 2=NH_4^+, 3=K^+, 4=Mg^{2+}, and 5=Ca^{2+}.

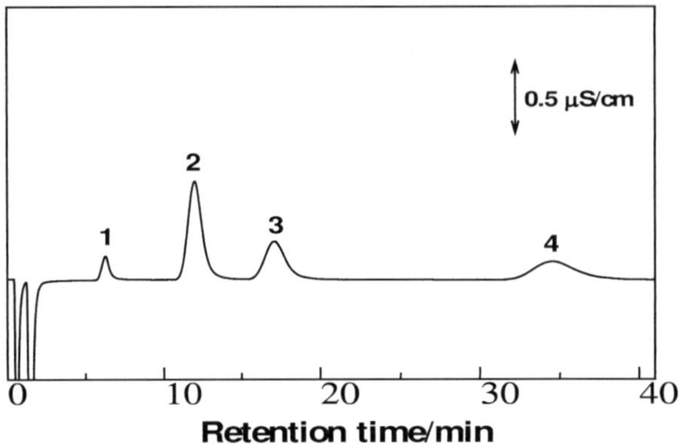

Figure 4. The representative chromatogram of inorganic anions using standard sample. Eluent: 10 mM tartaric acid. Column: IC-Anion-SW (50 mm x 4.6 mm I.D.). Eluent flow-rate: 0.5 mL/min. Column temperature: 35°C. Injection volume: 20 μL. Analytes (concentration, 1.0 mM each): 1=PO_4^{3-}, 2=Cl^-, 3=NO_3^-, and 4=SCN^-.

3.4 Validation of the system

Table 2 shows the linear relationships between the peak height of each ions and molar concentration were observed up to 1 mM with the correlation coefficients, $R^2 \geq 0.99$ for all anions and cations, and indicated as a good linearity.

The detection limit (LODs) for 20 μl injection volume of sample was calculated at a signal-to-noise ratio (S/N) of 3:1. As in Table 2, the LODs obtained were 3.10-12.14 ppb (μg/l) for cations and 16.00-88.10 ppb for anions.

Table 2. Summarized data for the detection limits, correlation coefficient, and retention time of ions under the optimum operating conditions, as in Figures 3 and 4

Analytes	**Detection Limit/ ppb**	**Correlation coefficient (r^2)**	**Retention times (t_R)/ min**
Cations			
Na^+	3.10	0.9956	5.51
NH_4^+	3.99	0.9921	6.01
K^+	12.14	0.9935	7.18
Mg^{2+}	4.87	0.9918	11.22
Ca^{2+}	10.80	0.9954	12.58
Anions			
PO_4^{3-}	88.10	0.9994	6.30
Cl^-	18.12	0.9971	12.02
NO_3^-	83.11	0.9963	17.07
SCN^-	16.00	0.9998	34.55

Table 3 shows the reproducibility of peak heights, peak areas, and retention times under the optimum condition in Figures 3 and 4. The relative standard deviations (RSDs, %) of these ions were less than 4.81% for peak height, less than 6.30% for peak area, and less than 6.22% for retention time, respectively.

Table 3. RSD of ions under the optimum operating conditions, as in Figures 3 and 4

	RSD (%), n=5[a]		
Analytes	**Peak height**	**Peak area**	**Retention time**
Cations			
Na^+	2.72	2.65	3.49
NH_4^+	3.28	2.79	3.37
K^+	3.79	3.25	3.58
Mg^{2+}	4.43	3.29	4.56
Ca^{2+}	4.36	4.81	4.73
Anions			
PO_4^{3-}	3.59	3.45	3.49
Cl^-	4.68	4.79	4.45
NO_3^-	4.59	5.43	4.06
SCN^-	4.81	6.30	6.22

[a]n=number of measurements

3.5 Application to urine and saliva samples

Good chromatograms for inorganic cations in both heavy smoker's urine and saliva sample were achieved, as in Figure 5.

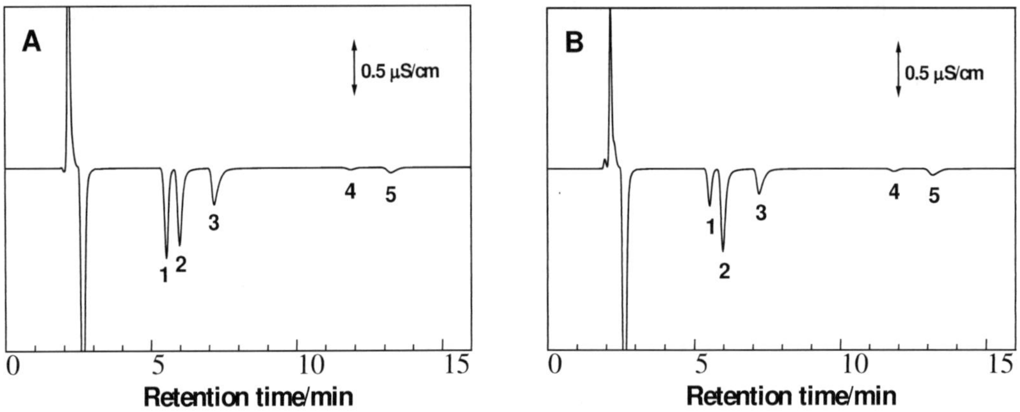

Figure 5. Determination of inorganic cations in heavy smoker's urine (A) and saliva (B) samples with 5-times of dilution. Ions: 1=Na^+, 2=NH_4^+, 3=K^+, 4=Mg^{2+}, and 5=Ca^{2+}. Other chromatographic conditions, as in Figure 3

The ammonium ion found as much more concentration, especially in saliva samples. Since ammonium is one of compounds contained in cigarettes, that can influence the physically and chemically composition of human physiological fluids such as urine and saliva.

The concentration of sodium, ammonium, potassium, magnesium, and calcium ions for heavy smoker's urine were determined to be 3.45, 5.02, 3.03, 0.18, and 0.42 mM, respectively, whereas for heavy smoker's saliva sample, were determined to be 1.12, 5.19, 1.24, 0.16, and 0.48 mM, respectively, as listed in Table 4.

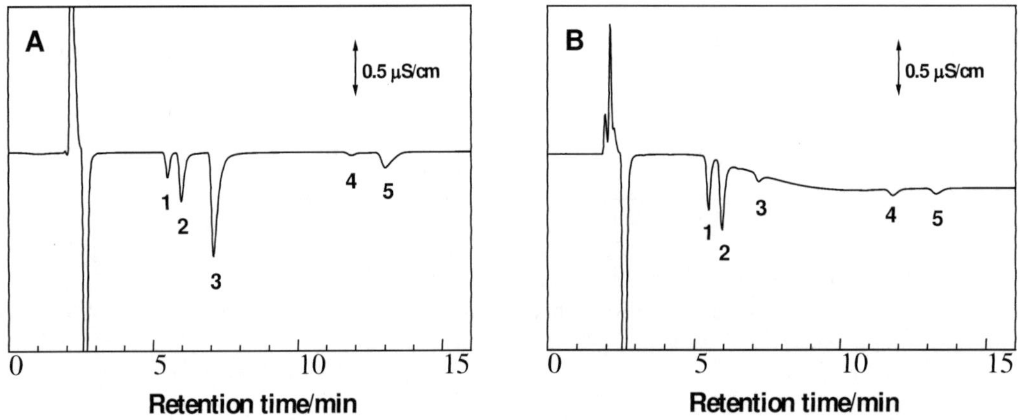

Figure 6. Determination of inorganic cations in non-smoker's urine (A) and saliva (B) samples with 5-times of dilution. Ions: 1=Na^+, 2=NH_4^+, 3=K^+, 4=Mg^{2+}, and 5=Ca^{2+}. Other chromatographic conditions, as in Figure 3

Figure 6 shows good determination for inorganic cations in both non-smoker's urine and saliva samples. An uneven baseline occurred when the saliva sample was injected to the IC system. It might be due to analyte chemical composition. However, it should be noted that the results did not influenced. Table 4, summarizes the concentrations of cations for all smoker's urine and saliva samples, and the composition of cations contained both types of samples, as shown in Figure 7. Moreover, the obtained results also show the concentrations of these cations in moderate smoker, light smoker, and passive smoker for urine and saliva, respectively.

Table 4. Determination results for inorganic cations in urine and saliva samples (n=5)

Smoker types	Sample types	Cations (mM)				
		Na^+	NH_4^+	K^+	Mg^{2+}	Ca^{2+}
Heavy smoker	o Urine	3.45	5.02	3.03	0.18	0.44
	• Saliva	1.12	5.19	1.24	0.16	0.48
Moderate smoker	o Urine	3.22	4.64	3.10	0.20	0.52
	• Saliva	1.62	4.66	1.22	0.22	0.42
Light smoker	o Urine	2.53	4.02	4.28	0.26	0.52
	• Saliva	2.39	4.00	0.82	0.19	0.68
Passive smoker	o Urine	2.40	2.61	4.54	0.19	0.60
	• Saliva	1.98	4.1	0.65	0.30	0.42
Non-smoker	o Urine	0.53	0.12	6.50	0.22	0.64
	• Saliva	1.55	3.22	0.26	0.34	0.38

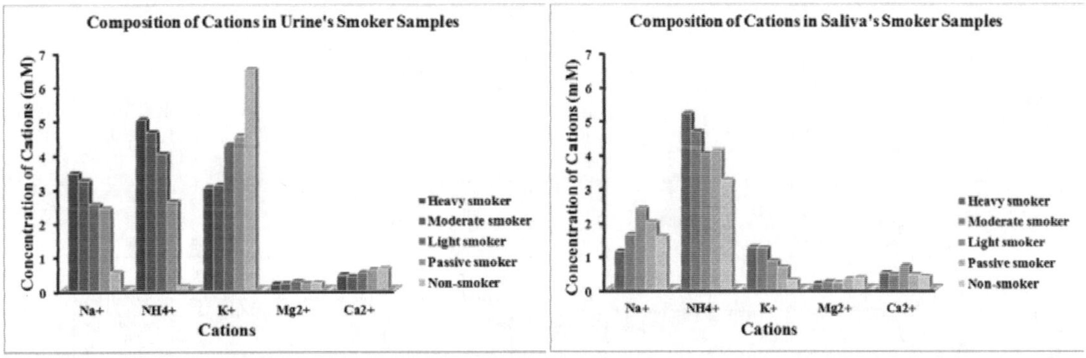

Figure 7. Composition of inorganic cations contained in smoker's urine and saliva samples. Other chromatographic conditions, as in Figure 3.

Good chromatograms for inorganic anions in both heavy smoker's urine and saliva samples as showed in Figure 8, and for non-smoker's samples as showed in Figure 9. From Figure 8, the four anions (particularly thiocyanate ion) commonly contained in these types of samples could be determined under this optimum conditions. The determination of SCN^- ion in urine and saliva samples is a great interest to monitor cyanide exposure from tobacco smoke. Therefore, in this work, SCN^- ion is the main target ion to be determined, and it can be evaluated in the zoomed chromatograms. Table 5 summarized the concentration of these anions. When the heavy smoker's urine sample was injected to the IC instrument, an uneven baseline occurred. It might be due to the chemically compositions contained in the samples. Therefore, the concentration of SCN^- ion could be calculated.

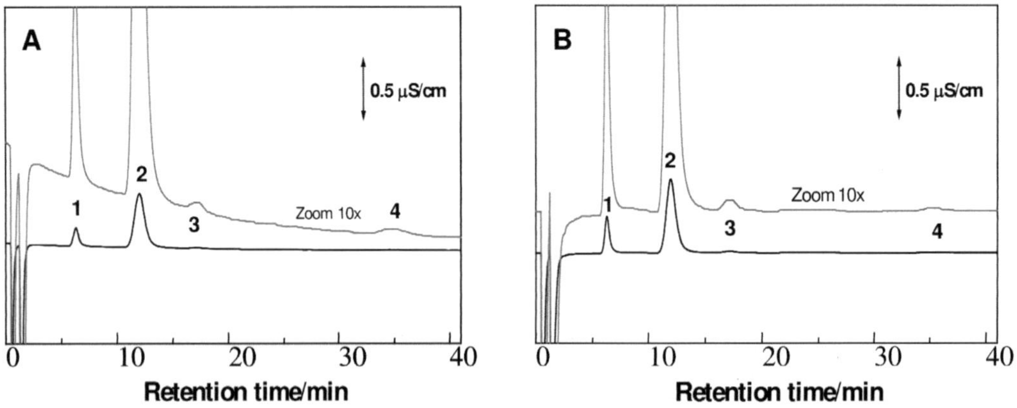

Figure 8. Determination of inorganic anions in heavy smoker's urine (A) and saliva (B) samples with 5-times of dilution. Ions: 1=PO_4^{3-}, 2=Cl^-, 3=NO_3^-, and 4=SCN^-. Other chromatographic conditions, as in Figure 4.

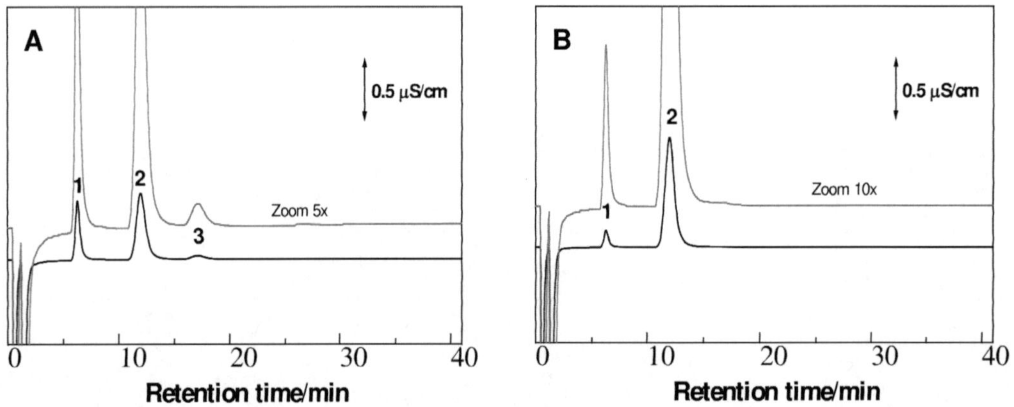

Figure 9. Determination of anions in non-smoker's urine (A) and saliva (B) samples with 5-times of dilution. Ions: 1=PO_4^{3-}, 2=Cl^-, and 3=NO_3^-. Other conditions, as in Figure 4.

IOP Conf. Series: Journal of Physics: Conf. Series **1095** (2018) 012022 doi:10.1088/1742-6596/1095/1/012022

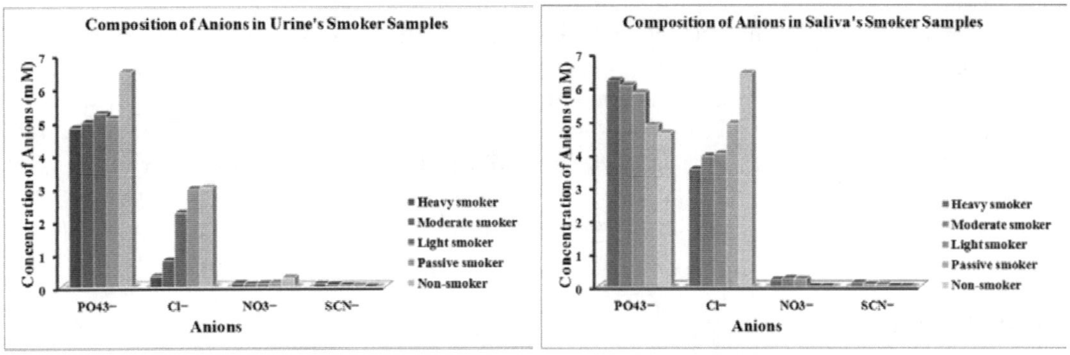

Figure 10. Composition of inorganic anions contained in smoker's urine and saliva samples. Other chromatographic conditions, as in Figure 4.

Figure 9 illustrates the chromatograms of anions for all non-smoker's urine and saliva samples. All anions could be detected except SCN^- ion. It assumed that the concentration of SCN^- was lower than the detection limit, so that it could not be calculated.

Table 5, summarizes the concentrations of anions for all smoker's urine and saliva samples, and the composition of anions contained in both type of samples, as shown in Figure 10. Moreover, the obtained results also show the concentrations of these anions contained in moderate smoker, light smoker, and passive smoker for urine and saliva samples, respectively.

Table 5. Determination results for inorganic anions in urine and saliva samples (n=5)

Smoker types	Sample types	Anions (mM)			
		PO_4^{3-}	Cl^-	NO_3^-	SCN^-
Heavy smoker	o Urine	4.82	0.32	0.11	0.09
	• Saliva	6.20	3.52	0.21	0.10
Moderate smoker	o Urine	4.98	0.80	0.08	0.06
	• Saliva	6.04	3.92	0.24	0.06
Light smoker	o Urine	5.25	2.24	0.10	0.03
	• Saliva	5.82	4.02	0.22	0.04
Passive smoker	o Urine	5.11	2.96	0.12	0.02
	• Saliva	4.86	4.90	ND	ND
Non-smoker	o Urine	6.50	3.10	0.28	ND
	• Saliva	4.62	6.41	ND	ND

Note: ND=not detected

4. Conclusions

The most interesting aspect of this dual-column switching technique is the fact that cation-exchange column and anion-exchange columns were connected serially via 10-port switching valve, a single pump, and a single detector for the determination of five important cations (Na^+, NH_4^+, K^+, Mg^{2+}, and Ca^{2+}) and the four anions (PO_4^{3-}, Cl^-, NO_3^-, and particularly SCN^-) in smoker's urine and saliva samples. The eluent of 10 mM tartaric acid showed good performance in term of selectivity and good peaks shapes for the determination of the above cations and anions. The described technique offered simple and convenient operation, and could be applied in physiological fluid such urine and saliva samples.

Acknowledgments:

This research was supported by funding of Incentive of National Innovation System Research (InSINAS) 2017, Ministry of Research, Technology, and Higher Education of the Republic of Indonesia, with grant Recipient No. 160.

References

[1] Small H, Stevens T and Bauman W C S 1975 *Anal. Chem.* 47 1801.
[2] Weiss J 1995 *Ion chromatography, 2nd ed* (Weinheim VCH).
[3] Fritz J. S and Gjerde D T. 1999 *Ion chromatography 3nd ed* (Weinheim VCH).
[4] Synder LR, Kinkland JJ and Glajch JL Practical HPLC Method Development, 2nd ed 1997 (New York – JW) *Journal of Chromatography* **602** 127.
[5] Amin M, Lim LW and Takeuchi T 2005 *Anal Bioanal Chem* **381** 1426.
[6] Amin M, Lim LW and Takeuchi T 2006 *Anal Bioanal Chem* **384** 839.
[7] Amin M, Lim LW and Takeuchi T 2007 *Talanta* **71** 1470.
[8] Amin M, Lim LW and Takeuchi T 2008 *J Chromatogr* A **1182** 169.
[9] Amin M, 2014 *Indonesian Joumal of Applied Chemiistry* **1**(6) 853.
[10] Amin M 2016 *Jurnal Techno* **5** (1) 978.
[11] David E.C. Cole, Evrovski J 1997 *J Chromatogr* **789** 221.
[12] Narkowicz S, Polkowska Z, Marc M, Simeonov V and Namiesnik J 2013 *Ecotoxicology and Environmental Safety* **96** 131.
[13] Sunden T, Lingren M and Cedergen A 1983 *Anal. Chem* **55** 2.
[14] Connolly D, Barron L and Paull B 2002 *Journal of Chromatography B* **767** 175.
[15] David E.C. Cole, Evrovski J and Pirone R 1995 *J Chromatogr B* **672** 149.
[16] Demkowska I, Polkowska Z and Namiesnik J 2008 *Journal of Chromatography B* **875** 419.
[17] Goldstein F 1950 *J. Biol. Chem.* **187** 523.
[18] Rubab Z and Rahman MA 2006 *J Pak Med Assoc.* **56** 323.

IOP Conf. Series: Journal of Physics: Conf. Series **1095** (2018) 012023 doi:10.1088/1742-6596/1095/1/012023

Production Biomethane from Palm Oil Mill Effluent (POME) with Truncated Pyramid Digester in Fed Batch System

Martha Aznury[1*], Jaksen [1], Abu Hasan[1], Arin Putri Dila [1]

[1] Department of Chemical Engineering, Politeknik Negeri Sriwijaya, Palembang Jl. Srijaya Negara Bukit Besar,Palembang 30139, Indonesia

Email : martha_aznury@polsri.ac.id

Abstract. Palm oil mill effluent (POME) production in Indonesia is estimated around 63 million ton/year [1]. Therefore it is a potential source of contamination when it directly discharged into the river. Even though, POME will be as raw resource biogas production.The aims of this research were to know the effect of fermentation time toward biogas production in truncated pyramid digester. The study was done in three combinations of feedstock with POME concentrations of 90, 80, and 70% and activated sludge concentrations of 10, 20, and 30%. A mixture of POME and activated sludge at 6 L/minutes for 30 days were fed into the digester. The result showed the fermentation time gave significant effect toward biogas production in the truncated pyramid digester. All combinations showed the upward trend in the volume of biogas during fermentation. The highest quantity of biomethane in biogas was 25% mol that the active microbes were added to POME in the ratio 1: 9 in a fed-batch system.

Keywords: biomethane, fed-batch, palm oil mill effluent (POME), truncated pyramid digester.

1. Introduction

Palm oil mill effluent (POME) is wastewater generated by palm oil mill mainly from condensate stew, hydrocyclones water and sludge separator. Every ton of fresh fruit bunches (FFB) processed formed about 0.6 to 1 m3 POME. POME is rich in organic carbon with chemical oxygen demand (COD) of 40 g/L and total nitrogen of approximately 0.5 g/L.

The fermentation process takes place in an anaerobic digester with the aid of bacteria. This process can be done in batch or continuous. In a continuous process, the substrate is inserted every day at certain flow rate corresponding to the retention period. The substrate is derived from a material that is rich in organic matter and mixed by activated sludge derived from manure as an activator. Nutrient content in manure varies depending on the state level of production, the type and amount of feed, as well as individual animals themselves [2]. Nutrient content in manure, such as nitrogen (0.29%), P2O5 (0.17%), and K_2O (0.35%) [3].

Biogas digestion consists of four stages of reaction that requires optimum conditions at each stage, if there are no optimum conditions, often the formation of biogas becomes obstructed and sludge still has the potential to produce biogas. In an effort to optimize the production of biogas, the fermentation process is done in two steps with two digesters interconnected. One of them can be done by assembling the digester interconnected in two stages.

Based on this, the authors conducted a study of wastewater treatment process of POME into biogas used by microorganisms as the activator. This research is expected to represent the real situation as possible so that it can be used to produce an environmentally friendly alternative energy. These technologies include the tank (biogas reactor) truncated pyramid-shaped. Truncated pyramid shape was chosen due to avoid channeling occurs in the bottom of the pool for their deposition. At the top of the reactor by the cover is that the methane gas trapped above the surface and can be converted to produce energy.

This technology generally avoids greenhouse gases, especially methane escape into the atmosphere. The subsequent processing stages can also be done with a truncated pyramid-shaped digester to treat of POME more effectively and efficiently. Methane gas as one of the products from oil POME can be accommodated and utilized to energy as compared to released into the atmosphere and added to greenhouse gas emissions. The other products, Sludge, can be used as a liquid fertilizer which is applied to palm oil plantations. Industrial liquid waste processing palm oil at the stage of sedimentation by using this design tool is one way that is more effective and efficient in terms of place, time, and cost of processing.

The purposes of this study include:

- Determine the influence of starter volume percentages given in POME in the digester to the quantity of biogas produced.
- Determine the influence of fermentation time on POME to the quantity of biogas by using a truncated pyramid-shaped digester on fed-batch system

2. Methods

2.1 Fermentation anaerobic for activator microorganism

The active microbes were added to POME in the ratio 1:4 L in a batch bioreactor with a volume of 8 L. Temperature and pH on anaerobic fermentation were not regulated, whereas the sampling time, each day with as much volume of 1 L and 1 L of water was replaced with POME.

2.2 Fermentation anaerobic for biogas production

Main raw material substrates of POME are pumped into the first tank to the brim. This tank serves as shelter bait. Feed that has been entered into the first tank will undergo deposition (sedimentation). The substrate will be divided into two layers. The bottom layer will be a rich substrate in the form of the slurry and the top layer of substrate that has been separated from the slurry.

The top of layer flowed into the second tank after deposition for 24 hours. In the second tank of POME mixed with a starter. Digester used with a truncated pyramid-shaped fed-batch type with a capacity of 18 L. The mixture of POME and starter varied by comparison wastewater and starter 10:90, 20:80, and 30:70. POME from the tanks streamed every day and also continuously added a starter. The fermented liquid in the tank to be also streamed every day two to three tanks to prevent in the second tank so on until the gas obtained in the variant designated day (Figure 1).

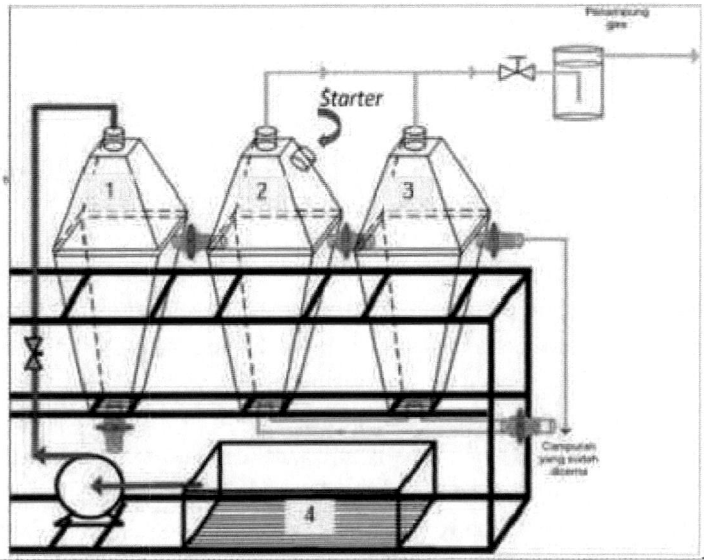

Figure 1. The design of the truncated pyramid-shaped measurement in processing of POME on biogas production

Note :

Tank 1 : Bioreactor for sedimentation

Tank 2 : Bioreactor for fermentation

Tank 3 : Bioreactor for collection

Tank 4 : Tank for collection of slurry

2.3 Analytical analysis

All the tests for the samples were analyzed according to the guidelines of the American Public Health Association [4] for the examination of POME and POME after fermentation. Biogas yield was measured with a wet gas meter (W-NK-O.SA, Shinagawa). Gas samples were obtained through an inverted funnel placed above baffles near the top of the reactor. Biogas composition was determined using a gas chromatograph (GC-8A, Shimadzu, Kyoto).

3. Results and Discussion

3.1 Raw materials analysis

POME is analyzed by nitrogen content, pH, COD, and BOD to get treatment in the sewage treatment pyramid-shaped anaerobic digester. POME as substrate was fermented to produce biogas. The results of the analysis of the substrate can be seen in Table 1

Table 1. Analysis Data from POME

Parameter	Result from this research	Kep Men LH 51
pH	4	6.0-9.0
BOD (mg/L)	1755	250
COD (mg/L)	26352	350
NH_3 Total (mg/L)	199,36	50

The results of the analysis of the chemical characteristics of POME from PT. Mitra Ogan (Table 1) shows that the waste has a pH of 4, BOD 1,755 mg/L, COD 26,352 mg/L, nitrogen 199.36 mg/L. The overall parameter is above the threshold value the quality standards set by Kep Men LH 51, 1995 (Table 1), so that POME which has a high organic matter content has the potential to be used as a substrate in the process of anaerobic fermentation. Biological treatment and in the process uses microorganisms to degrade organic pollutants [5] worthy recommended as the appropriate processing method. According to Davis and Cornell (1991) [6], the content of BOD in excess of 1,000 mg/L makes the anaerobic process to be applied as a pretreatment.

COD from POME is 26,352 mg/L. COD value is much lower when compared with studies Lang, (2007) [7] in the amount of 50,000 mg/L and is still smaller than research Hasanah, (2011) [8] is 47105.26 mg/L. COD is a measure of the quantity of organic material in the wastewater that showed high content in the form of lipids, carbohydrates, and proteins. According to Zhang et al (2008) [9], the processing potential of anaerobic fermentation is done for handling POME due to high organic material characteristics. Additionally, effluent BOD value of palm oil mill is 1755 mg/L is still the very high content of organic material so that the value of TDS and TSS is high anyway.

Based on the analysis, the total nitrogen content of POME is in the range approaching the total of nitrogen. According to Simamora et al. (2006) [10], C/N ratio that is optimum for the microorganisms is 20-25. Therefore, the liquid waste used as a substrate in the process of anaerobic fermentation. Sources C and N necessary microbes involved in the anaerobic process as a source of nutrients for growth and development of the microbes. If the N content in the substrate slightly, then the bacteria cannot produce the enzymes needed to synthesize the compound (substrate) containing carbon. Meanwhile, if the substrate is too much to contain N, bacterial growth is inhibited due to the ammonia contained in large quantities [11].

3.2 Aanalysis of Chemical Oxygen Demand (COD) and Biological Oxygen Demand (BOD)

COD is used as a measure of the quantity of organic material in the wastewater industry-and the potential to produce biogas [12]. In the reform process of anaerobic, which is biodegradable COD present in the organic material and the resulting end product called methane and past the newly formed bacteria. Measurement of COD value in the wastewater end of the palm oil industry after the fermentation process in the digester is shown in Figure 2.

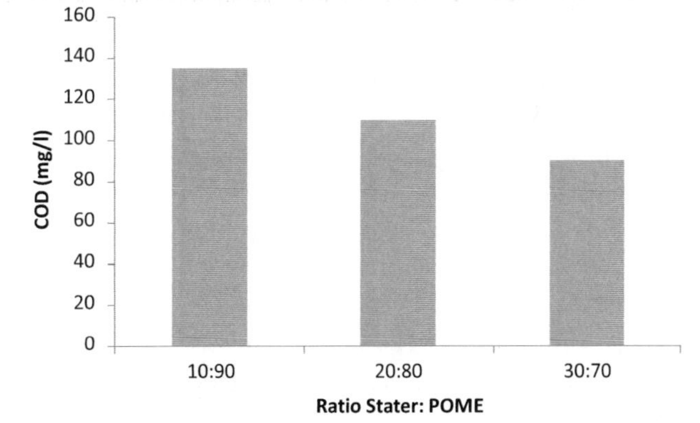

Figure 2. The value COD of POME in Fermentation Anaerobic Process

Based on the analysis, a combination of POME and starter has a significant influence on the COD value in the end. It can be seen that the value of COD in waste water mixture of palm oil and cow dung after undergoing a process of anaerobic fermentation decreased from 26,352 mg/liter before treatment to 135 mg/liter for the first sample and 110 mg/liter and 90 mg/liter for each of the second and third samples. This indicates that the POME waste containing high pollutant loads and can be dropped using the process of anaerobic fermentation processing.
The content of organic substances in the effluent POME has been degraded by microbes active from cow dung so that the organic material has been changed and its COD value decreases. According Nugrahini et al (2008) [13], COD is an important variable that indicates success or failure of the process of degradation. Overall COD measurement detects organic compounds, both organic as well as simple organic compound [14]. According Kresnawaty (2008) [15] due to impairment COD has become the hydrolysis process. At this stage of organic materials utilized by microorganisms as nutrients and convert it into a form simpler compounds.

As for the measurement BOD value in POME after the fermentation process in the digester is shown in Figure 3.

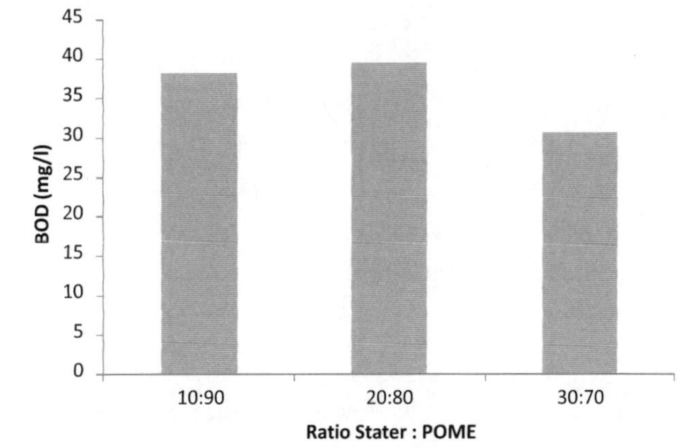

Figure 3. Concentration of BOD from Ratio of Stater and POME

The Figure 3, it can be observed that the reduction in BOD loadings occur in POME occurs due to the influence of the reduction in COD where the value of the initial BOD content before getting treatment in the digester pyramid. This reason is POME contains high pollutant loads and can be dropped using the process of anaerobic fermentation processing.

POME is dumped directly into the environment will cause environmental pollution, therefore, to reduce COD and BOD concentration required treatment in the form of waste treatment in the pyramid-shaped digester with a sedimentation process. The decrease of COD and BOD happens after treatment because some organic substances in the waste has been degraded into sludge and others have been degraded into other products such as methane [16]. Although at the sewage treatment of COD and BOD has decreased, but still need to do the processing for the reduction of COD as low as possible by regulations made by the government of Kep. MEN-LH No. Kep-51 / MENLH / 10/1995 dated October 23, 1995, where the quality of raw palm oil industry waste ready to be released into the receiving water body or the environment should have a maximum COD value of 500 mg / liter and maximum BOD of 250 mg/L.

POME has a high organic matter content and containing macro nutrients such as nitrogen (N), phosphorus (P) and potassium (K). Analysis of total nitrogen content before and after getting treatment by sedimentation processes in the pyramid-shaped digester is shown in Fig. 4.

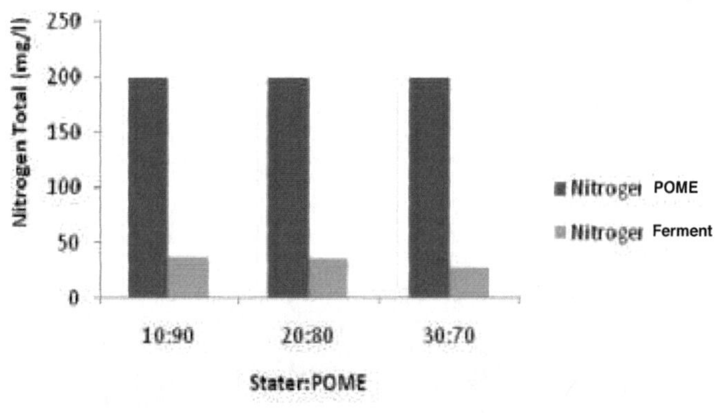

Figure 4. Total Nitrogen Levels Before and After Anaerobic Fermentation Process

After analysis of sludge produced, sludge wastewater treatment results of POME by means of sedimentation has a total nitrogen content of the previous 199.36 mg/L to Rp 38.08 mg/L for the first sample while samples in second and third respectively are 36.84 mg/L liter and 28 mg/L. This figure is already approaching the levels nitrogen of POME, so it could potentially be used as a liquid fertilizer for palm oil plants, which can automatically replace the use of chemical fertilizers price is too expensive. In addition to reducing waste disposal to the environment and can save processing costs.

3.3 Effect of time and ratio on methane production

Gas production from POME and starter using truncated pyramid digester system result fed laboratory-scale batch volume of 18 litres shows the whole combination showed the trend of increasing volume of biogas fermentation. The highest quantity of CH4 is 24.95865% mole in ratio 10:90, while the lowest quantity of CH4 is 9.4782% in ratio 30:70. Each variation of the mixture composition shows fermentation time significantly affected the production of biogas. Wherein an increase in CH4 composition during the 30 days of fermentation, as shown in Figure 5.

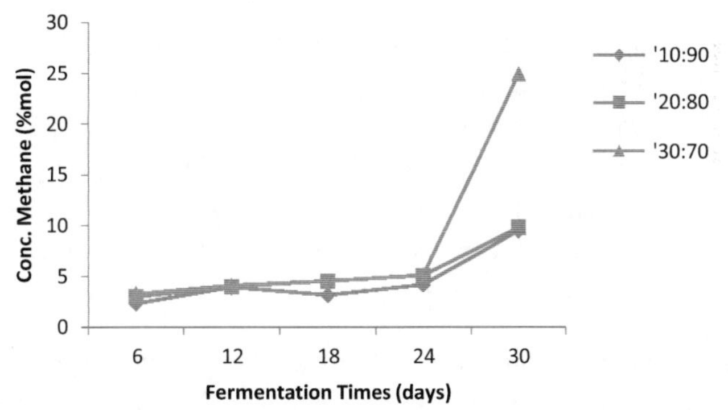

Figure 5. Effect on Time and Ratio on Methane Production

The optimumpercentage of methane concentration in optimum condition 30 days with a volume ratio of starter and liquid waste, 10:90, is 9.4782%. The percentage of gas in the ratio of 10:90 raw material are the compared with a ratio of 20:80 and 30:70 raw materials. However, the percentage of CH_4 always rises until the 30th day. The percentage of gas for 30 days of fermentation will experience enhancement with increasing fermentation time. According to research [17], the longer the fermentation time the increased activity of microorganisms to use a substrate so that it will affect the resulting product. On the 6th day of the first quantity produced was 2.9824%. The lower the percentage of CH_4 in this comparison due to the many factors that lead to microbial activity stops, such as pH and temperature factors so that microbes do not work optimally in overhauling the substrate. Such data has been shown in Table 1, the initial pH of POME that is 4. After the addition of the mixture to starter, the pH will increase. The reason is the pH value on anaerobic processes will decline with volatile acids produced and will increase with volatile acids consumed by methane forming bacteria [18]. If the pH drops below 6.5, the organic acid began to take shape with the help of hydrolytic bacteria and fermentation stage begin to stop. Methanogenic bacteria are very sensitive to changes in environmental pH. The pH value of 10 best in producing biogas ranged from 7.0. If the pH value below 6.5, the activity of methanogenic bacteria will decrease and pH below 5.0, the fermentation activity will stop [11].

The combination 20:80 of gas production in the digester produces truncated pyramid CH4 percentage is 9.8200%. Gas production on a combination was higher than 10:90 combination. It was explained earlier that the factors that influence differences in gas production in each treatment are the total N content of the substrate, the pH value and the content of nutrients and organic matter content in each mixture. The graph in Figure 5 above shows that in addition there is also the CH4 content of O2 and N2. The content of O2 and N2 looks quite higher than the content of CH4. This affects the fermentation process in the formation of biogas because microbes contained in the substrate cannot work properly if a lot of air contained within the digester. The presence of O2 and N2 content of that magnitude in the digester can occur due to the less closed digester, causing air can enter easily into the digester. The content of O_2 and N_2 causes a state in no aerobic digester while the methanogenic bacteria are living in anaerobic or minimal O_2.

In Figure 5, note that the content of CH_4 volume ratio of 30:70 starter and liquid waste is the highest, amounting to 24.95865%. This is the most appropriate composition of the three variations are done. This occurs because the starter, written more than 30% of the remaining mixture is POME with rich organic matter content so that the bacteria were able to remodel perfectly. The organic material will change the protein, cellulose, and fats into amino acids, glucose, and fatty acids. Level of methane in the biogas is affected by the ability of microorganisms to decompose organic substrate and the carbon and nitrogen in the digester.

The graph above shows the relationship quality of biogas against time to the comparison of the volume of each starter and POME. Quantities indicated a significant increase in the ratio of 30% and 70% starter liquid waste. This is because the ability of the bacterial hydrolysis of each comparison is different. Where the organic material from the starter ratio of 30% and 70% of POME is greater than the other resulting in the formation of methane by bacteria more leverage. Besides the comparison between batch and fed-batch systems are presented in Table 2.

Tabel 2. Comparison of percentage methane in batch and fed batch system

System	Stater:POME	Ferm. time (days)	Conc Methane (%mol)	Ref.
Batch	30:70	4	1.7972	Saputri (2015)
Fed Batch	30:70	30	24.9586	This research (2016)

4. Conclusions

Based on the results of research that has been done is the treatment of POME into biogas fed-batch system, it can be concluded of the three variations of combinations of substrates that starter and POME are 10:90, 20:80, 30:70 shows that the more volume starter is added to the liquid waste then the percentage of biogas produced will be higher. It can be observed in the percentage of methane generated is directly proportional to the volume of starter added.

Time of fermentation in each variation of the substrate, namely a combination of POME and starter 10:90, 20:80, 30:70 by fed-batch system provides real influence on the production of gas, with the upward trend in gas production during the 30 days of fermentation. Fermentation time is long lead time live feed in digesters longer so that the bacteria can remodel more organic material into biogas.

In further research is necessary to add other nutrients to the substrate so that the result can be a maximum of biogas. In addition, variations in the other livestock manure activators could be in with a longer retention time again.

Acknowledgments
The authors would like to acknowledge the financial suport of Penelitian Stategis Nasional, Directorate General of Higher Education provides funding research project grants NOMOR SPPK : 189/SP211/LT/DRPM/III/2016, Date: 7 Desember 2015, entitled Rancang Alat Biodigester Untuk Pengolahan Air Limbah Industri Minyak Kelapa Sawit Untuk Memproduksi Biometan Dan Pupuk. PT. Perkebunan Mitra Ogan was a su0port of POME.

References

[1] Direktorat Jenderal Perkebunan.2015.*Statistik Perkebunan Indonesia 2014-2016*. Jakarta.
[2] Abdulgani, I. K., 1998. *Seluk Beluk Mengenai Kotoran Sapi serta Manfaat Praktisnya. Fakultas Peternakan,* Institut Pertanian Bogor, Bogor.
[3] Hardjowigeno S 2013. *Ilmu Tanah*. (Jakarta Akademika Presindo)
[4] American Public Health Association (APHA). 1995. *Standard Methods for the examination of Water and Wastewater,* 19thEd. Washington, D.C.
[5] Hermana J 2016.*Pengolahan Biologis Aerobik Sistem Tersuspensi dan Terlekat*.
[6] Davis, M L and Cornell DA 1991. *Introduction to Environmental Engineering*. 2nd edition. (New York: Mc Graw Hill International).
[7] Lang, LY 2007 *Treatability of Palm Oil Mill Effluent (POME) using Black Liquor in an Anaerobic Treatment Process*. Tesis (Malaysia University Sains Malaysia)
[8] Hasanah, Hilda. 2011. *Penurunan Beban Pencemar Limbah Cair Kelapa Sawit Melalui Proses Fermentasi Anaerob Menggunakan Digester Anaerob Dua Tahap*. Skripsi. Fakultas Teknologi Pertanian. (Bogor-IPB).
[9] Zhang Y, Yan L, Chi L, Long X, Mei Z and Zhang Z. 2008. *Journal of Environmental Science*. **20** 658.
[10] Simamora S, Salundik, Wahyuni S, and Surajudin. 2008 *Membuat Biogas: Pengganti Bahan Bakar Minyak & Gas dari Feses Ternak*. Jakarta: Penerbit Agromedia.
[11] Yani M, and Darwis AA. 1990.*Diktat Teknologi Biogas*. Bogor: Pusat Antar Universitas Bioteknologi- IPB.
[12] Zeeman. 2003.*Pengolahan Kelapa Sawit*. Pusat Antar Universitas, Universitas Gadjah Mada, Yogyakarta.
[13] Nugrahini. 2001 *Peningkatan Zat-Zat Tercemar Mengakibatkan* Pemanasan Global. Fakultas Kesehatan Masyarakat: Universitas Sumatera Utara.
[14] Syamsudin; Purwati, S.,and Andri, T. R. 2008, Bandung . *Berita Selulosa* 43(2) 83.
[15] Kresnawaty, I., I. Susanti, Siswanto and Panji, T. 2008. *Optimasi produksi biogas dari limbah lateks cair pekat dengan penambahan logam, Jurnal Menara Perkebunan*. Balai Penelitian Bioteknologi Perkebunan Indonesia: Bogor 76(I),
[16] Deublein, D. and Steinheuster A 2008.*Biogas from Waste and Renewable Resourcers* (Weinheim Wiley-VHC Verlag GmBH & CO)

IOP Conf. Series: Journal of Physics: Conf. Series **1095** (2018) 012023 doi:10.1088/1742-6596/1095/1/012023

[17] Saputri, Hardina Apri, 2015. *Pengaruh Persentase Volume Starter Dan Waktu Fermentasi Pada Pome Terhadap Kuantitas Biogas Menggunakan Digester Balok Sistem Batch. Laporan Akhir*. Jurusan Teknik Kimia. Politeknik Negeri Sriwijaya. Palembang.
[18] Gerardi, M.H. 2003. *The Microbiology of Anaerobic Digesters*. (New Jersey - John Wiley dan Sons. Inc)

Preparation And Characterization of Bio-\Polymeric Nano Feed Incorporating Silage-Derived Organic -Acids And The Polar Fraction of Papaya Leaf Extract

[1]Sofia Sandi, [2]Miksusanti*, [3]Mardiyanto, [4]Fitra Yosi, [5]Meisji Liana Sari
Sofiasandi_nasir@yahoo.com; miksusanti@gmail.com*, dianto72@yahoo.com, fitrayosi@unsri.ac.id, meisji@yahoo.com
[1]Depatment of Animal Science Faculty of Agriculture, [2]Department of Chemistry, Faculty of Mathematics and Natural Sciences, [3]Pharmacy Study Program Faculty of Mathematics and Natural Sciences, [4,5]Depatment of Animal Science Faculty of Agriculture, University of Sriwijaya,.

Abstract: This research has aimed to incorporate the fermented organic acids and papaya leaf extract into chitosan-alginate composites using an ionic -gelation method. The concentration of chitosan was 35 mg/7.5 mL, sodium alginate was 20 mg/7.5 mL, and cross-linking the $CaCl_2$ cross linker was 20 mM. Variations in the formula Different formulations, are carried out on organic acids with the code of referred to as kumpai-ZnO, legum-ZnO, and kumpai-legum-ZnO. These sample using various organic acids, and the formulations contained as much as 50 mg/7.5 mL of the organic acid. The results of the identification of the chemical contents of the fermented product was lactic acid derivatives as a major component, and there were no alkaloids, flavonoids, terpenoids, and or poly phenols content present. The particle characterization results showed they were a homogeneously distributed in size with PDIs from 0.199 to 0.218 with a sizes between of 253.6 to and 286.8 nm. The potential Zeta potential is was found to be -20 to -15 mV by using a particles size analyzer (PSA). Particles were spherical based on the images of transmission electron microscopy (TEM) images. The best formula formulation was is the formula 3, containing with the kumpai-legum-ZnO content, which was demonstrated by to be 85% encapsulation percentage encapsulated (%EE). The anti-*E. coli* activity test showed an increase of the particles were 40% of activity the particles compare to more active than the same substances not in particle. The ionic -gelation method successfully created a homogenous particle suspension loading loaded with the active compounds, such as the fermented organic acids and the papaya leaf extract. This system could deliver the active compounds to the bacterial cell of *E. coli* and *S. typimurium* bacterial cells.
Key words: Organic acid, chitosan, alginate, ionic -gelation, antibacterial activity

1. Introduction

Silage is one of the bioprocessing technique used for fresh materials of with fresh high- water contents in a places like (silos) under anaerobic conditions. In The process (ensilage) formed generates secondary metabolite products such as the organic acids of the fermentation process products from the conversion of carbohydrates into lactic acid by lactic acid bacteria. The more greater the amount of lactic acid bacteria, the more acid form formed in the silage. The acids that are formed are organic acids such as lactic acid, acetic acid, and butyric acid. The content of organic acids in silage derived from silage grass is higher than the acid content in the legume silage, because legumes contain higher protein contents, which can inhibit the acidification process. Previous research studies have reported that the copper-based silage resulted in higher organic acid profile content of silage made from legumes [1]. Organic acids are feed ingredients that can be used as feed additives in animal feed as a substitutes for antibiotic function because it they have antibacterial properties, which can allow them to penetrate the bacterial cell walls and disrupt the normal physiology of some types of pathogenic bacteria. Pathogenic bacteria such as *Escherichia coli* and *Salmonella typhimurium* can compete with microflora of animal in obtaining nutrients in the gastric intestinal tract, and can produce harmful metabolites products that are harmful to the host, thus which can result in impaired livestock growth and increasing the chances of contracting the disease [2]. A disadvantage of the direct administration of organic acids in to livestock in liquid form has a disadvantage is that is not optimal in the acceptance absorption of organic acids up to the small intestine because of its their volatile nature, and they may at first initially be absorbed in the upper gastrointestinal tract (cache, gizzard and proventriculus), thus reducing their effectiveness in the small intestine and caecum. This weakness can be overcome by changing the organic acid into its salt, and then it can be delivered by a biopolymer. Organic acid salts are obtained by mixing the liquid organic acid liquid with a base to form crystals, that can which prevent evaporation and reduce organic acid metabolism in the upper gastrointestinal tract. Subsequently, Organic acid salts derived from complete maize-based corn waste material with dose of 12.5% were more effective in inhibiting *Escherichia coli* (0.33 cm) than *Salmonella typhimurium* (0.34 cm).

Papaya leaf, can be consumed as livestock feed. Papaya leaves contains flavonoids, which has antibacterial properties [3] [4] [5] [6]. Attempts have been made to extract the beneficial constituents from Papaya leaf, but the extract is less stable than the leaf so far has been tried to be extracted but the stability is less obtaining. With particle technology, papaya leaf can be diversified and made more practical the empowerment and diversification of papaya leaf use could be realized as livestock feed for preventing the growth of pathogenic bacteria. This particle technology uses polymers such as chitosan that can be destroyed in by living organisms (biodegradable) . Chitosan is not expensive biopolymer; therefore, it can be used for livestock rather than other biopolymers that more expensive pharmaceutical products. Chitosan could form particles with the presenting of in the presence of alginate and cross linking with $CaCl_2$ as a crosslinker by ionic -gelation method [7] [8] [9].

Based on the above background, the purpose of this study was to prepare, and characterize, the antibacterial activity of particles chitosan-alginate loaded with organic acids and ethanol extract of papaya leaf. The chemical content of the organic acids was performed the evaluated qualitatively test of chemical content. Against the particle was characterized the the size, distribution pattern, and homogeneity, also as well as the zeta potential (using a PSA). Particle morphology was determined with TEM images. Percentage encapsulation (% EE) was determined using UV -spectrophotometer tool. The antibacterial activity test was performed by the challenge analyzed by testing the particles against the growth of bacteria on nutrient-agar media.

2. Materials and Methods

2.1. The process of making silage

The stages of in making the silage refer to the described in a previous method report [10]. In the first step, grass of kumpai tembaga (Hymenachne acutigluma) and legume of kemon air (Neptunia oleracea Lour.) were cut with the size 2-5 cm and stored for 24 hours for the process them to withering. After that, it was weighed as many as 500- g batches of withered material were weighed out for each treatment. Each sample was mixed with molasses, which had been dissolved in water, as many as at as much as 3% by weight of forage, and then the mixture was stirred until blended. Furthermore, each treatment forage was put into transferred to a 3-layered plastic bag of 3 layers and then compacted to air compressed to remove all air. The bags were then tied tightly and stored for 21 days in a dry place and not exposed to protect from direct sunlight. After 21 days, the bags of silage were opened, and taken to be tested for testing in the laboratory.

2.2. Isolation of organic acids from the silage

The process of isolating the organic acids carried out using maceration extraction method with methanol as the solvent [11]. First of all, 50-g sample of swamp forage weighed 50 g and then was immersed into 250 ml of methanol. The sample covered with plastic wrap and coated with aluminum foil. Furthermore, it was stored on in a fume hood. Maceration was done for 3 days. Every 24 hours, the methanol extracts obtained from the silage were combined. The combination of the methanol extract was evaporated using a rotary evaporator at 50°C, with rotation speed of 60 rpm to form a viscous extract.

2.3. Dispersion of the polymer

To prepare the the chitosan preparation solution, 35 mg of chitosan powder was dissolved in 7.5 mL of 1% acetic acid solution in Beaker and then homogenized with magnetic stirer at 75 rpm for 30 min at room temperature until solution was completely dissolved. Every 20 mL of chitosan solution pipeline contains 12 mg of chitosan for each formulations. To prepare the sodium alginate solution, the 9.6 mg of sodium alginate preparation was dissolved within 60 mL of API in a glass beaker and glass then homogenized with magnetic sttirer at 75 rpm for 30 min at room temperature until sodium alginate solution was completely dissolved. Every 20 mL of sodium alginate solution pipette contains 3.2 mg for each formulation [12].

2.4. Preparation of the particles

A 50-mg sample of papaya leaf extract 50 mg was added to 7.5 mL of the chitosan solution in the a vial using a magnetic stirrer and magnetically stirred at 750 rpm [6] [8] [9] [10], and this mixture was used for each formulation and to submicron particles [13]. Sodium alginate was dispersed in aqueduct using magnetic stirrer at a rate of 750 rpm 60 ^{o0}C, and taken 20 mL of this mixture was taken as mass 2 for each formulation and as mass 2. The mixture mixing of the two masses was performed by dropping the solutions from with a 50- µL micro pipet above the a solution being magnetically stirred and then continuing to stir the solution at 750 rpm for 1 hour. Calcium chloride solution was added to avoid the rapid agglomeration [14] [15].

2.5. Purification and determination of encapsulation (%EE)

Purified sub micro particles of chitosan- sodium alginate was carred out by using 300 kDa viva spin filters and centrifugation for 15 min until 2 phases. The phases was absorbed by the Viva-spin 300 kDa filter [16]. The supernatant phase was separated, with 30 mL of Aqua Injection (API) into in the sample holder and centrifuged again. Determination of maximum wave length scanning was done at the lowest concentration solution possible using UV-Vis spectrophotometer. After that, the absorbance of solutions measurement at all concentration of a standard (quercetin). Determination of the EE% values were determined using calibration curve with concentration each from 0.10; 0.20; 0.30; 0.40 and 0.50 µg/mL solutions of standard prepared from the stock solution with a concentration of 10 µg/mL stock solution. Measurement of absorbance was done using standard solutions and the supernatant of purification were measured. EE% values were then determined based on the amount of free substance present in the solutions.

2.6. Antibacterial Activity

One Ose of each culture of test bacteria (*E. coli and S. typhimurium*) was, inoculated on the surface of the nutrient agar medium in a Petri dish by a scratch method. After that, the samples were incubated for 24 hours at 37°C. Bacterial suspensions were made using rejuvenated bacteria. Furthermore the bacteria were inoculated into 10 mL of nutrient broth medium and homogenized. After that it the homogenates were then incubated for 24 hours at 37°C. The concentrations of the extract used refers to the optimum formula obtained after the based on the percentage of efficiency of the encapsulation (% EE). Antibacterial testing was performed by diffusion method using 6- mm discs paper. Antibacterial activity was reported as shown by the diameter of the clear zone formed around the paper disc paper.

A total of 1- mL aliquot of each the suspension of the each test bacteria was inoculated into a petri dish which contained 15 mL of nutrient media. Approximately 50 µL of the test solutions were dropped on to the discs 50 L. The supply of nanoparticles was placed in a nutrient agar medium inoculated with bacteria. In each petri dish there were also negative controls (chitosan and alginate), a positive control (tetracycline (30 ppm), and papaya leaf ethanol extract (2% (w/v). The dishes were then incubated for 24 hours at 37°C. The clear zones were measured using a sliding term with total

IOP Conf. Series: Journal of Physics: Conf. Series **1095** (2018) 012024 doi:10.1088/1742-6596/1095/1/012024

area calculations reduced by the diameter of the paper disc diameter. Three replicates were performed for each treatment consists of three repetitions.

3. Results

The acids formed during the this process include lactic acid, acetic acid and butyric acid. Leguminous feed such as *C. calothyrsus*, with their high protein contents, such as *C. calothyrsus* can be stored using an appropriate method such as silage. However, these legumes generally have low salt contents and low buffer capacity, which cause low silage quality. In this research study, we already evaluated the contents of organic acids produced by during the fermentation of kumpai and legumes. Precipitation of the organic acids was facilitated by the hydroxides of base of Na, Ca, and Zn. The results in Table 1 shown that kumpai has more potential than the legume. Because of source of our system is based on the herbs, we also determined found that other secondary metabolites such as alkaloids, flavonoids, terpenoids, and polyphenols another compound that might could also be entrapped present in the fermentation result products such secondary metabolites alkaloid, flavonoid, terpenoids, and polyphenol. These secondary metabolites had influence to the growth of normal flora in the digestive tract of livestock. The results was that only lactic acid derivatives were only available as the precipitation salts of the organic acids.

Papaya plants produce fruit that is with known to benefits, but the leaves can still be increased its more efficiently used use. Papaya leaves are commonly used consumed as vegetables and are also used as animal feed. Papaya leaves have a high fiber and flavonoid contents that can counteract the growth of pathogenic microbes. Papaya leaves contain semi polar to polar active compounds, that are in the fraction of semi polar to polar while the nonpolar fraction contains terpenoids compounds and waxes. The results regarding of to the screening photochemistry phytochemical studies on the of extract papaya leaf extracts was shown in Table 1 and Figure 1. Based on previous report, besides so far papaya leaf is in addition to being a fodder, papaya leaf is also then a good nutrient delivery system required by the farm for livestock. So that the This means that organic acids from the fermentation process can be used as a probiotics for the support of good beneficial bacteria, such as *Lactococcus*, in the digestive system, that is *Lactococcus* and the acids can be then followed by papaya extract, to which can handle the pathogenic bacterial pathogens. Livestock have the Healthy digestive systems ensure the propertion there for the supply of nutrients will be guarantee livestock.

Table 1. Identification of The Fermentation Products and Phytochemical Tests of papaya leaves

Identification	Result
Protein test	-
Lipid test	-
Organic acid test	+++
Disacharide test	-
Polysaccharide test	-
$CHCl_3 + H_2SO_4$ -	-
$FeCl_3$ test	+
Dragendorff test	++
Cyanidin test	++

IOP Conf. Series: Journal of Physics: Conf. Series **1095** (2018) 012024 doi:10.1088/1742-6596/1095/1/012024

Described mark : The symbols indicate the following: - (revealed showed no reaction), , + (revealed color by reaction showed a color change in the reaction), +++ (revealed color by very strong reaction showed a very strong color change)

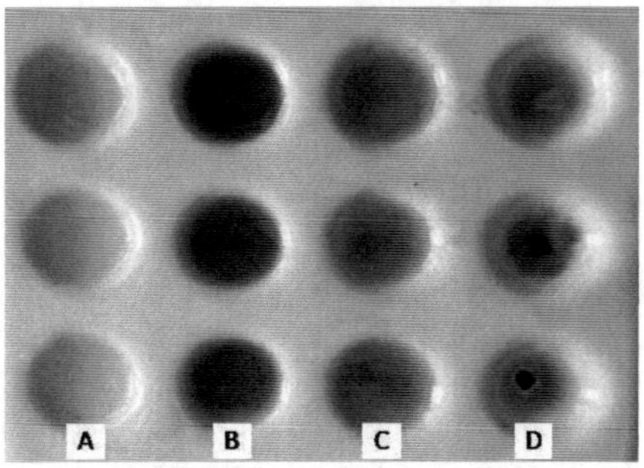

Figure 1. Image of Colors_Observed in the Phytochemistry Tests Steroids Test (A) Polyphefol Test (B) Alkaloid Test (C) and Flavonoid Test (D)

Table 2. Inhibition Zone Nanoparticle Feed From Silage

Bacteria	Green Silage	*Inhibition Zone (cm)* ----------------*Doses ZnO (%)*---------------- 12,5	17,5	22,5	*Avarage*
E. coli^{tn}	P1	0,39 ± 0,01	0,31 ± 0,06	0,24 ± 0,04	$0{,}32^{b}$
	P2	0,32 ±0,08	0,33 ± 0,08	0,28 ± 0,10	$0{,}31^{b}$
	P3	0,22 ± 0,03	0,27 ± 0,06	0,23 ± 0,03	$0{,}23^{a}$
	Average	0.30	0,31	0,25	
S. typhi*	P1	0,22 ±0,03	0,32 ±0,03	0,15 ±0,01	$0{,}22^{b}$
	P2	0,08 ±0,03	0,14 ±0,04	0,14 ±0,02	$0{,}12^{a}$
	P3	0,10 ±0,05	0,12 ±0,03	0,12 ±0,04	$0{,}11^{a}$
	Average	$0{,}13^{a}$	$0{,}19^{b}$	$0{,}12^{a}$	

Explanation : Silver Kumpai (P1), Silver Kumpai + Water Kemon (P2), Water Kemon (P3). Significant Effect of Interaction (*). Unsignificant Effect of Interaction (^{tn}). The same Superscript in the same colum and row showed significance difference (P<0,05).

From statistical data, showed that interaction between green silage and ZnO dose effect unsignificantly (P<0,05), aganits inhibition zone of *E. coli*. ZnO concentration in this research can not give good impact on inhibition zone aganist *E. coli.* Concentration organic acid from silage significantly enhance inhibition zone against *E. coli.* Organic acid has known as antimicrobial which is potential to inhibit patogen [17]. P3 showed the lowest result compare with other treatment. P3 was composed by water kemon legum which is contain hight protein concentration. Hight protein concentration can effect to the BAL growth due to ensilase process and gave negatif impact on organic acid production. According to Ali *et al.* (2012), water legum kemon contain crude protein about 28,02%. Hight protein could declaine acidify process, and inhibit the growth of BAL [18].

Due to variance analyses, there was significant interaction (P<0,05), between green silage and ZnO concentartion against avarage inhibition zone against *S. typhimurium.* From further test result P1 has give the best result from other treatment, and with concentration 17,5% organic acid and the existence of ZnO, can inhibit the growth of *S. typhimurium.*

Combination of low pH with organic acid (especially lactic acid) act as important aspect to inhibit patogen bacteria [19]. Lactic acid can inhibit *S. typhimurium* because of low pH and intraselluler activity of bacteria sel [20]. According to Sharma et al (2017), Zn as micro mineral play as antibacterial mineral. ZnO can disrupt bacterial membranne, alter the permeability of membrane and can induce oxidative stress on bacteria and inhibit cell growth [21].

The polymers that support the ionic-gelation loading papaya leaf extract and acid derivatives through-derivated ionic gelation are chitosan and alginate. Positive charges on the chitosan many benefits can beneficially can be mainly interacted interact with the negatively charged alginate. First, chitosan is interacts with the papaya leaf extract, and then this interaction the newly formed complex is coated with alginate. Alginate can make bonds with calcium chloride. Therefore, on the outer layer of the particles, we can also put append the fermented organic acids from the fermentation for the legumes and kumpai as probiotics to form NPs loading loaded with 85% of probiotics and papaya extract. The size, PDI, and zeta potential also as well as the TEM images of the particles was shown in Figures 2 and -3. The conclusions of the arrangement of ultimately, materials in the particles we designed are, in the surface layer there is a probiotic, and the coating contains coated in there substances that can inhibit the growth of pathogenic bacteria. In some Previous research, it studies have indicated that papaya leaves did not cause a problem to inhibit the growth of normal flora, so that the cattle eat well can these leaves with no negative side effects. Probiotics could also suppress the growth of pathogenic bacteria as shown in Table 2 and Figure 4. In the results of this study, when it the materials were incorporated into became particles, there was a notable increase in activity a visible activity increase compared to nanoparticles extract papaya leaf extract not in particle form.

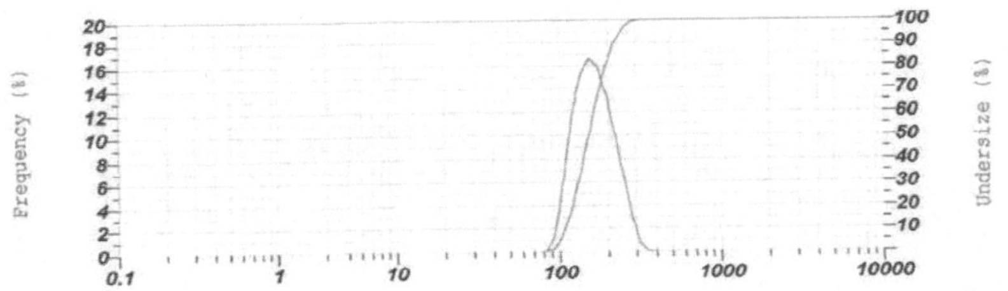

2A

IOP Conf. Series: Journal of Physics: Conf. Series **1095** (2018) 012024 doi:10.1088/1742-6596/1095/1/012024

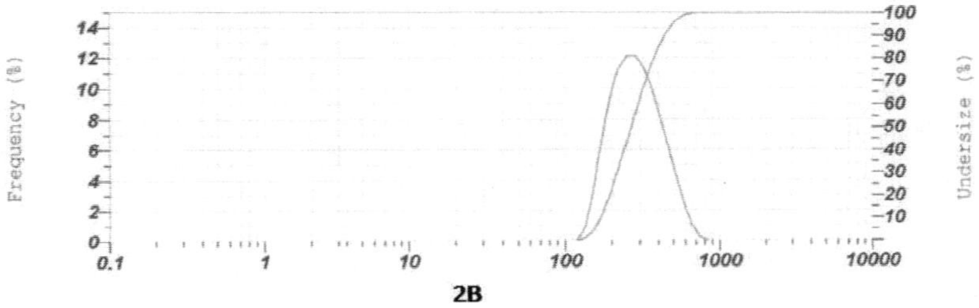

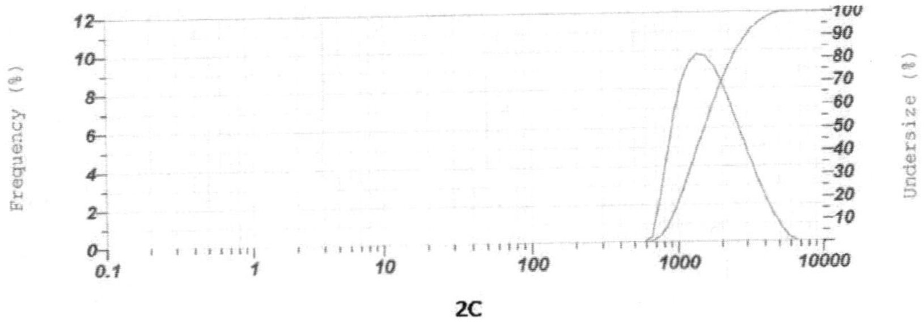

Figure 2. Size of the Particles and Their PDIs with at concentrations of 12.5 (A) 17.5 (B) and 22.5 (C).

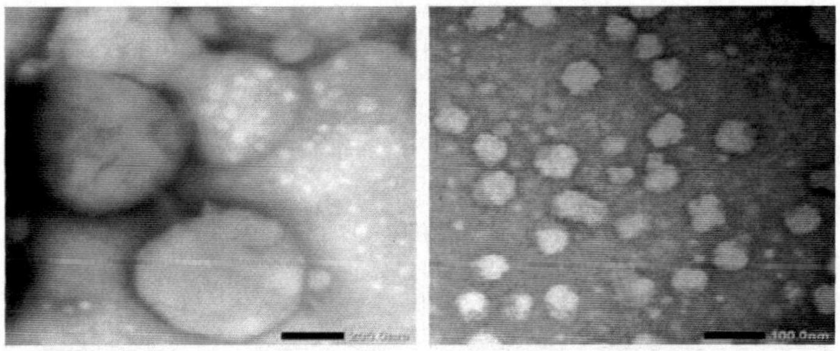

Figure 3. Transmission electron microscopy (TEM) Images.

Table 3. Trials Tests of Pathogenic Bacteria Against Organic Acid Salts from Swamp Forage Silage

Type of silage	Base	*concentration*	*Escherichia coli* (cm)				*Salmonella typhimurium* (cm)			
			1	2	3	average	1	2	3	average
Kumpai	NaOH	12.5	0.1	0.2	0.1	0.13	0.2	0.15	0.12	0.16
		17.5	0.2	0.15	0.15	0.17	0.13	0.1	0.12	0.12
		22.5	0.1	0.2	0.1	0.13	0.3	0.3	0.2	0.27
	$Ca(OH)_2$	12.5	0.1	0.3	0.1	0.17	0.2	0.3	0.15	0.22
		17.5	0.15	0.1	0.3	0.18	0.15	0.1	0.25	0.17
		22.5	0.1	0.15	0.1	0.12	0.1	0.13	0.12	0.12

	ZnO	12.5	0.38	0.4	0.39	**0.39**	0.23	0.27	0.23	0.24
		17.5	0.25	0.37	0.3	0.31	0.15	0.15	0.1	0.13
		22.5	0.28	0.25	0.2	0.24	0.3	0.35	0.3	**0.32**
Legume	NaOH	12.5	0.1	0.05	0.1	0.08	0.1	0.1	0.2	0.13
		17.5	0.1	0.15	0.1	0.12	0.25	0.22	0.2	0.22
		22.5	0.1	0.1	0.1	0.10	0.13	0.1	0.15	0.13
	$Ca(OH)_2$	12.5	0.15	0.15	0.1	0.13	0.12	0.2	0.2	0.17
		17.5	0.2	0.1	0.1	0.13	0.1	0.2	0.1	0.13
		22.5	0.15	0.2	0.2	0.18	0.1	0.12	0.2	0.14
	ZnO	12.5	0.25	0.4	0.3	0.32	0.1	0.1	0.05	0.08
		17.5	0.35	0.4	0.25	0.33	0.14	0.12	0.15	0.14
		22.5	0.25	0.2	0.4	0.28	0.1	0.13	0.18	0.14
Kumpai-Legume	NaOH	12.5	0.15	0.1	0.1	0.12	0.2	0.2	0.25	0.22
		17.5	0.1	0.1	0.1	0.10	0.15	0.2	0.15	0.17
		22.5	0.1	0.1	0.1	0.10	0.25	0.25	0.15	0.22
	$Ca(OH)_2$	12.5	0.1	0.25	0.1	0.15	0.2	0.15	0.2	0.18
		17.5	0.2	0.15	0.15	0.17	0.15	0.15	0.1	0.13
		22.5	0.1	0.1	0.1	0.10	0.15	0.2	0.15	0.17
	ZnO	12.5	0.2	0.2	0.25	0.22	0.15	0.1	0.05	0.10
		17.5	0.3	0.3	0.2	0.27	0.15	0.08	0.12	0.12
		22.5	0.25	0.25	0.2	0.23	0.15	0.1	0.1	0.12

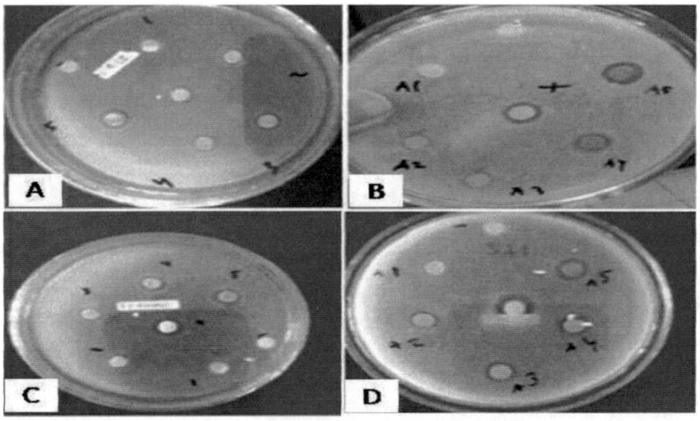

Figure 4. The growth of *E. coli* by NPs-eks activity (A). *S. Typhimurium* by NPs-eks activity (B). *E. coli* by NPs-org Ac-eks activity (C). and *S. Typhimurium* by NPs-org Ac-eks activity (D)

4. Discussion

Silage is a fodder produced by the fermentation of materials with high water contents fermentation process. The High-quality of the silage is achieved when lactic acid as the dominant acid is produced as the dominant acid; a rapid decrease in pH indicates, an efficient acid fermentation is occurring when the pH silage decrease occurs rapidly. The faster the fermentation occurs, the more nutrients can be retained contained in the silage can be obtained. *Enterococcus faecium* EF9296 type can be used as an inoculum because it can be used to suppress the work of pathogenic bacteria such as *Salmonella spp*. It can also be used to maintain the environmental ecosystem of the silage and the quality of microbiological quality during the fermentation process [27].

There are two ways of making silage, by chemically and biologically. The chemical way is carried out by method involves the addition of acids as preservatives such as formic acid, propionic acid, hydrochloric acid and sulfuric acid as preservatives. The addition is required for to rapidly decrease the pH of the silage to descend immediately (to about 4 approximately 4.2). This condition will inhibits respiration, proteolysis and prevents the activation of Clostridia bacteria. While in biologically method, by fermenting the ingredients are fermented until the naturally formed acid is formed that decreases the pH of the silage. The identification of the chemical contents of the fermented product was indicated the presence of lactic acid derivatives, and there were no alkaloids, flavonoids, terpenoids, and poly phenols content present. The polar substances, such as lactic acid derivatives, are is suitable for interacting with the chitosan encapsulated by alginates. This lactic acid derivatives are a food source that fully nourish normal flora in cattle rumen. If normal flora is fertile leads to the livestock digestion will also be being optimal for fiber processing because of normal flora will produce hydrolytic enzymes and decompose the fiber breaking regime into nutrients that enter the body of the livestock. These nutrients will have a direct effect on muscle weight and livestock health. [22]

Nutrition in livestock is generally absorbed through a mechanical process by involving the organs of the livestock because it depends on the size of the substance to be being eaten and requires diminution of in size. Many creations prepared nanoparticles production have been tried tested primarily for targeted drug delivery, but that were not many few of them have been applied to increase livestock productivity. This research is the first step to develop more broaden the potential from applicability of an extracts from Indonesia to for enhancing the productivity of livestock. Natural botanical extracts is herbal ingredients that, they must have been proven safe for humans, will also be safe for livestock. For nutrient delivery systems, the effects of mechanical processes need to be reduced minimized by reducing the length duration of the process of chewing and rumen destruction processes [23]

With the an emphasis of on the overlap between nanoparticle technology that also has touched and the world of farms agriculture, then this there is an opportunity to create the ideal particles form and certainly that are also biodegradable. Nano encapsulation systems of for polar compounds such as lactic acid derivatives have long been experimented studied and are often formed by researchers to be made into small objects particles for rapid absorption in the intestines. Nano precipitation Starting from the is a precipitation technique in requiring two mixed solvents with the but the polymer is being more soluble in one than the other, and the product precipitates out precipitation in the form of nanoparticles, this method is known as Nano precipitation [27]. There is another method that uses a water-insoluble polymer as that is soluble in oil and, therefore may be able to form it might be an emulsion form. The emulsion occurs when one or even two phases are by emulsified one or even two stages. When the oil solvent evaporates, the polymer will be suspended as into a nanometer- sized particles;, this method is known as emulsion solvent evaporation. The opposite thing is In contrast to this is a the simple method such as ionic gelation; because the polymers used in this method are polar, and most natural materials such as acid derivatives, flavonoids, and saponin are more polar or semi polar, so it they are better suited to fits with this system. In addition, polymers such as chitosan that can be used for in ionic gelation systems are polymers that are numerous common and easy to find in Indonesia. For the foreseeable future, the assurance of bbioactivity test and toxicity test of the prebiotic and extra-unnatural materials in the nanoparticles needs to be done conducted soon to gain a full and become the material of its own understanding of the ingredients in the various extracts of Indonesian natural ingredients products [25]

As the In conclusions the production produced Nano feed has been successfully characterized and showed to be homogeneous in distributed size with PDIs between 0.199 to and 0.218 with a size and sized of 253.6 to 286.8 nm. The potential Zeta potential was found to be between -20 to and -15 mV by using particle size analyzer (PSA). Particles were spherical based on the images of transmission electron microscopy (TEM) images. The best formulation is the formula 3, which contains with the kumpai-legum-ZnO content, which was demonstrated by and was found to be 85% encapsulation percentage encapsulated (%EE) by using based on UV -spectrophometers. The anti-*E. coli* activity test showed an increase of the particles were 40% of activity the particles compare to nor active than the same active substances not in particle active substances [26]

Acknowledgements
This work was financially supported by the Ministry of Research, Technology, and Higher Education through a competence grant in 2016-2017. We certify that there is no conflict of interest with any financial organization regarding the results had that have been discussed in the manuscript.

References

[1] Fitra Yossi, Sofia Sandi and Miksusanti, 2017. *American Journal of Animal and Veterinary Sciences* **12** 143

[2] Negara W. 2009. *Production of organic acid salt from silage complete ransum as growth enhance for broiler to protec from Escherichia coli and Salmonella typhimurium.* Disertation (Bogor/IPB) Indonesia.

[3] Seigler DS, Pauli GF, Nahrstedt A, and Leen R. 2002 *Phytochemistry* **60** 873.

[4] Kafshgari MH, Khorram M, Khodadoost M, and Khavari S 2011. *Polymer J* **20** 445.

[5] Paques JP, van der Linden E, van Rijn CJM, and Sagis LMC. 2004 *Colloid and Interface Sci* **209** 163

[6] Kawashima Y, Yamamoto H, Takeuchi H, and Kuno Y. 2000, *Pharm Dev and Tech* **5** 77.

[7] Kumar DP, Subas D, Subrata C, and Soumen R. 2012. *Int. Res. J. Pharm.* **3** 132.

[8] Gao C, Liu M, Chen J, and Zhang X. 2009. *Polymer Degr and Stab* **94** 1405.

[9] Gazori T, Khoshayand MR, Azizi E, Azizi E, Yazdizade P, Nomani A, and Haririn I. 2009 *Carbohydrate Polymers J* **77** 599.

[10] Gyliene O, Razmute I, Tarozaite R, and Nivinskiene O. 2003. *J Chem* **14** 121.

[11] Li P, Dai YN, Zhang JP, Wang AQ, and Wei Q. 2008. *Int J of Biomed Sci* **4** 221.

[12] Luo Y, Zi T, Qin W.2012. *J of Agric and Food Chem* **5** 561.

[13] Mardiyanto, 2013. *PhD, Saarland University, Saarbrucken,* Germany,

[14] Pillai CKS, Paul W, Sharma CP. 2009. *Progr in Polymer Sci* **34** 641.

[15] Rasheed H, Hari Babu R, Mohidin MK, Vineela J, Raviteja A, and Kishore PR *et al.* 2011. *J. Pharm. Biol. Chem. Sci.* **2** 233.

[16] Rajeshwari A, Prathna TC, Balajee J, Chandrasekaran N, Mandal AB, and Mukherjee A.2013 *Int. J. Pharm. Sci.* **5** 619.

[17] Ricke, SC. 2003. *Journal of Poultry Science* **82** 632.

[18] Sandi S, F.Yossi, and M Miksusanti 2016. *Pakistan J. Nutr*, **15** 268

[19] Desniar., Rusmana, I., Suwanto, A. dan Mubarik, NR. 2012.. *Jurnal Akuantika.* ISSN 0853-2523. Vol. III, No. **2** 135.

[20] Yesillik, SN., Yildirim, A., Dikici, A. dan Yildiz. 2011. *Asian Journal Animal Veteriner.* **6** 189.

[21] Espitia, PJP., Soares, NFF., Coimbra, JSR. Andrade, NJ., Cruz, RS., Medeiros, EAA. 2012.. *Food Technology*, Federal University of Vicosa. **5** 1447.

[22] Ali, A.I.M., Sandi, S. Muhakka. dan Riswandi. 2012. *Quality of swam grass Lebak Padang Development of Pampangan Buffalo.* Prosiding INSINAS

[23] Halim A, Abdullah NR, Afzan A, Abdul Rashid BA, Jantan I, and Ismail Z. 2011. *J. Med. Plant.* Res. **5** 1867.

[24] Moradhaseli S, Mirakabadi AZ, Sarzaeem A, Kamalzadeh M, Hosseini RH. 2013 *Iranian J of Pharm Res*, **12** 155.

[25] Sharma, N., Jandalik, S., Kumar, S. Chitkara, M. dan Shandu, IS. 2015. *Nanoparticles. Journal of Experimental Nanoscience.* **5** 1745.

[26] Sumihe G, Runtuwene MRJ, and Rorong JA. *2014. J Ilmiah Sci* **14** 125 .

IOP Conf. Series: Journal of Physics: Conf. Series **1095** (2018) 012024 doi:10.1088/1742-6596/1095/1/012024

[27] Wang W, Liu X, Xie Y, Zhang H, Yu W, Xiong Y, Xie W, and Ma X. 2006 *J of Mat Chem* **16** 3252.

IOP Conf. Series: Journal of Physics: Conf. Series **1095** (2018) 012025 doi:10.1088/1742-6596/1095/1/012025

Synthesis of Sorbitan Oleate from Sorbitol as Iron Adsorbent and Comparative Capacity of Adsorption on Pectin

Muhammad Arif Darmawan[1], Agustino Zulys[1, a)], Misri Gozan[2]

[1]Department of Chemistry, Faculty of Mathematics and Sciences, Universitas Indonesia
[2]Department of Chemical Engineering, Faculty of Engineering, Universitas Indonesia

[a)] Corresponding Author: zulys@ui.ac.id

Abstract. Heavy metal pollutant such as iron species is harmful to the environment and the health of living things. The method used in heavy metal adsorption was adsorbent such as bio-sorbents and synthethic adsorbents. In this research, sorbitan oleate was used as an iron (III) adsorbent which aims to determine the adsorption capacity of iron(III) with sorbitan oleate and its ratio to pectin. The synthesis of sorbitan oleate was carried out in two stages: dehydration of sorbitol into sorbitan and esterification of sorbitan with oleic acid to form sorbitan oleate. The greater the synthesis time, the acid number of sorbitan oleic synthesis is smaller. Adsorption of iron (III) ions with sorbitan oleate and pectin was carried out at varied temperature (20 °, 30 °, 40 °, 50 ° Celsius), time (30-180 min), and pH (4, 5, 6, 8, 9). There was found an optimum condition for sorbitan oleate at 20 °C, 30 minutes, and pH 4; While for pectin at 20 °C, 90 min and pH 5. Adsorption isotherms iron(III) with sorbitan oleate and pectin both follow Freundlich adsorption isotherms with adsorption capacity of iron (III) with sorbitan oleate and pectin respectively are 1.193 and 0.8304. The interaction of iron (III) ions with sorbitan oleate has a stronger interaction than pectin.

Introduction

Environmental pollution of heavy metals becomes a serious problem as the heavy use of heavy metals in the industrial field [1]. Heavy metals are toxic substances and generally carcinogenic. As a marine pollutant, heavy metals at high concentration are very harmful to living things. One of the heavy metals that has an adverse effect when the excess amount is the ferrous metal. At high concentrations, iron can form radical iron species due to non-binding iron species with proteins that can cause cell damage to living things [2]. Therefore, the concentration of ferrous metal ions in the environment must be maintained so as not to overdo in the environment.

An effort to overcome the problem of heavy metal waste pollution is applying an adsorbent to adsorb heavy metal ions. Many researchs have been done to find out and use different types of adsorbents which are effective in reducing the amount of heavy metals such as using activated charcoal, nanoparticles and other types of adsorbents [3].

Pectin is one of the most widely used adsorbents for heavy metal adsorbents. Pectin can be obtained through an extraction process from natural materials such as banana peel, apple and others [4].

Sorbitan Oleate is one of the organic compounds belonging to the sorbitan ester group and is the result of esterification of sorbitan and oleic acid. The use of sorbitan oleate itself is widely used as emulsifier or surfactant due to sorbitan oleate having hydrophilic (like water) and lipophilic (like lipid) groups [5]. However, there are currently no studies using sorbitan oleates as heavy metal ion adsorbents.

IOP Conf. Series: Journal of Physics: Conf. Series **1095** (2018) 012025 doi:10.1088/1742-6596/1095/1/012025

Therefore, in this research will be studied whether the sorbitan oleate can be used as metal adsorption of iron and the ratio of its adsorption capacity by using pectin.

Method

Sorbitan oleates compound were synthesized by two stages, that is dehydration of sorbitol and esterification of sorbitan with fatty acids. Sorbitol was dehydrated using concentrated sulfuric acid for 45 minutes at 120 °C to produce sorbitan. Sorbitan esterification with oleic acid at a temperature of 180 - 200 °C and the presence of alkali catalyst. Synthesis time varied at 2, 4, 6 and 8 hours. Sorbitan oleate of synthesized product was tested its acid value with acid-base titration. Sorbitan oleate synthesized and pectin are reacted with iron metal ions in the presence of variations of pH (4, 5, 6, 8, 9), temperature (20, 30, 40, 50 °C), and time (30; 60; 90; 120; 180 minute). To identify the rest of iron metal contents at optimum condition, FAAS instrumentation was . After the optimum conditions are known, sorbitan oleate and pectin are respectively reacted with variations in the concentration of iron metal ions (50, 100, 150, 200 and 250 ppm) to determine the type of adsorption isotherm. The interaction between iron metal ions with sorbitan oleate and pectin is done with FTIR.

Result and Discussions

The synthesis of sorbitan oleate was carried out in two stages: dehydration of sorbitol and esterification of sorbitan with oleic acid. The first stage was dehydration of sorbitol by using concentrated sulfuric acid as a catalyst. Sorbitol was reacted to remove a mole of water by heating and the presence of an acid catalyst for the dehydration process. Acids which was used in this process are non-oxidizing acids such as sulfuric acid, phosphoric acid, hydrochloric acid or aryl sulphonic acids (such as p-toluene sulfonic acid). In this study we used sulfuric acid easily to obtain and a hygroscopic acid. The dehydration process was carried out for 45 minutes due to the conversion of sorbitan into isosorbide at higher temperatures which is the final stage product of the sorbitol dehydration process. The research conducted by Yabushita et al produced the time data to obtain optimum levels of the sorbitol dehydration process is a reaction of 45 minutes [6]. When the reaction time gets longer, it will decrease the sorbitan content formed due to change to isosorbide. Figure 1 is a reaction scheme of sorbitol dehydration.

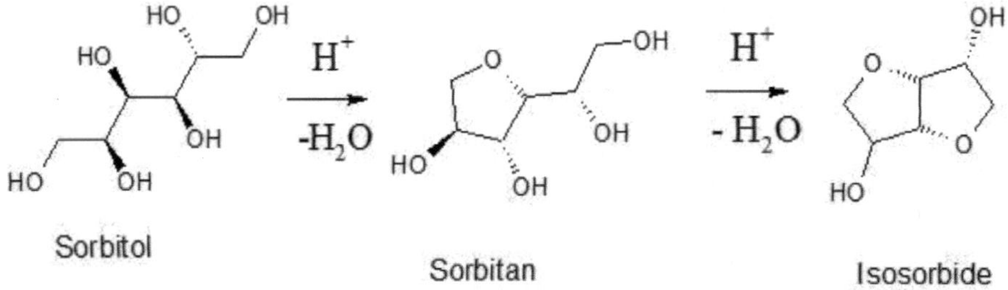

Figure 1. Sorbitol Dehidration Reaction

Sorbitan synthesis then conducted esterification reaction with oleic acid and added NaOH to neutralize excess acid in dehydration sorbitol process. The reaction was carried out with the reflux system during the time variation of 2, 4, 6 and 8 hours. Sorbitan oleate synthesis then taken some parts and done the titration to measure the value of acid numbers. The acid number is the number of free fatty acids (oleic acid) that do not react with sorbitan to form sorbitan oleate. The lower the acid number, the less oleic acid because it has formed to form sorbitan oleate. Figure 2 shows the value of the acid number of sorbitan oleic synthesis. The greater the synthesis time, the acid number will be smaller because more oleic acid reacts with sorbitan to form sorbitan oleate. The acid value obtained in the study with synthesis time 2, 4, 6 and 8 hours were 147.26; 72.93; 50.49 and 24.289 mg KOH / g samples.

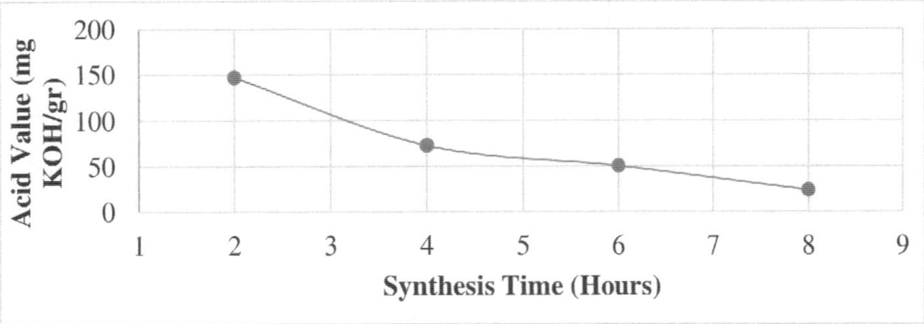

Figure 2. Graphic of Acid Value from Sorbitan Oleate

Sorbitan oleate that has been synthesized for 8 hours then reacted with iron ion solution to know the adsorption interaction that occured and pectin as comparison. Some parameter variations were performed to determine the optimum conditions of the reaction, temperature variations (20 °, 30 °, 40 ° and 50 ° C.), time (30, 60, 90, 120, 150, and 180 min) and pH (4, 5, 6 , 8 and 9). Figures 3 and 4 represent the optimization graph of iron ion interaction time with sorbitan oleate and pectin. Where Q value is the unit of adsorption efficiency used with units of mg of metal / gram adsorbent (sorbitan oleate or pectin) (mg / g).

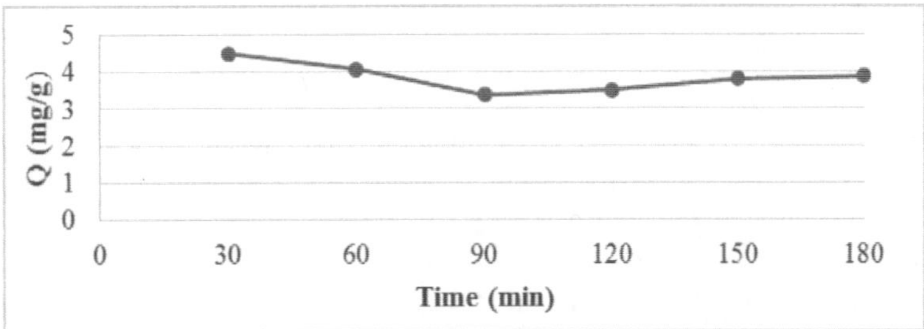

Figure 3. Time Variation of Iron(III) and Sorbitan Oleate

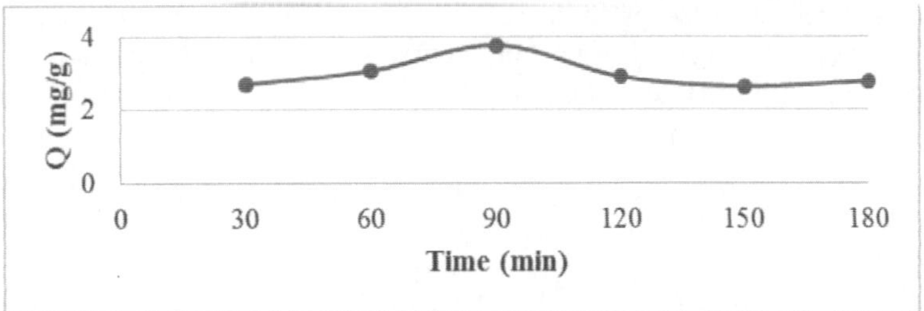

Figure 4. Time Variation of Iron(III) and Pectin

In the research using sorbitan oleate, the optimum adsorption time of iron was 30 minutes with adsorption efficiency of 4.5 mg / g. While using pectin obtained optimum time for 90 minutes with adsorption efficiency of 3.76 mg / g. The difference phenomena is probably due to differences from the active site found in sorbitan oleat and pectin that serves as a place of interaction. In the sorbitan oleate, the amount of adsorbate (in this case the iron (III) ion) had reached equilibrium to interact with the oleic sorbitan at 30 minutes so as not to bind more iron (III) ions. Meanwhile, pectin adsorption took a longer time as 90 minutes to reach equilibrium, possibly because pectin is a polymer so many sites to bind iron (III) ions.

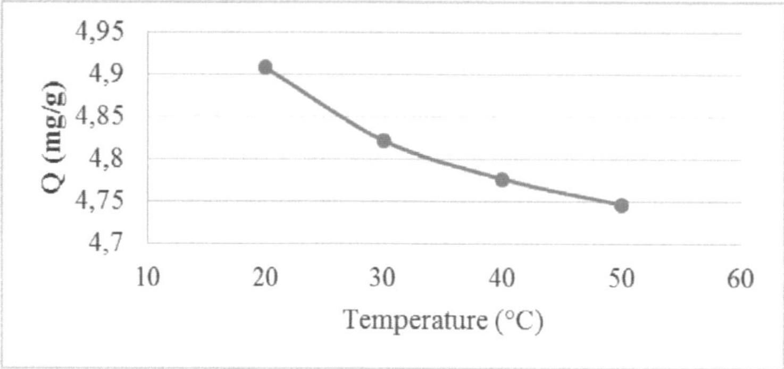

Figure 5. Temperature Variation of Iron(III) and Sorbitan Oleate

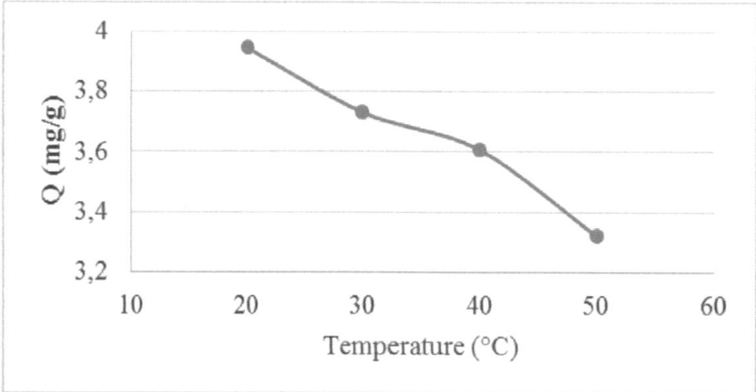

Figure 6. Temperature Variation of Iron(III) and Pectin

Figures 5 and 6 show temperature variations of iron ion (III) reactions with sorbitan oleate and pectin. Based on the graph data from temperature variation of iron adsorption (III) with sorbitan oleate and pectin there is a similarity of phenomenon, that is decreasing adsorption with increasing temperature. The increased temperature causes the energy and ion reactivity to increase so that more ions are able to pass energy levels to interact with sites on the surface. The greater ionic reactivity will also increase ion diffusion in the pores of the adsorbent. However, an increase in temperature that causes increased ionic reactivity will interfere with the interaction that has been formed. The optimum temperature obtained from both adsorption was 20 °C.

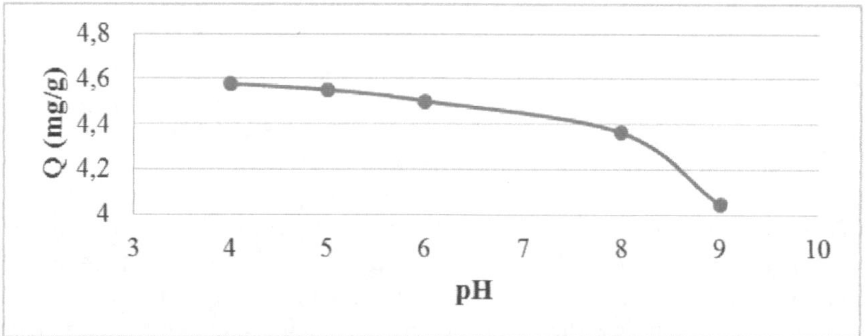

Figure 7. pH Variation of Iron (III) and Sorbitan Oleate

In Figure 7, it is known that the greater pH value in the adsorption process using sorbitan oleate, the smaller the Q value. This indicates the less number of iron (III) ions that interact with the sorbitan oleate. One of the biggest factors is due to the alkaline pH (pH> 7), sorbitan oleate will undergo saponification reaction or generally called hydrolysis. Sorbitan oleates which are fatty acid esters will undergo hydrolysis under alkaline conditions and reshape sorbitan and salts of oleic acid. Since some sorbitan oleates undergo hydrolysis into sorbitan and fatty acid salts, the less iron (III) ion interacts with the remains sorbitan oleates that do not hydrolysis.

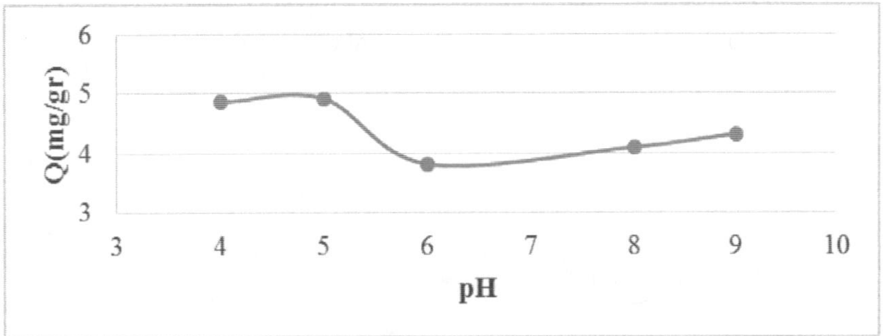

Figure 8. pH Variation of Iron(III) and Pectin

A different phenomenon occured in Figure 8 when pectin was used as an adsorbent. Initially adsorption decreases as it approaches the neutral pH, because the iron (III) ions form the insoluble precipitate $Fe(OH)_3$. The reason is the iron (III) ion has a very small solubility that is 2 x 10^{-18} M, so that the iron (III) ions will begin to precipitate at pH 7 and will more precipitate when the pH increases. At an alkaline pH (pH> 7), the carboxylic group (-COOH) in pectin becomes negatively charged and it is capable of binding to a positively charged iron metal (III) ion. Although the carboxyl group (-COO-) in pectin has better interacting ability, the remaining amount of iron (III) ions is less due to forming $Fe(OH)_3$ deposits so that the adsorption value at base pH is no greater then at the acidic pH.

Determination of adsorption isotherm between iron(III) ions with sorbitan oleate and pectin was conducted using the optimum conditions (pH, temperature, and time) that had been obtained before. Tests of adsorption isotherms were performed using two types of isotherms that is langmuir isotherms and freundlich isotherms. Figures 9 and 10 are the langmuir isotherm model of iron interaction (III) with sorbitan oleate and pectin.

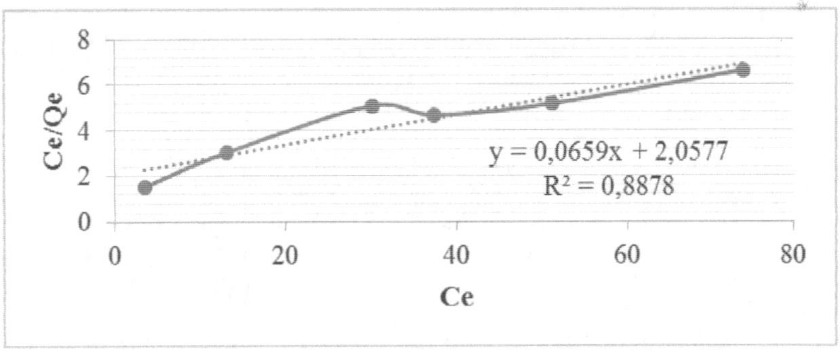

Figure 9. Langmuir Isotherm of Iron(III) and Sorbitan Oleate

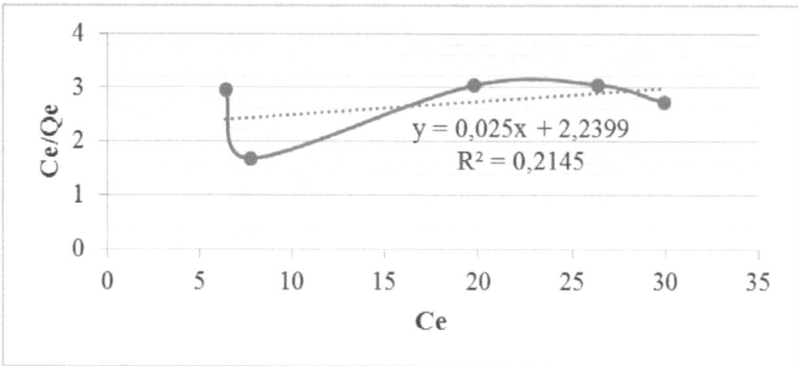

Figure 10. Langmuir Isotherm of Iron (III) and Pectin

From the equation of line can be determined from the value of Qo that is the maximum adsorbate coverage value of the monolayer layer (State et al. 2012). The Qo value of iron adsorption (III) with sorbitan oleate was 15.174 mg / g while the Qo value of iron adsorption (III) with pectin was 40 mg / g. Figures 11 and 12 are freundlich isotherm models between iron (III) ions with sorbitan oleate and pectin.

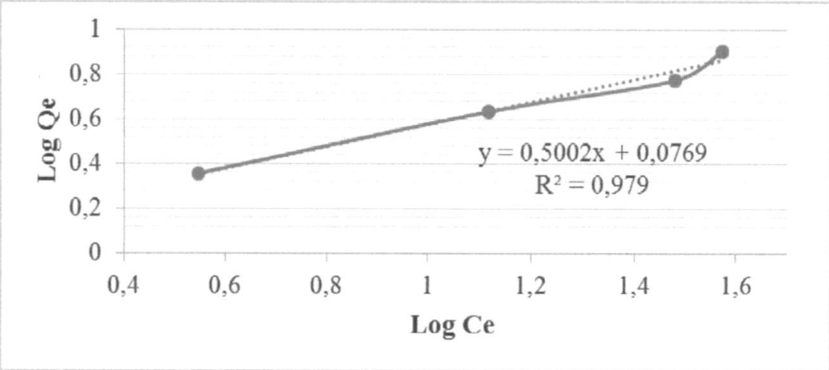

Figure 11. Freundlich Isotherm of Iron(III) and Sorbitan Oleate

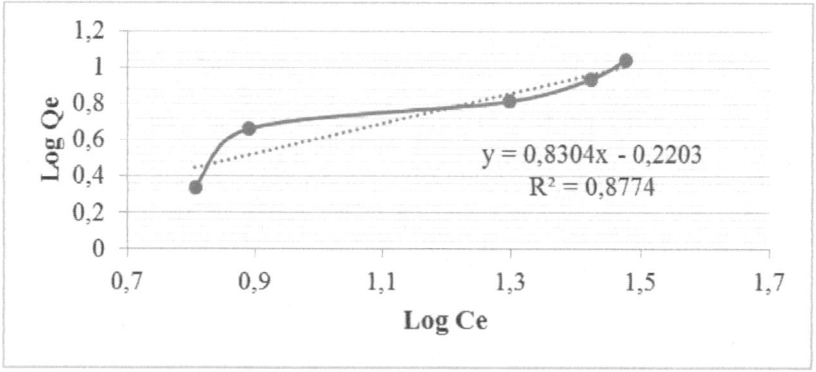

Figure 12. Freundlich Isotherm of Iron(III) and Pectin

From the langmuir and freundlich isotherms model, the greatest R^2 value is found in the freundlich isotherm model so that the adsorption of iron (III) with sorbitan oleate and pectin follows the freundlich isotherm model. From the linear equation of freundlich isotherm model can be determined the value of Kf ie Freundlich characteristic constant and 1 / n heterogeneity factor of absorption [8]. The value of 1

/ n can be changed to a value of n that is the intensity of adsorption. The Kf value or adsorption capacity between iron (III) ions with sorbitan oleate and pectin are 1,193 and 0.8304, respectively.

The Fe (III) interactions of sorbitan oleate and pectin can be seen from the spectrum form of Sorbitan oleic and pectin FTIR respectively before and after mixed with iron (III) solution. Figs 13 and 14 are FTIR spectra between Fe (III) with sorbitan oleate and pectin.

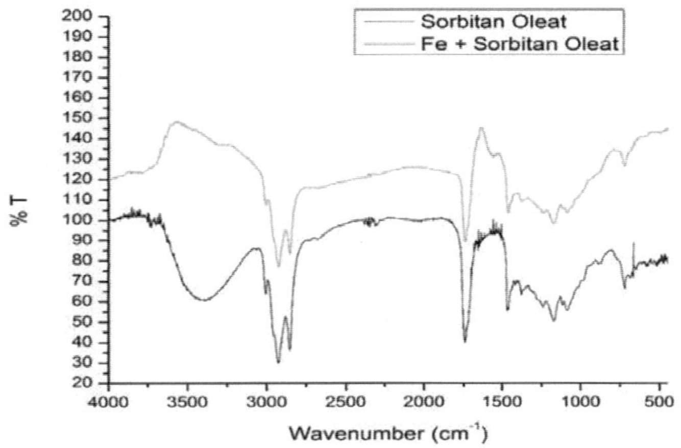

Figure 13. Interaction of Iron(III) and Sorbitan Oleat

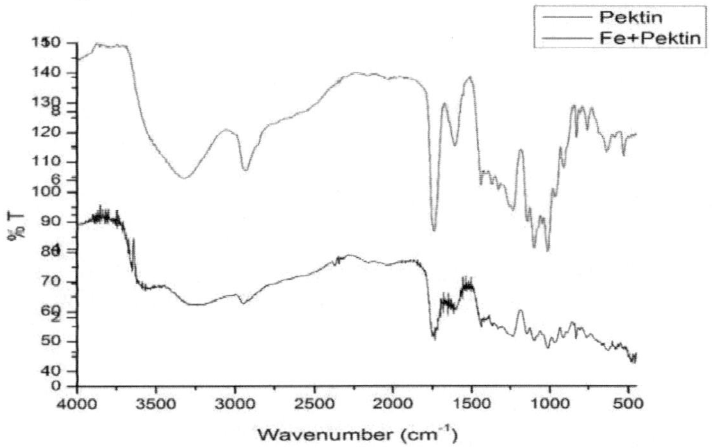

Figure 14. Interaction of Iron(III) and Pectin

From the data in Figures 13 and 14 it is shown that the iron(III) metal ions are related to the present of O-H group in the sorbitan oleate. In the FTIR spectrum of sorbitan oleate + iron(III) there is a decrease in thc intensity of hydroxyl groups when compared with the spectrum of sorbitan oleate without iron(III) ion. Reduced intensity of the O-H spectrum due to iron(III) interactions with O atoms thus reducing the stretching of the O and H bonds on the hydroxyl group of sorbitan oleate. The O atoms of the O-H group will share with their free electron pairs to the iron (III) ions for co-ordinate covalent interactions. One of the reasons why O atoms only interact covalently in coordination with iron(III) ions is that iron(III) ions have greater electropositive value than H^+ ions so that O atoms have a greater tendency to bind covalently to H^+ ions than iron(III). The same phenomenon also occurs between the interaction of pectin and iron (III) ions, but the interaction is weaker compared to the interaction of iron (III) ions with sorbitan oleate.

Conclusion

Sorbitan oleates were synthesized in two stages: dehydration of sorbitol into sorbitan and esterification of sorbitan with oleic acid into sorbitan oleate. The acid value obtained in the study with synthesis time 2, 4, 6 and 8 hours were 147.26; 72.93; 50.49 and 24.289 mg KOH / g samples. There was found an optimum condition for sorbitan oleate at 20 °C, 30 minutes, and pH 4; While for pectin at 20 ° C, 90 min and pH 5. The iron adsorption process (III) with pectin and sorbitan oleate both followed the Freundlich adsorption isoterm model. The iron adsorption capacity (III) with sorbitan oleate is higher than that of pectin. The interaction between sorbitan oleate and iron (III) is higher than that of iron (III) interaction with pectin.

Acknowledgments

This experiment supported by grant of PUPT program from Indonesia Minister of Research, Technology and Higher Education (002/SP2H/LT/DRPM/IV/2017).

References

[1] N. Nirmala, V. M. A. Tiwow, and S. Suherman, "Adsorpsi Ion Tembaga (Cu) Dan Ion Besi (Fe) Dengan Menggunakan Arang Hayati (Biocharcoal) Kulit Pisang Raja (Musa sapientum)," *J. Akad. Kim.*, vol. 4, no. 4, p. 189, 2017.

[2] M. Jaishankar, T. Tseten, N. Anbalagan, B. B. Mathew, and K. N. Beeregowda, "Toxicity, mechanism and health effects of some heavy metals," *Interdiscip. Toxicol.*, vol. 7, no. 2, pp. 60–72, 2014..

[3] J. L. Gong *et al.*, "Copper (II) removal by pectin-iron oxide magnetic nanocomposite adsorbent," *Chem. Eng. J.*, vol. 185–186, pp. 100–107, 2012.

[4] I. W. Sutapa, V. Petrus, D. Siahay, and M. F. J. D. P. Tanasale, "Adsorption Cu^{2+} Metal Ion of Pectin from Tongka Langit Banana's Crust (Musa Species Van Balbisiana)," pp. 72–77, 2014.

[5] L. Peltonen, J. Hirvonen, and J. Yliruusi, "The Behavior of Sorbitan Surfactants at the Water – Oil Interface : Straight-Chained Hydrocarbons from Pentane to Dodecane as an Oil Phase," vol. 276, pp. 272–276, 2001.

[6] Yabushita, M., Kobayashi, H., Shrotri, A., Hara, K., Fukuoka, A., "Sulfuric Acid-Catalyzed Dehydration of Sorbitol: Mechanistic Study on Preferential Formation of 1,4-Sorbitan," 2015.

[7] A.O, Dada., A.P, Olalekan., A.M, Olatunya., O, Dada., "Langmuir , Freundlich , Temkin and Dubinin – Radushkevich Isotherms Studies of Equilibrium Sorption of Zn^{2+} Unto Phosphoric Acid Modified Rice Husk," vol. 3, no. 1, pp. 38–45, 2012.

[8] A. K. Agarwal, M. S. Kadu, C. P. Pandhurnekar, and I. L. Muthreja, "Langmuir , Freundlich and BET Adsorption Isotherm Studies for Zinc ions onto coal fly ash," vol. 3, no. 1, pp. 64–71, 2014

Synthesis Alumina-Activated Carbon Composite Using Sol-Gel Method As Adsorption for Methylene Blue Dye

Poedji Loekitowati Hariani [1], Muryati [2], Fatma [1]

[1] Department of Chemistry, Faculty of Mathematicts and Natural Science, Sriwijaya University
[2] Magister Program of Chemistry, Faculty of Mathematicts and Natural Science, Sriwijaya University

Email ; muryatiengzlem@gmail.com

Abstract. The research about synthesis alumina-activated carbon composite using the sol-gel method as adsorption for methylene blue has been done. Activated carbon is made from durian shell. The alumina-activated carbon composite was characterized by FTIR (Fourier Transform Infrared), SEM-EDS (Scanning Electron Microscopy-Energy Dispersive X-Ray Spectroscopy. The FTIR characterization for activated carbon indicated that the functional groups O-H, C=C, C-H, and C-O while the alumina-activated carbon composite has increased of the functional group of Al-O. SEM analysis of the surface of alumina-activated carbon composite showed that alumina sticks to the activated carbon surface. EDS results showed a decrease of the element C from 64.60% to 20.87% and the increase of Al from 0.86% to 23.02%. The optimum condition adsorption of methylene blue using activated carbon obtained at an initial concentration of 25 mg/L, contact time of 75 minutes and a temperature of 55°C, while the composite alumina-activated carbon obtained at an initial concentration of 30 mg/L, contact time of 90 minutes and the temperature of 75°C. The ability of activated carbon and alumina-activated carbon composite for adsorption methylene blue were 10.7205 mg/g and 14.3662 mg/g, respectively.

Keywords : alumina-activated carbon composite, Adsorbtion, Methylen blue

1. Introduction

Nowadays research on composites is growing up. Composites are a macroscopic combination of two or more different components. The purpose of composite synthesis was to obtain new materials that have better properties than their constituent components. These properties include the ability to absorb a compound. Composite synthesis has been widely done from various materials one of them using activated carbon

Activated carbon is an amorphous carbon material having a large surface area of 300 to 2000 m^2/g. This large surface area is due to having a pore structure. These pores result in activated carbon having the ability to absorb compounds [1]. Activated carbon can be synthesized from a variety of basic materials such as gelam wood [2], cassava shell [3], coconut shell [4] and durian shell [5].

Durian shell contains cellulose content of about 50%-60% (carboxymethylcellulose) and 5% lignin [6]. Durian shell is very potentially as a base material to make activated carbon because it has a high enough carbon content of 60.31% [5]. Durian shell activated carbon has micropore and mesoporous structures. The presence of micropores and mesopores in activated carbon increases the absorbency, especially for large adsorbate molecules such as dyes. The synthesis of activated carbon from durian shell has been successfully done to absorb the methylene blue dye [7].

The ability of the activated carbon depends on the surface area, the internal structure of the pore, the surface characteristics and the functional group on its porous surface [7]. Addition alumina for activated carbon to composite is expected to increase the ability [8]. Alumina (Al_2O_3) is a porous material with macropore and mesoporous structures and has an active site in surface and thermostable [9] so that the adsorption process can be done at high temperature. The alumina surface active site is expected to interact with the methylene blue dye. The alumina-activated carbon composite can increase absorption because the adsorption process occurs physically and chemically. Activated carbon has pores that can absorb while alumina has functional groups that can interact with adsorbate [8].

Many methods can be used to synthesize composites, one of which is sol-gel method. The sol-gel method has several advantages: lower used temperature, shorter process, and lower pollution [10]. The alumina-activated carbon composite synthesized by the sol-gel method has a high absorbency to acetone vapor [11]. The sol-gel method has been successfully used to make Nanocomposite TiO_2-Carbon Nanotubes [12] and Nanocrystalline Metal Oxide [10]. Iriani [2] also succeeded in synthesizing the alumina-activated carbon composite from the gelam wood using the sol-gel method as adsorption the methylene blue dye.

Methylene blue is a toxic dye [13]. Methylene blue can cause irritation of the gastrointestinal tract if ingested, causing cyanosis if inhaled, and irritation of the skin if touched by the skin [14]. Methylene blue is often used in the process of staining in various industries such as textile, ceramics, paper, printing, and plastics industries. In the process of coloring, the industry uses a lot of water and many produce liquid waste containing dye [7].

In this study, the activated carbon of durian shell was modified by alumina using sol-gel method. Activated carbon with alumina is applied for adsorption of methylene blue dyes. Characterization of alumina-activated carbon composite includes functional group analysis using FTIR and morphology using SEM-EDS.

2. Materials and Methods

The chemicals used are analytical grades such as KOH, HCl, NaOH, $Al_2(SO_4)_3$, $NaNO_3$ (was supplied from Merck), methylene blue dye, and aquadest. Durian shell was collected from Pasar Kuto Palembang. Prior to the process, durian shell was repeatedly washed with distilled water in order to remove dust and other inorganic Impurities. The size of these materials was reduced to ±2x2 cm and then dried at 120°C for 24 h to reduce the moisture content. Subsequently, as dried durian shell was grounded until it became powder (140 mesh) and then stored in desiccators [5].

2.1 Characterization

Analysis instrument were used Spectrophotometer UV-Vis Spectronic 20. The pore structure characteristics of composite and activated carbon were determined using, SEM-EDS 6510-LA (operated using argon gas with a current of 6mA for 4 minutes and observed at a voltage of 15 Kv) and FTIR Shimadzu 5000 (sample was measured as a mixture on KBr pelleted samples were scanned over the wavenumber range 400-4000 cm^{-1}). Characterization moisture content and ash content of activated carbon using SNI No. 06-3730-1995 method.

2.2 Synthesis of Alumina-Activated Carbon Composite
Preparation of activated carbon was conducted according to [5]. A 25 g of durian shell in the form of powder was mixed with KOH solution (50%). During the impregnation period, the mixture was stirred at 200 rpm for 5 h at room temperature. The resulting homogeneous slurry was dried at 110°C for at least 24 h.The dried product carbonized at 500°C for a 1 hour. These activated carbons were washed sequentially with a 0.1 M HCl solution. Consecutively, carbon powders were repeatedly washed with hot distilled water until the pH of the solution reaches 6.8 and finally washed with cold distilled water. After that, these powders were dried at 110°C for 2 h and stored in a desiccator.

Synthesis of alumina-activated carbon composite was conducted according to [11]. Synthesis of alumina-activated carbon composite the sol-gel method was used by adding the activated carbon into 1 M $Al_2(SO_4)_3$ 50 mL, later 6 M NaOH by portions. Wet alumina activated-carbon derivates were granulated by extrusion, then dried at 120°C for 4 h and heated at 420°C for 4 h. The amount of activated carbon in the adsorbents was ~5% (wt).

2.3 pH_{PZC}
The procedure of pH_{PZC} was conducted according to [15]. The determination of pH_{PZC} using 50 mL of 0.01 M $NaNO_3$ solution was included in 10 Erlenmeyer respectively. Initial pH (2, 3, 4, 5, 6, 7, 8, 9, 10, 11, and 12) was performed by adding a 0.1 M HCl or 0.1 M NaOH solution. 0.1 g of composite is added to each further Erlenmeyer in the shake for 2 hours. The mixture was allowed to stand for 2 days and the final pH was measured.

2.4 Adsorption Experiments
Optimum condition adsorption of methylene blue dye using adsorbent o,1 g composite and activated carbon by durian shell each added into 50 mL methylene blue pH 7.25 by varying the concentration (10, 15, 20, 25, 30, and 35 mg/L), contact time (60, 75, 90, 105, and 120 min) and temperature variations (35, 45, 55, 65, 75, and 85°C). The mixture was stirred at 120 rpm for 60 minutes. The solution of the methylene blue substance was separated by filtration and then measured its concentration by using UV-Vis spectrophotometer.

The amount of dye adsorbed is calculated based on the following equation:

$$q_t = \frac{(C_o - C_t)\,V}{m}$$

3. Results and Discussion

3.1 Water Content and Ash Content
Based on Table 1 it can be seen that the activated carbon of durian shell has good water content because it has fulfilled the requirement of active carbon quality according to SNI No.06-3730-1995, but the ash content has not fulfilled the requirement because it exceeds the maximum limit.

Table 1. Characteristics of Activated Carbon

Characteristics of Activated Carbon	Research Result	SNI No. 06-3730-1995 (powder)
Water Content (%)	3.6494	Max. 15
Ash Content (%)	13.5877	Max. 10

The amount of content ash is caused by the amount of mineral content in activated carbon. The process of heating when the manufacture of activated carbon is not flowed by nitrogen gas, this is caused inside the furnace there is still oxygen. This oxygen causes the formation of oxide minerals on the activated carbon produced. This metal oxide causes high ash levels. The metal is according to EDS results (Table 3) ie Mg, Si, K, Ca, and Fe. The concentration of HCl used for washing affects the amount of ash content. Washing process should use HCl with concentration more concentrated, this is like done by [5] using HCl 0.5 N in the washing process.

3.2 Characterization

The result of the synthesis of active alumina-carbon composite using sol-gel method with 5% carbon content showed smaller size and uniform [2]. Figures 1a and b show the activated carbon and alumina-activated carbon composite. The alumina-activated carbon composite is lighter than the activated carbon. This is due to the presence of alumina on the activated alumina-carbon composite.

(a) (b)

Figure 1. Image of (a) Activated Carbon and (b) Composite Alumina-Activated Carbon

FTIR. FTIR spectroscopy is used to identify functional groups. The functional group identification is used to determine the chemical interaction between the active functional groups in the alumina, the activated carbon, and the alumina-activated carbon composite. The FTIR spectra alumina, activated carbon, and alumina-activated carbon composite were shown in Figure 2. The characteristic band of alumina, activated carbon, and alumina-activated carbon composite at 3450.4, 3446.5, and 3448.5 cm^{-1} is the peak of O-H. The functional group of O-H on alumina comes from the remaining H_2O. Activated carbon and composite showed functional groups C=C at wavenumbers 1627.8 and 1639.4 cm^{-1} respectively. Alumina shows peak C=C at wave number 1639.4 cm^{-1} caused by impurity on the alumina. The activated carbon C-O functional group appears at 1022.2 cm^{-1} wavenumbers.

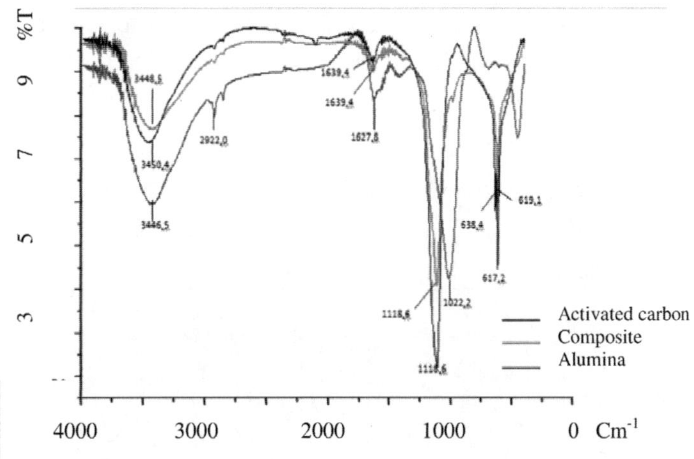

Figure 2. FTIR spectra of Alumina, Alumina-Activated Carbon Composite

Alumina and composite alumina-activated carbon have peaked at 1118,6 cm^{-1} indicating the SO_4^{2-} free functional group. this is indicated by the presence of a peak at 617.2 cm^{-1} in alumina and 619.1 cm^{-1} in the composite [16].The SO_4^{2-} functional group is derived from $Al_2(SO_4)_3$. The other broad spectra showed at 620 cm^{-1} to 850 cm^{-1} is the peak of Al-O group [11]. Alumina and alumina-activated carbon composite show Al-O functional group at wave number 638,4 cm^{-1}. The FTIR spectra of alumina, activated carbon and alumina-activated carbon composite can be seen in Figure 2.

Table 2. Functional Group and Wavenumber of Functional Group

Functional Group	Wavenumber of Functional Group(cm^{-1})		
	Alumina	Activated Carbon	Alumina-activated carbon composite
O-H	3450.4	3423.4	3448.5
C-H	-	2922.0	-
C=C	-	1627.8	1639.4
C-O	-	1022,2	-
SO_4^{2-}	1118.6	-	1118.6
	617.2		619.1
Al-O	638.4	-	638.4

Shifts occur only in adjacent areas (Table 2). It indicates an interaction of the activated alumina-carbon composite, i.e. the physical interaction. The active alumina-carbon composite does not appear new peak. The resulting peak indicates a functional group derived from activated carbon and alumina.

SEM-EDS. Figures 3a and b show the morphology of activated carbon and alumina- activated carbon composites with 30,000× magnifications. The composite synthesis results show alumina attached to the surface of the activated carbon and some enter into the pores of activated carbon. The white granules on the composite surface show alumina, Figure 3b. The entry of alumina causes some of the pore activated carbon is closed. The surface of the activated carbon initially has a larger pore but after modification, the alumina closes part of the pore so that the pore becomes smaller.

(a) (b)

Figure 3. Morphology (a) Active Carbon and (b) Alumina-Activated Carbon Composite

The adsorbent EDS spectra showed percent of active carbon element and active alumina-carbon composite (Table 3). Activated carbon and activated carbon each contained C of 64.60% and 20.87% respectively. The composition of the alumina-activated carbon composite changes the elemental composition. The presence of Al in the activated carbon shows the presence of alumina adsorption into the activated carbon [17].

Table 3. Element Analysis

Element	Activated Carbon (%)	Alumina-Activated carbon Composite (%)
C	64.60	20.87
O	13.39	39.09
Al	0.86	23.02
S	-	9.86
Na	-	4.41
Mg	0.86	0.18
Si	7.59	1.81
K	4.35	0.77
Ca	1.15	-
Fe	7,18	-

a. pH_{PZC}

pH Point Zero Charge (pH_{PZC}) is a state when the surface of the adsorbent is neutral. The pH_{PZC} value is determined by the point of intersection of the initial pH curve and the final pH. The pH_{PZC} data was used to estimate pH conditions suitable for the adsorption process. Based on Figure 4, pH_{PZC} of activated carbon of durian shell is 7.10 (surface of the activated carbon is neutral). At pH <7.10 the surface of the activated carbon is positively charged to form $R\text{-}OH_2^+$ so it is easy to absorb the anions, whereas at pH> 7.10 the surface of the activated carbon is negatively charged to forms $R\text{-}O^-$ so it is easy to absorb the cation [2].

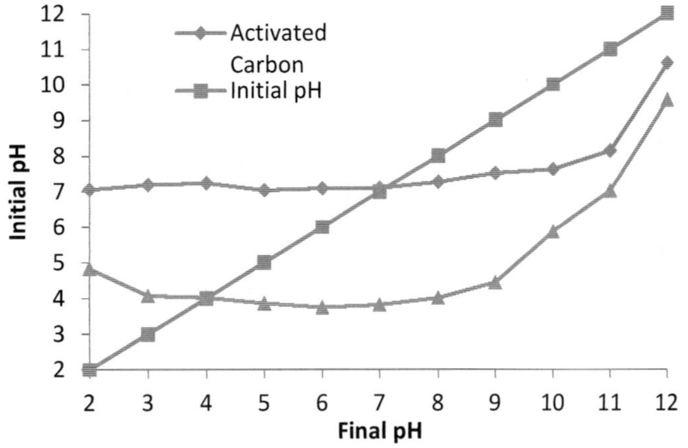

Figure 4. Curve of pH_{PZC}

Composite pH_{PZC} was obtained at pH 4.02. This pH indicates that the composite surface is neutral with the presence of a neutral Al-OH group and a zero charge. When the pH <4.02 hydroxyl group on the alumina can experienced protonation to form $Al\text{-}OH_2^+$ causes the composite surface to be positively charged so it easily absorb the anion. At pH> 4.02 hydroxyl group on the alumina can experience deprotonation to form $Al\text{-}O^-$ so the composite surface is negatively charged so it easily absorbs the cations [17].

pH_{PZC} measurements show that there is no pH adjustment during the adsorption process. The pH of methylene blue (7.25) is already above the activated carbon pH_{PZC} as well as the alumina-activated carbon composite. At the pH the adsorbents are negatively charged and easily absorb the positively charged methylene blue.

b. Methylene Blue Dye Adsorption

Figure 5 shows the optimum concentration of methylene blue dye on the activated carbon and composites 25 mg/L and 30 mg/L, respectively. During the adsorption process, the color molecules first diffuse into the surface of the adsorbent and then diffuse into the adsorbent pores [14]. The composite has an active site of alumina so that the adsorption process is not only played by pore but also played by alumina. This results in higher composite absorption than activated carbon. This alumina will bind to the methylene blue dye ions.

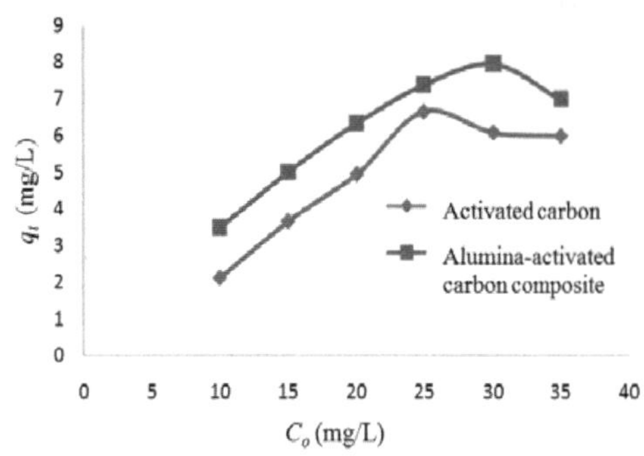

Figure 5. Effect of Concentration

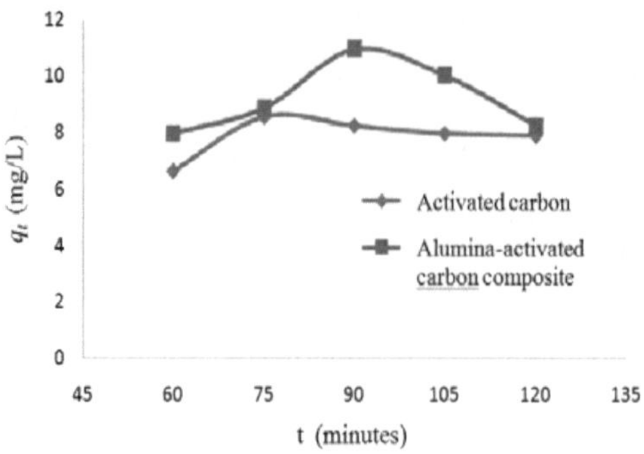

Figure 6. Effect of Contact Time

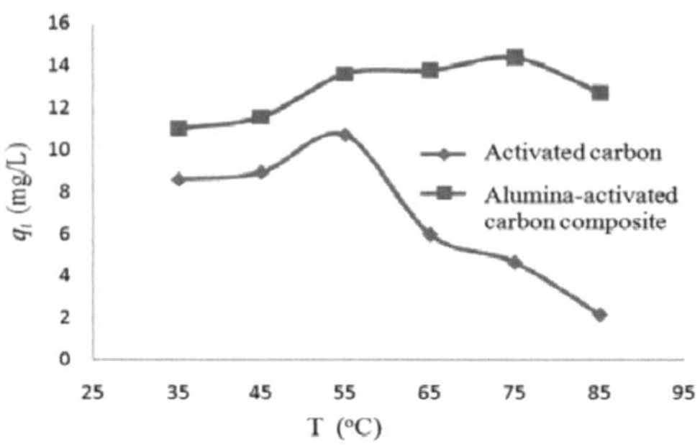

Figure 7. Effect of Temperature

Figure 6 shows the optimum contact time of activated carbon obtained at 75 minutes at 8.5823 mg/g while in the composite at 90 minutes of 10.9826 mg/g. Composites are able to absorb more dyes because the composite has an active site of alumina. Increased contact time causes more adsorbent particles to be in contact with dye ions. This causes more and more dyes ions to be absorbed by the adsorbent. This condition will continue until it reaches saturation condition or optimum contact time [18]. If the adsorbent has been saturated by the dye ions, the increased contact time does not increase adsorption, it will be desorption. This desorption is influenced by continuous stirring so that the dye ions is released again.

Figure 7 shows the optimum temperature of activated carbon and composite absorption obtained at 55°C of 10.7205 mg/g at 75°C of 14.3662 mg/g respectively. The alumina contained in the composite results in a more thermostable composite than the activated carbon. This results in higher optimum absorption temperature composite than absorption using activated carbon. Adsorption of organic components (including dyes) involves a physical bond that will decrease with increasing temperature. Increased temperatures involve greater kinetic energy so that collisions between particles are more frequent. This collision causes the dyes ion to be desorbed.

4. Conclusions

Characteristics of using FTIR showed that the active alumina-carbon composite was physically interacting. SEM analysis showed that alumina sticks to the activated carbon surface. EDS results showed a decrease of the C element and the increase of Al element. The results of the optimum condition analysis show that the alumina-activated carbon composite absorption is higher than the activated carbon.

References

[1] Dewi TK, Nurrahman A and Permana E 2009 *Jurnal Teknik Kimia* **1** **24** 30
[2] Iriani F 2015 *Sintesis dan karakterisasi komposit alumina-karbon aktif dari kayu gelam menggunakan metode sol-gel* (Palembang: Sriwijaya University)
[3] Santoso R H, Susilo B and Nugroho WA 2014 *Jurnal Keteknikan Pertanian Tropis dan Biosistem* **2** **(279)** 86
[4] Gratuito M K B, Panyathanmaporn T, Chumnanklang RA, Sirinuntawittaya N and Dutta A 2008 *Bioresource Technology* 99 (4887).95
[5] Chandra T C, Mirna M M, Sunarso J, Sudaryanto Y and Ismadji S 2009 *Journal of Taiwan Institute of Chemical Engineers* **40** 457.
[6] Soekardjo 1990 *Kimia Anorganik* (Jakarta: Rineka Cipta)
[7] Chandra T C, Mirna M M, Sudaryanto Y and Ismadji S 2007 *Journal of Chemical Engineering* **127** 121.
[8] Fatma and Hariani P L 2014 *Modifikasi karbon aktif dari kayu gelam dengan alumina untuk menyerap zat warna procion dari limbah cair industri songket* (Pelambang: Sriwijaya University)
[9] Chartterje M, Naskar M K and Ganguli D 2003 *Journal of Sol-Gel Science and Technology* **28** 217.
[10] Widodo S 2010 *Teknologi sol gel pada pembuatan nano kristalin metal oksida untuk aplikasi sensor gas* Semarang
[11] Kirpsaite E, Kudzmiene G D and Kitrys S 2010 *Journal of Materials Science (Medziagotyra)* **16** 353.
[12] Subagio F A A and Nurhasanah I 2011 *Telaah Jurnal Ilmu Pengetahuan dan Teknologi* **29** 63.
[13] Ozer A and Dursun G 2007 *Jornal of Hazardous Materials* **146** 262.
[14] Hamdaoui O and China M 2006 *Journal Acta Chimica Slovenia* **54** 407.
[15] Dewi S H and Ridwan 2012 *Jurnal Sains Materi Indonesia* **13** 136.
[16] Roto, Tahir I, and Sholikhah U N 2008 *Indo. J. Chem* **8** 307.
[17] Hariani P L, Fatma and Zulfikar 2015 *J.Pure App. Chem. Res* **4** 25.
[18] Farda E and Maharani D K 2013 *Journal of Chemistry* **2** 19.
[19] Marlinawati, Yusuf B and Alimuddin 2015. *Jurnal Kimia Mulawarman* **13** 23.

The Effect of Natural Based Oil as Plasticizer towards Physics-Mechanical Properties of NR-SBR Blending for Solid Tyres

Nasruddin[1] and Tri Susanto[1]

[1] Palembang Institute for Industrial Research and Standardization, Ministry of Industry, Indonesia

E-mail: nas.bppi@gmail.com

Abstract. Petroleum derivatives oil has been used dominantly as a primary plasticizer in rubber compounding for tyre industries. Despite its harmful effect to the environment, it is also a nonrenewable resources that need to be replaced. This research would study the effect of natural based oil such as castor oil, palm oil, fried - palm oil as an alternative plasticizers to replace petroleum based oil like white oil and minarex oil in Natural Rubber and Styrene Butadiene Rubber (NR-SBR) composites for rubber solid tire. The rubber compounding applied the same amount of each plasticizers which was 8,75 phr in 90/10 phr of NR-SBR composites filled by 64 phr Carbon Black N330 and 34 phr $CaCO_3$. Physical mechanical properties of 5 formulas were examined, the hardness values range from 68 - 74 ShoreA; tensile strength 118 - 136 kg/cm^2; modulus 200% 19,25 – 31,16; density 1,22 - 1,29 g/cm^3, abrasion resistance 102,37 - 135,64 mm^3, compression set (tested in 25% defl, 70 ^{O}C, 22 hours) range from 65,44 to 72,35; and there were no crack for ozone resistance (tested in 50 pphm, 20% strain, 24 hours, 40 ^{O}C). To sum up, it is indicated that there is no significant difference in physical mechanical properties of composites using either natural or petroleum based oil as plasticizers.

Keywords: natural based plasticizers, petroleum based plasticizer, NR-SBR compoistes, solid tyres.

1. Introduction

Plasticizers for rubber industry primarily used petroleum or mineral oils derivates. On the one hand, the mineral resources has depleted recently, while the petroleum based plasticizer experienced constant competition with many other petroleum consumers [1, 2]. Mineral oils are made from crude petroleum which are effective, tightly controlled for quality, inexpensive and easy to use for rubber compounding. The necessity of releasing the polymer industry from its dependence on depleting resources represents a major concern. On the other hand, plant oils represent a renewable resource that can be used as reliable starting material to access new polymer products [3]. Its availability and relatively low cost becomes a competitive advantages of an industrial material for plastic or polymer industries [4]. The largest sources of vegetable oils are annual plants such as corn, castor oil, cottonseed, peanut, linseed, palm oil, soybean as well as oil bearing perennials such as palm, oil or coconut.

Plasticizers are added in rubber compounding process in order to soften or as lubricant in rubber milling, mixing, and extrusion process and increase dispersion of fillers as well as extend the rubber giving a larger volume of elastomer [5]. The type of plasticizers such as naphthenic, paraffinic petroleum-based, and natural oils could affect the physical properties namely hardness, fleck life, low temperature performance, abrasion resistance, and tensile strength [1, 3].

SBR has good wear resistance and traction performance, it is the primary rubber for passenger car tire tread compounds. While Natural rubber has good properties in tensile properties [6]. Various types of oil can be added into the rubber compound in order to reduce its viscosity, get better processing properties, increase low temperature flexibilities and decrease cost [5]. Producing of oil-extended rubbers has been industrialized since 1950s in the United States using petroleum based extender oils such as aromatic oil, naphthenic oil and paraffinic oil. Aromatic/ Mineral oil is the most widely used extender oil in modern tire manufacturing due to its good compatibility with tire rubbers, such as SBR and natural rubber. However, in 1994 a report by Swedish National Chemicals Inspectorate showed that polycyclic aromatic hydrocarbon (PAH), which is the main constituent of aromatic oil, is highly carcinogenic. Hence, plasticizer from mineral oil also not eco-friendly, corrosive, and irritate stated that natural / plant oils could be developed in polymer industry due to its availability, low cost, eco-friendly and less toxic for nature and human being [4].

The research on utilize the commercial brown factices of castor oil at temperature 170°C, then use it in rubber compounding of natural rubber and EPDM 6250, and the results showed that the compounding energy and modulus of torque were decreasing [7]. The addition of castor oil as a plasticizer with a 5 phr ratio has a positive effect towards the mechanical properties which is increasing tensile strength, elongation at break, and fracture resistance. The other vegetable oil utilizations were the vegetable oil and epoxidized palm oil (EPO), the addition of 2-3 phr vegetable oil and EPO in natural rubber compounding and yielded the increasing of cross linking density, hardness, modulus but decreasing tensile strength and elongation at break [8]. While, EPO improved the carbon black dispersion in the matrix polymer as well as increase its heat resistance. However, research on the utilization of natural based oil as plasticizers for Natural Rubber and Styrene Butadiene Rubber composites filled with Carbon Black and Calcium Carbonates is still limited. Hence, the effect of using natural based oil as a greener plasticizer towards its properties of rubber products is an urgency to be investigated.

Bearing in mind the importance of searching new alternative mineral plasticizers and with regard to some weakness of using petroleum based oil for rubber compounding in solid tires industries, it is a necessity to do this research. The objective of this research was to study the effect of natural plasticizers namely castor oil, palm oil, fried-palm oil towards it physic-mechanical and tensile properties. Comparison those properties with commercial extended petroleum based oil such as white oil and minarex oil was evaluated to determine the feasibility using natural based plasticizers. Consequently, the research would contribute new information about the characteristics of natural plasticizers that potentially could replace the petroleum oil in rubber compounding of NR-SBR composite in solid tyres industries

2. Materials and Methods

2.1 Materials

*Natural Rubbe*r/ NR (Standard Indonesian Rubber/SIR 20) and Styrene Butadiene Rubber/ SBR were used as polymer. Other chemicals used were Zinc Oxide (NC 105 Global Chemical Co, Ltd), Stearic Acid

(SA 1806, Oleochemicals Industry, PT Sumi Asih), Carbon Black (CB N330 or HAF VULCAN, CABOT), Calcium Carbonate (400 mesh, SK Product). While, MBTS, Diphenyl Guanidine (DPG), Paraffin Wax, N-(1,3-dimethylbutyl)- N'-phenyl-p-phenylenediamine 6 PPD, Sulphur were bought from Brataco Chemical. Three types of natural based oils namely castor oil, palm oil, fried-palm oil, while the petroleum based oils were industrially commercial white oil, and minarex oil.

2.2 Methods

2.2.1 Rubber Compounding for Solid Tyre
The rubber compounding process were conducted based on the standard procedure of ASTM D3182 using laboratory two roll mill XK-160, Shanghai Rubber- Machine Worker, China. The formulations are given in Table 1. The five different types of plasticizes oils were in the same quantity of 8.75 phr. The composites were vulcanized and pressed for 14 minutes at 160 °C. Three replication for each formulas were compounded in order to obtain an accurate data.

Table 1. Formulation of NR-SBR Composites for Rubber Solid Tyre

Ingredients	Formula				
	A	B	C	D	E
NR (SIR 20)			90		
SBR			10		
ZnO			4.75		
Stearic Acid			2.25		
Carbon Black N330			62		
$CaCO_3$			34		
Fried Palm Oil	8.75	-	-	-	-
Palm Oil	-	8.75		-	-
Castor Oil	-	-	8.75	-	-
Minarex Oil	-	-	-	8.75	-
White oil	-	-	-		8,75
MBTS			1.65		
DPG			0.87		
Parafin Wax			1.55		
6PPD			1.37		
Sulfur			2.65		

2.2.2 Physical-Mechanical Properties Testing
The five formulas of solid tyres were tested for parameters as follows hardness Shore A (ASTM D 2240); tensile strength (ASTM D 412); modulus 200% (ASTM D 412); density (ASTM D 624); abrasion resistance (ASTM D 5963); compression set, 25% defl, 70°C, 22h (ASTM D 395); and ozon resistance, 50 pphm, 20% strain, 24 h, 40°C (SNI.7655-2010). Each replication was tested for those parameters, the data presented in this results are the average of 3 replications.

3. Results and Discussion

3.1 Physical Properties

Table 1 shows the physical properties of SBR-NR composites filled by CB and $CaCO_3$ using castor oils, palm oil, fried-palm oil, minarex oil, and white oil. It shows that composites using fried-palm oil as plasticizer yield the lowest hardness of 68 ShoreA, while minarex oil could yield 74 ShoreA. It could be said that the natural oil based could plasticize the composites well, even though the value of hardness smaller than the petroleum based oil. Natural based oil has different fatty acid that could soften the interaction between NR, SBR, CB, and $CaCO_3$ during the mixing process. However, the compatibility between the types of plasticizers could effect the cross linking reaction in rubber matrix [9]. Thus, different plasticizer influence the speed of insertion filler into rubber matrix and the homogeneity between reinforcing and non-reinforcing filler [10]. Thus, it could yield the different level of hardness.

Table 1. Physical Properties of NR-SBR Composites filled by CB and $CaCO_3$ using Various Plasticizer

No	Parameter	Unit	Testing Results				
			A Castor Oil	B Palm Oil	C Fried-Palm Oil	D Minarex Oil	E White Oil
1.	Hardness	ShoreA	71	70	68	72	74
2	Density	g/cm^3	1,28	1,27	1,22	1,26	1,29

The effect of using different natural and petroleum based plasticizers on the density of vulcanized rubber is not significant. The density ranges from 1,22-1,28 g/m^3. The composites using free palm oil has the lowest density among others. It is stated that the hardness and density is directly influeced by the volume of reinforcing filler, but in order to obtain homogeneous and fine dispersion, plasticizer plays an important roles along with the type and particle size of the fillers [11]. Furthermore, the rheological analysis would be beneficial in order to know the effect of different plasticizers on rubber mixing and vulanization. Based on the data as listed in Table 1, it is indicated that hardness have linear relationship with density. Similarly, those relationship also connected to the abrasion resistance [12].

3.2 Tensile Properties

The value of tensile strength and modulus of each composites using various plasticizer depicted in Figure 1. Compared to other formulas, minarex oil could give the highest tensile strength 145 kg/cm^2, while the lowest is formula C (fried-palm oil) 118 kg/cm^2. Aforementioned, the phenomena could be explained as follows. Each plasticizer has different ability to stretch rubber matrix during the chemicals dispersion and also compatibility between polymer matrix and filler [5]. Its ability depends on the fatty acid and other components contained in the palsticizers [11].

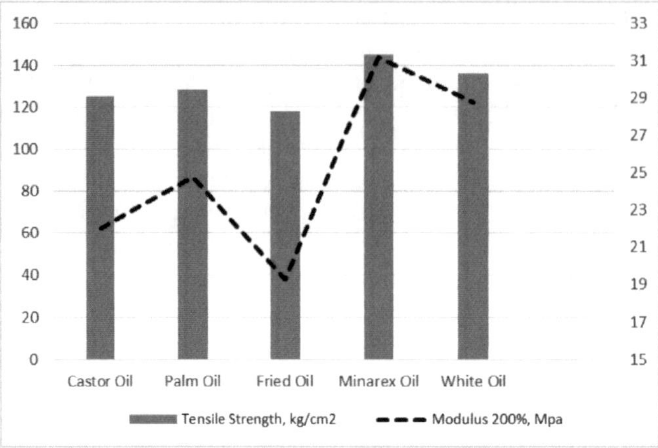

Figure 1 Tensile Properties of NR-SBR Composites filled by CB and $CaCO_3$ using Various Plasticizer

It is stated that increasing of plasticizer loading gradually reduces the bounding rubber contents of the compounds, but the bound rubber content is insignificantly influenced by the types of plasticizers [1]. The more saturated rubber shows less bonding to the same type of carbon black than unsaturated rubber. The composites in this research used 90 phr NR and 10 phr SBR. NR that possesses a higher un-saturation than SBR shows a higher bound rubber content than SBR at the same carbon black loading. In comparison with the un-plasticized compound, the addition of plasticizer remarkably decreases the bound rubber content for the filled SBR compounds, whereas the filled-NR and NR/SBR compounds show a narrower range of reduction in bound rubber content with increasing plasticizers loading [13]. This indicates a different level of interaction strength in NR and SBR because of their difference in chemical structure. In addition to their different unsaturation levels, the chain break-down of NR during mixing generates free radicals which may lead to strong interactions with the filler surface [6]. Similarly, the bounding rubber could affect the tensile strength properties.

Modulus 200% for the given formulas range from 19 - 31 MPa. Similarly, the minarex oil could give indirectly insignificant effect to the tensile properties rather than the natural based oil. As expected, an increase of plasticizers volume reduces the hardness and modulus of the vulcanizes since the rubber molecules can deform easier due to an increase of the free volume and the lubrication effect of the each plasticizers [14]. Thus, at the same plasticizers content, NR/SBR composites filled CB and $CaCO_3$ display the slightly same modulus irrespective of the plasticizer types.

3.3 Physical-Mechanical Properties

Similarly with the tensile properties minarex oil is able to soften the rubber mixing and yield the highest abrassion resistance as shown in Figure 2. The 90/10 phr of NR/SBR composites would be beneficial for the rubber product since NR has a good tensile, tear strength, flexing and fatigue resistance, and elastic properties [9]. In addition, rubber compounding has advantage such as reduced rolling resistance, low temperature flexibility and enhances adhesion between components [15]. However, filler are practically added into NR composites to enhance properties like stiffness, abrasion and skid resistance [16]. Hence, the plasticizers type in the same volume could not give significant different of abrasion resistance of each composites. This results was highlighted that particle size of filler influence the 3 dimensional networking in rubber polymer matrix in a very fine dispersion, not the plasticizers yet.

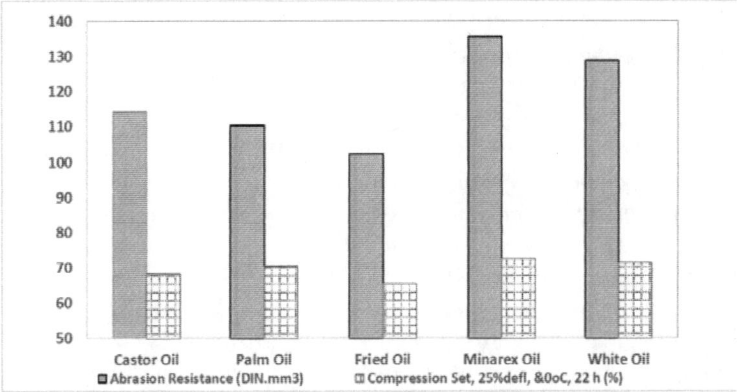

Figure 2. Physical-Mechanical Properties of NR-SBR Composites filled by CB and $CaCO_3$ using Various Plasticizer

The compression set (25% deflation, temperature 70 °C, 22 hours) for each formulas shown in Figure 2. It depicts that using fried-palm oil yield the compression set approximately 65%, while the about 72% using minarex oil. Some researchers concluded that the compression set is dominantly influenced by the presence of reinforcement fillers within the rubber polymer matrix. The incorporation of fillers (silica – quartz) into rubber increases both the dynamic storage modulus and the phase angle [17]. The filled-rubber compounds are highly complex systems in which various solid and liquid ingredients are dispersed in a rubber matrix. In such a system, the reinforcing filler develops strong interactions that lead to the development of a structural material and the associated specific flow properties [14]. Filler-rubber interactions give rise to additional cross links in the network structure, and the immobilized elastomeric layers on the filler surface have an influence on the dynamic response of the material [3]. Thus, the plasticizers type doesn't give any massive impact on the physical mechanical properties like compression set as well as modulus.

The test results showed that there is "no crack" after expose of 50 pphm ozone, 20% strain for 24 hours at 40°C. It is implied that the plasticizers type does not influence the abrasion resistance of the vulcanizates. The abrasion resistance is closely related to the chemical of anti-degradation quantity and types in rubber compounding. Aforementioned, the plasticizers activity dominantly play as lubricant role not for protect rubber polymer matrix from ozone, oxidant attack or even flex cracking. Therefore, the same amount of MBTS (1.65phr) is in appropriate to yield the solid tyres which has a good ozone resistance.

4. Conclusion

In summary, the research showed that natural based oil namely castor oil, palm oil, fried palm oi were a potentially alternative plasticizer to substitute naphthenic/ petroleum basedoil in NR-SBR composites filled with CB and $CaCO_3$ for rubber solid tyres industries. It is indicated that the physic mechanical properties of rubber vulcanizates using castor oil, palm oil, fried-palm oil are not significantly different from using minarex oil and white oil as plasticizers. However, using minarex and white oil yielded a little bit better physical properties for tensile strength, modulus, and abrasion resistance.

Acknowledgement

The authors would thank the Chief of Baristand Industri Palembang for the financial support, Crumb Rubber Factory for supplying the NR, and PT Shima Prima Utama for the providing machines and other equipment. We are grateful to research team in laboratory for the hard work and technical support in conducting the research.

References

[1] Kukreja T 2003 *Journal of applied polymer science,* 87(**10**) 1574.
[2] Kukreja, T.2002 *European polymer journal,* **38(7)** 1417.
[3] Kundu, P 2000 *Journal of applied polymer science,* **75(6)** 735.
[4] Vieira, M.G.A 2011 *European Polymer Journal,* **47(3)** 254.
[5] Cataldo, F., O. Ursini, and G. Angelini, 2013 *ISRN Polymer Science,*
[6] Iqbal, M., 2008 *Journal of the Chemical Society of Pakistan,* **30**(4) 507.
[7] Puspitasari, S. and A. Cifriadi,2013 *Jurnal Tanaman Industri dan Penyegar,* **4**(2) 99.
[8] Chandrasekara, G. 2011 *Journal of the National Science Foundation of Sri Lanka,* **39(3)**.
[9] Nandanan, V., R. Joseph, and K. George, 1999 Journal of Applied Polymer Science, **72(4)** 487
[10] Puspitasari, S.,2014 *Jurnal Penelitian Kare* **32**(2) 189.
[11]Lapig, J 2016 *International Journal of Research in Pharmacy and Chemistry* **6(3)** 604.
[12]Joshi, R 2015 *To Study the Effect of Various Additives on Vulcanized Rubber–A.*
[13]Tanrattanakul, V. and P. Bunkaew 2014 *Express Polymer Letters,* **8(6).**
[14]Basfar, A., M. Abdel-Aziz, and S. 2002 MoftiRadiation Physics and Chemistry **63(1)** 81.
[15]Samaržija-Jovanović, S 2014 The Scientific World Journal.
[16]Sun, J 2012 *Effects of the Synergy of Hardness and Resilience on the Akron Abrasion Properties of SBR Vulcanizates.* Journal of Macromolecular Science, Part B **51(8)** 1658.
[17]Akcakale, N 2010 *Scientific Research and Essays,* **5(8)** 758.

Functionality Analysis of Carbon Nanosheet, Oxidized Carbon Nanosheet and Reduced Carbon Nanosheet Oxide by Using Fourier Transform Infra Red and Boehm Titration Method

N F Syabania [1], W Sudarsono [2], D Rohendi [3] and N Syarif [3]

[1]Master Student of Chemistry, Universitas Sriwijaya, Indralaya, Sumsel, Indonesia
[2]Master Student of Fuel Cell Institute, Universiti Kebangsaan Malaysia, Bangi, Malaysia
[3]Department of Chemistry, Universitas Sriwijaya, Indralaya, Sumsel, Indonesia

E-mail: nsyarif@unsri.ac.id

Abstract. Functionality analysis of carbon nanosheet (CNS), oxidized carbon nanosheet (o-CNS) and reduced carbon nanosheet oxide (r-CNSO) has been done by using fourier transform infra red (FTIR) and boehm titration methods. The results showed that the functional group differences between CNS, o-CNS and r-CNSO were variated in intensity value for FTIR spectrogram. CNS spectrogram showed a wide peak about 3400 cm^{-1} for hydroxyl group, bond C=C vibration in 1500 cm^{-1}, epoxy and alkoxy in 1200-1100 cm^{-1}. o-CNS spectrogram shows the carboxilyc acid about 2300 cm^{-1}. Peak at 1700 cm^{-1} showed for C=C bond of epoxy and alkoxy functional groups were shown in 1250 cm^{-1} and 1125 cm^{-1}. Hydroxyl group of r-CNSO were showed in the 3441 cm^{-1}. The highest intensity of IR absorbance for hydroxyl group can be observed in r-CNSO. Carboxylic functional group in three carbon type was observed in 2337 cm^{-1}. Moderate peak at 1627 cm^{-1} showed C=C group. Epoxy and alkoxy were shown in 1172 cm^{-1} and 1056 cm^{-1}. The Boehm titration method showed that three carbon have differences amounts of funtional groups, except for the carboxylic and lactone groups. The CNS have funtional group concentration for fenolic, lactonic and base were 0.089, 0.069 and 0.082 mmol g^{-1}, respectively. The o-CNS have funtional group concentration for fenolic, lactonic and base were 0.089, 0.078 and 1.146 mmol g^{-1}, respectively. r-CNSO have funtional group concentration for fenolic, lactonic, carboxylic and base were 0.05, 0.1, 0.025 and 0.05 mmol g^{-1}, respectively.

Keywords : carbon nanosheet (CNS), boehm titration methods, o-CNS, r-CNS

1. Introduction

The performance of electode is highly related with the performance of lithium battery and supercapacitor [1,2]. One of key-factor for the performance of electrode is functionality of electrode material [3] . One of common material used as an electrode is a carbon nanosheet (CNS). CNS can be made of gelam wood bark (GWB) [4]. It showed that the high electronic and ionic conductivity can be achieved after its surface modification [5]. Surface modification has been done by oxidation of the

CNS (CNSO) by Hummer method [6] and followed by the reduction of CNS-oxide by using liquid exfoliation [7] to have reduced carbon nanosheet oxide (rCNSO). The changing in CNS, CNSO and rCNSO functionalities can be detected by using FT-IR method and Boehm titration [8].

2. Methods

2.1 Preparation of Carbon Nanosheet (CNS)

GWB was cleaned by soaking in water for 12 hours and then dried under the sunlight for one day. GWB was cut into small pieces using scissors and put into blender to make powder. The mixture of powder 40 g GWB and 0.08 g KOH in 100 mL water was stirred for 3 minutes, placed in the hydrothermal reactor and heated in the electric baking oven at 200 °C. Under this condition, the pressure inside the reactor reached 20 – 30 bars. After 16 hours in the oven, the reactor was cooled to room temperature and atmospheric pressure. Hydrothermal treatment resulted torrefaction material [9].

Microwave pyrolysis was applied to the mixture of torrefaction material powder 20 g and graphite rod 4 g as microwave absorber in a crucible and microwave furnace. The mixture was put into microwave oven and heated for 20 minutes so that the temperature can reach 800°C. The result of this process was carbon nanosheets (CNS).

2.2 Oxidation of Carbon Nanosheet (CNS) and Reduction of Carbon Nanosheet Oxide (CNSO)

3 g of CNS was mixed with 3 g of $NaNO_3$ in erlenmeyer, and dissolved with 80 ml of H_2SO_4, followed by homogenizing with shaker for 1 hour. The addition of 9 g $KMnO_4$ into mixture were done after homogenizing and was shaked for 1 hour, in ice bath for maintained the temperature. The mixture was deployed from ice bath and allowed to reach 35 °C. The mixture was then stirred for 12 hours to form brown paste. The paste was added with 10 mL 32% of H_2O_2 and followed by destilled water as the mixture temperature reached room temperature. The precipitated was washed with distilled water and 3% HCl to netralized. Finally filter sediment and ultrasound for 30 minutes and oven for 1 hour with the temperature of 120°C.

0.01 g of powdered graphite oxide was added to 10 mL of deionized water and sonicated for 24 hours to make reduce carbon nanosheet oxide (r-CNSO). The sonicated solution was centrifuged at 15.000 rpm for 1 hour. Light brown supernatant was precipitated resulted rCNSO. Liquid part was removed gently form the tube by using pippete.

2.3 Characterization functionality of CNS, o-CNS and r-CNSO

The functionality of CNS, o-CNS and r-CNSO are characterization with FTIR methode and Boehm titration. The Boehm titration method was applied to determine the acid and basic sites in the activated carbons [10]. The total acid sites were neutralized with 0.1 M NaOH solutions, and the basic sites were neutralized with a 0.1 M HCl solution. The reactions between the reagents and acidic oxygenated functional groups on the surface were based on the difference strength of the acid/base. The strength order of acidic and basic groups is as follow: carboxyl > lactone > phenol. The carboxylic and lactonic sites were titrated with a 0.1 M Na_2CO_3 solution, the carboxylic sites were titrated with a 0.1 M $NaHCO_3$ solution, and the phenolic sites were determined by the difference. For each determination, 50 mL of the solution was added to one gram of activated carbon in a glass container. The container

IOP Conf. Series: Journal of Physics: Conf. Series **1095** (2018) 012028 doi:10.1088/1742-6596/1095/1/012028

was inserted into vortex shaker for over one day. A 10 mL sample was titrated with 0.1 M of HCl or NaOH solutions [11].

3. Results and Discussions

Surface modification has been done by the oxidation of the CNS and followed by the reduction of o-CNS to get r-CNSO. The characterization steps were proceed by using FTIR and boehm titration to analyzed functional groups that featuring CNS, o-CNS and also r-CNSO. Preparation of CNS, o-CNS and r-CNSO can be visualized as discoloration of carbon. Figure 1 show visible color changes from green to black in the form of slurry.. The changing of color is an indicator for making such carbons [12]. In general, the color of the solution of carbon change from greenish solution (figure 1a) gradually, to blackish that suggests the phase formation in o-CNS was occurred (figure 1b).

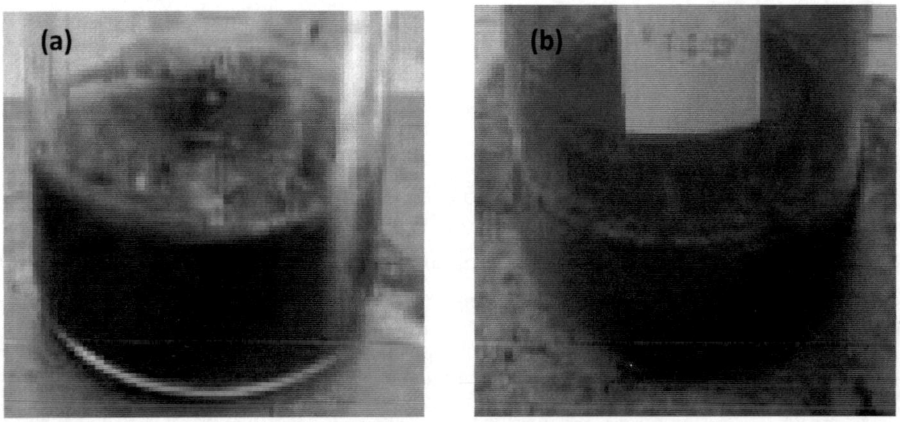

Figure 1. (a) Greenish solution after the addition of $KMnO_4$, initial condition (b) the black, final condition, the formation of CNSO

Black solid slurry was dried at 120 °C produced has the texture of soft, fragile and very hygroscopic consistency. Preparation of r-CNSO was done by using sonicator for o-CNS within 1 hour. This step was done because o-CNS reduced so that the product while have difference functional groups from the other.

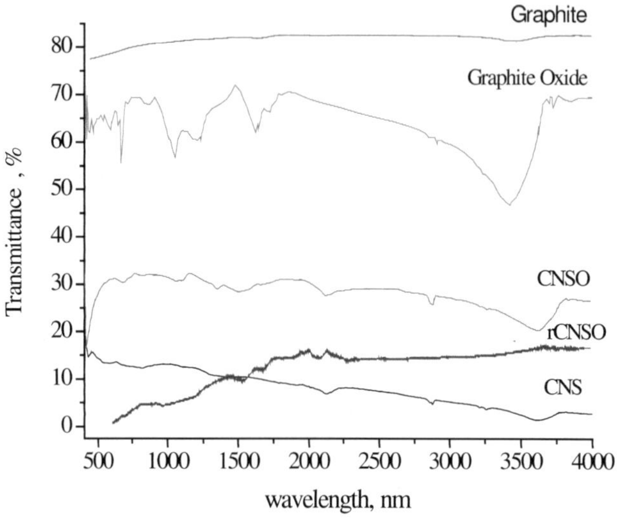

Figure 2. Spectrogram FTIR of CNS, o-CNS and r-CNSO

The result shows that the differentiation of functional group between CNS, o-CNS and r-CNSO is in intensity of FTIR spectrograma. Figure 2 showed that spectrum CNS, o-CNS and r-CNS. At which the CNS shows a wide peak in area 3400 cm^{-1} for hydroxyl group, bond C=C vibration in area 1500 cm^{-1}, epoxy and alkoxy in area 1200-1100 cm^{-1}. Spectrogram of o-CNS showed that the carboxylic acid appear in 2300 cm^{-1}; C=C in 1700 cm^{-1}. Epoxy and alkoxy are shown in 1250 cm^{-1} and 1125 cm^{-1}. The r-CNSO shows a hydroxyl group in the 3441 cm^{-1}. The hydroxy group is in highest intensity at r-CNSO. Carboxylic acid can be shown in 2337 cm^{-1}. Moderate peak at 1627 cm^{-1} is for C=C bond. Epoxy and alkoxy were shown in 1172 cm^{-1} and 1056 cm^{-1}.

Table 1. Functional group concentration

Samples	**Functional group concentration, mmol g^{-1}**			
	Base sites	**Carbo-xylic**	**Phenolic**	**Lacto nic**
CNS	0.082	0.003	0.089	0.069
o-CNS	1.146	0.003	0.089	0.078
r-CNSO	0.050	0.025	0.005	0.100

The Boehm titration method showed that three carbon have differences amounts of funtional groups, expept for the carboxylic and lactone groups. The CNS have funtional group concentration of fenolic is 0.089 mmol g^{-1}, lactonic is 0.069 mmol g^{-1} and base is 0.082 mmol g^{-1}. The o-CNS have funtional group concentration of fenolic is 0.089 mmol g^{-1}, laconic 0.078 mmol g^{-1}, and base is 1.146 mmol g^{-1}. r-CNSO have funtional group concentration of fenolic is 0.05 mmol g^{-1}, lactonic is 0.1 mmol g^{-1}, carboxylic is 0.025 mmol g^{-1} and base is 0.05 mmol g^{-1} as shown in Table 1. It can be shown that the oxidation treatment gives significantly the addition of base and lactonic groups concentration. There are no change in concentrations of carboxylic and phenolic. The concentration of functional groups decrease after reduction treatment except for lactonic. This can be interpreted that the treatment applied runs properly. This character appears with the additional oxygen bonded onto the surface of carbon in the form of C–O polar functional group that is responsible for increasing of the surface energy [13].

4. Conclusions

The conclusion of this research is that all carbons have the differentiation each other. The spectrograms show that the functionality of carbocilyc function for their CNS and o-CNS, while r-CNSO showed the difference functionality of each other.

Acknowledgments

Authors wishing to acknowledge assistance or encouragement from colleagues, special work by technical staff or financial support from SHERA Grant of Bandung Instititute of Technology - USAID for clustering work in Universitas Sriwijaya.

References

[1] Bleda-Martinez M J, Macia-Agullo J A, Lozano-Castello D, Morallon E, Cazorla-Amoros D and Linares-Solano A 2005 *Carbon* **1** 1
[2] Nian Y-R and Teng H 2003 *J. Electroanal. Chem.* **540** 119
[3] Flandrois S and Simon B 1999 *Carbon* **37** 165
[4] Syarif N and Prasagi M 2016 *Carbon – Sci. Technol.* **8** 35
[5] Yamada H, Tsunoe D, Shiraishi S and Isomichi G 2015 *J. Phys. Chem. C* **119** 5412
[6] Hummers W S and Offeman R E 1958 *J Am Chem Soc* **80** 1339
[7] Shih C-J, Lin S, Strano M S and Blankschtein D 2010 *J. Am. Chem. Soc.* **132** 14638
[8] Boehm H P 1994 *Carbon* **32** 759
[9] Kambo H S and Dutta A 2015 *Energy Convers. Manag.* **105** 746
[10] Allwar A 2012 *IOSR J. Appl. Chem.* **2** 9
[11] Fidel R B, Laird D A and Thompson M L 2013 *J. Environ. Qual.* **42** 1771
[12] Mijangos F, Varona F and Villota N 2006 *Environ. Sci. Technol.* **40** 5538
[13] Cvelbar U, Markoli B, Poberaj I, Zalar A, Kosec L and Spaić S 2006 *Appl. Surf. Sci.* **253** 1861

IOP Conf. Series: Journal of Physics: Conf. Series **1095** (2018) 012029 doi:10.1088/1742-6596/1095/1/012029

Structural study of the ordering processes of cold drawn *trans*-1,4-polyisoprene samples in the heating process on the basis of wide- and small-angle X-ray scattering measurements

P J Ratri[1] and K Tashiro[2]

Abstract. The regularization processes of the TPI glasses drawn at the various low temperatures were investigated by performing the temperature-dependent simultaneous measurements of the 2-dimensional wide-angle and small-angle X-ray scattering patterns. The glassy sample drawn below the cold crystallization temperature but above the glass transition temperature (-70°C) showed the disordered β form, which was found to transform to the regular β form at around -30°C. On the contrary, when we decreased the stretching temperature furthermore; for example -60 and -70°C, it gave the disordered α form. This disordered form transforms to the regular α form by heating to around 30°C. The small-angle X-ray scattering patterns were found to change in parallel with the wide-angle X-ray diffraction pattern changes, revealing a correlation between the crystalline phase transition and higher-order structure change in these regularization processes. These paracrystalline are packed in the smaller crystal cell than the regular form.

1. Introduction

Strain- or stress-induced crystallization phenomenon of some polymers has been reported in several papers, for example; polyethylene [1], natural rubber or *cis*-1,4-polyisoprene [2], *isotactic* polypropylene [3, 4], *syndiotactic* polypropylene [5, 6], polyethylene terephthalate (PET) [7, 8], poly(L-lactic acid) (PLLA) [9, 10], and so on. This method is quite important to study the strain effect during the crystallization process. In a lucky case, we may observe the existence of mesomorphic phase or mesophase before the appearance of crystalline phase. However, in some polymers, the mesophase was easier traced by rapid quenching over the large temperature intervals followed by applying some strain near the glass temperature point (Tg) than near the melting point (Tm) at high temperature. For example, in the case of PET and PLLA, nodular structure was reported as a mesophase, which is essentially amorphous with some degree of chain orientation order, by cold drawn of glassy sample above Tg [7 - 10]. Furthermore, this mesophase transforms to crystalline phase. In many cases, the crystalline phase, which formed at lower temperature, has some disordered of orientation (paracrystalline phase).

In the present paper, we will report the structural regularization of cold drawn sample of *trans*-1,4-polyisoprene (TPI) on the heating process. High molecular weight of TPI, well known as Gutta percha, exhibits properties that are attractive for various application fields such as root canal system, splint,

[1] Chemistry Department, Universitas Pertamina, Jl Teuku Nyak Arief, Simprug, Jakarta Selatan, 12220, Indonesia

[2] Department of Future Industry-oriented Basic Science and Materials,Toyota Technological Institute, Tempaku, Nagoya 468-8511, Japan

golf ball covers, and so on according to its mechanical and physical properties [11 - 18]. Since the discoveries of TPI content in *Eucommia ulmoides Oliv.* (Tochuu cha tea) trees, this polymer become one of the best candidates for supporting the green plastic industry by applying this polymer in the wider application fields [19-22]. In order to do so, we have to collect all of the basic information about the crystal structure study, the structure formation process and its related physical and mechanical properties [23 - 33]. In the bulk sample, TPI shows two crystal modification; α and β forms [23,24]. According to these two forms, series number of papers have been reported which related to our research theme to clarify the not-yet-solved structure-property relationship of this polymer through the crystal structure analysis and microscopically-viewed crystallization behaviour including phase transition mechanism of these structures [34-36].

Figure 1 shows the summary of our study in the crystallization process of the α and β forms formation [34-36]. As we see from figure 1, the melt-quenched sample crystallizes into the β form with no orientation at such a low temperature of -50°C [35]. The orientation of β form is easily performed by drawing the unoriented β sample below 40°C, before the phase transition occurred. On the other hand, when the melt-quenched sample is stretched at about -70°C near *Tg*, similar X-ray pattern of α form was observed. By heating up, it changes to the high temperature phase of α form with higher crystallinity [34]. When the X-ray diffraction patterns are compared carefully between low temperature phase and high temperature phase of α form, they are found to be different from each other. In this way the structure regularization process is quite complicated. It is needed to clarify these structural changes in detail. In the present paper the temperature dependent measurements of WAXD and SAXS were performed in the TPI samples stretched at the various low temperatures to investigate the complicated structural evolution process. However, for the TPI case, it was no report about the structural regularization in the cold drawn sample at low temperature. It was unclear whether the structural regularization of strain induced crystallization grows *via* a mesophase or not.

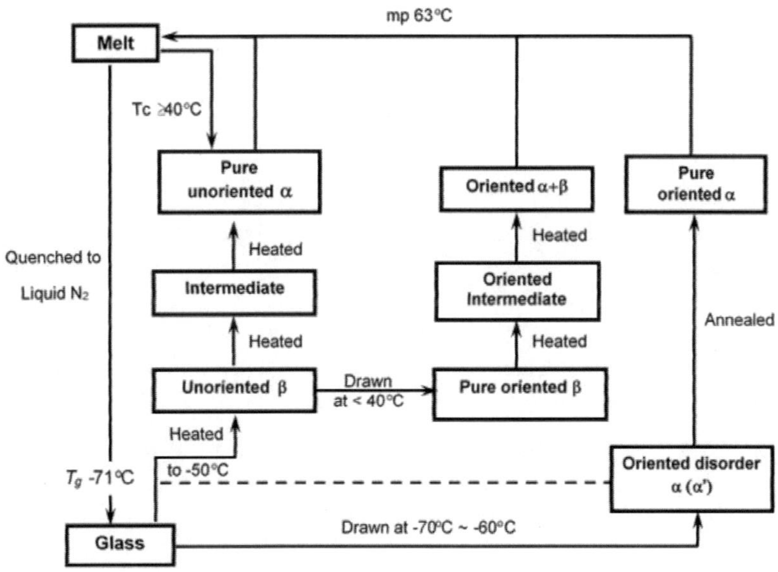

Figure 1. Schematic diagram of the formation process of the TPI α and β forms [34 - 36].

2. Experimental

2.1 Sample Preparation

Synthetic TPI sample with 99% *trans* content was purchased from Polyscience Inc. As described in previous paper, the glassy sample was prepared by quenching the melt into liquid N_2 temperature (-196°C) [34]. The three different temperatures were chosen for stretching the glassy sample near *Tg*;

-55°C (above *Tg*), -60°C, and -70°C (at *Tg*). The stretching was carried out in the cooling bath of the mixture of liquid N_2 and methanol. The stretching temperature was monitored by inserting a thermocouple into the cooling bath with temperature fluctuation of about 3°C. The thus-stretched sample was mounted on a goniometer head, during which the sample temperature was kept constantly by blowing the cooled nitrogen gas directly to the sample [37, 38].

2.2 WAXD and SAXS Measurements

The 2-dimensional WAXD patterns of cold-drawn samples were measured using a Rigaku RAPID-II X-ray diffractometer in the heating process. A cylindrical imaging plate of 127.4 mm radius was used as a 2-D detector and an incident X-ray beam was graphite-monochromatized Mo-Kα line (λ = 0.7107 Å).

The simultaneous measurements of 2-dimensional SAXS and WAXD patterns were performed using a Rigaku Nanoviewer X-ray diffractometer, where the WAXD pattern was detected using an flat imaging plate set at 97 mm from the sample position and the SAXS pattern was detected using a Pilatus 100k detector (Dectris, Switzerland) set at 761 mm distance from the sample [37, 38]. The temperature of the sample was monitored by inserting a thermocouple into a sample.

The 001 reflections profile of the cold-drawn samples were measured in the heating process using a Rigaku RINT-TTR III diffractometer in a transmission mode. An incident X-ray beam was a graphite-monochromatized Mo-Kα (λ = 0.717 Å) in the diffraction angle (2θ) range of 5° - 60° with the 1-D scintillation counter.

2.3 SAXS Data Analysis

The 2D-SAXS data were analyzed by transforming to the 1D-SAXS profiles by the circular integration an function of scattering vector *q*, where $q = (4\pi/\lambda)\sin\theta$ and 2θ is a scattering angle. The correlation function for the stacked lamellar structure *K(z)* was calculated using eq 1 [39].

$$K(z) = \langle [\eta(z') - \langle\eta\rangle][\eta(z+z') - \langle\eta\rangle] \rangle = 2\int_0^\infty (\pi)^{-1} q^2 I(q) \cos(qz) dq \qquad (1)$$

where $\eta(z)$ is an electron density at a position *z* along the direction normal to the stacked lamellar planes and $<\eta>$ is the ensemble average. The various points on the *K(z)* curve give the various structural parameters: an invariant *Q*, a mean lamellar thickness $<d>$, a mean transition zone thickness d_{tr}, a mean core thickness d_o, and a long period *LP* [39].

3. Results and Discussion

3.1. Stretching temperature above Tg (-55°C)

According to figure 1, as we pointed above, that β form starts to crystallize at -50°C. Therefore, we choose stretching temperature at -55°C. When the glassy sample was stretched at -55°C, the X-ray pattern was that of the β form as shown in figure 2a. The 1D-profile of the WAXD pattern is shown in figure 2b. However, the diffuse scattering of layer line was observed in the WAXD pattern. By annealing above -30°C, the diffuse scatterings became more spot like. As described in previous work [36], the β-to-α form occurs at around 50°C *via* the intermediate phase. From the SAXS pattern, the lamella of β form stood vertically along the draw axis. The scattering profiles were calculated from the 2-D SAXS patterns in figure 2a and are given in figure 3a. The SAXS pattern was of the meridional 2-points pattern, which increased in intensity steeply above 40°C, where the α form started to appear.

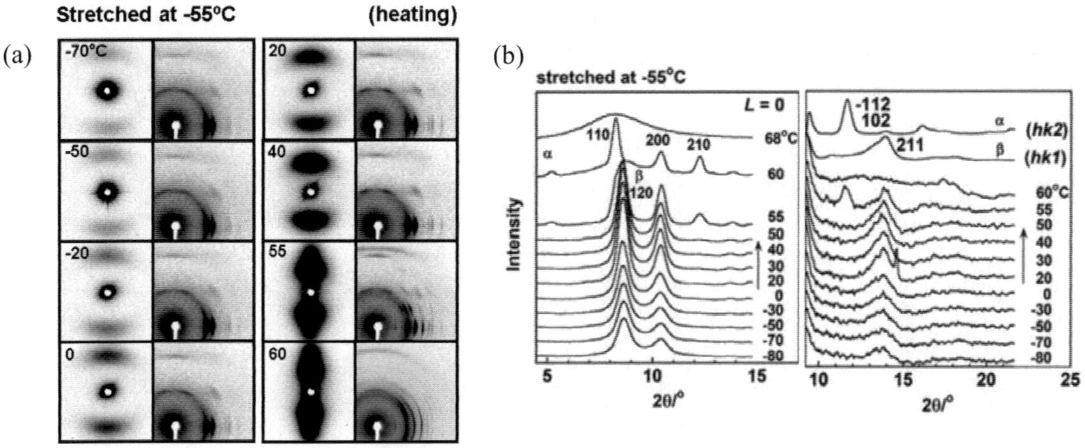

Figure 2. (a) The 2D WAXD and SAXS pattern of the oriented sample of TPI stretched at -55°C during heating, (b) The 1D-WAXD diffraction profiles of the oriented sample of TPI stretched at -55°C during heating process.

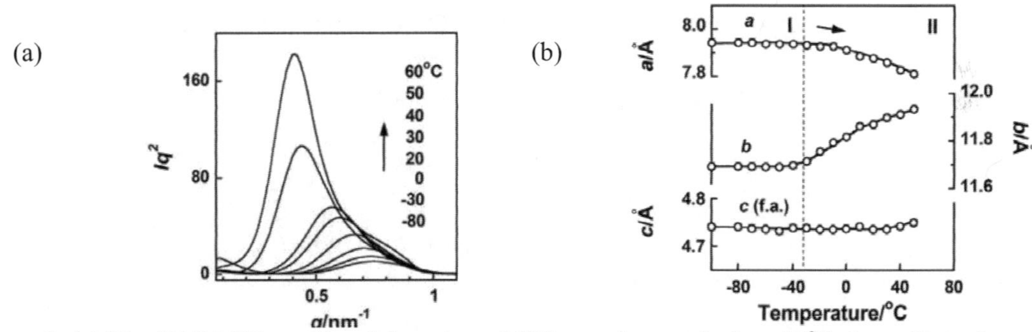

Figure 3. (a) The 1D SAXD profile of the oriented TPI sample stretched at -55°C above *Tg* in the heating process (refered to figure 2a). (b) Temperature effect to lattice cell of the oriented β sample of TPI stretched at -55°C (referred to figure 2b).

The unit cell parameters of the β form were estimated and plotted against temperature in figure 3b. Within the temperature region I, the parameters were almost constant. When temperatures reach at around -30°C, the *b*-axial length started to increase and the a-axial length decreased at the same time. These behaviours indicate that the β form in the region I is different from the β form in the region II. Let us call them the disorder β (low temperature) and regular β (high temperature), respectively. These two kinds of the β form take essentially the same chain conformation judging from the c-axial length. But, these chains are packed in a little more compacted manner in the regular β form as known from the cross-sectional area (a × b).

Figure 4 shows the temperature dependence of the *00l* reflection profile measured in the transmission mode. The 002 reflection intensity of disordered β form increased and became sharper in the regular β form. According to the space group of the β form, $P2_12_12_1{-}D_2^4$, only the *00l* reflections with l = even should be detected [40]. However, in the low-temperature phase of β form, the 001 reflection was observed, indicating the disorder in chain conformation from the regular 2_1 screw symmetry. The 001 reflection disappeared above 0°C and the disordered chain conformation was regularized there.

Figure 5 shows the intensity changes of the SAXS and WAXD data. In the region I (< -30°C) the β form was in the disordered state. By heating to the region II, the amorphous phase decreased and the 002 reflection of the β form increased in intensity in the region II. In this region the intensity of 120 reflection, the crystallinity (or Q), and long period gradually increased. In the region III the β-to-α transition occurred. The crystallinity (or Q) and long period significantly increased, where the transition to the α form was detected.

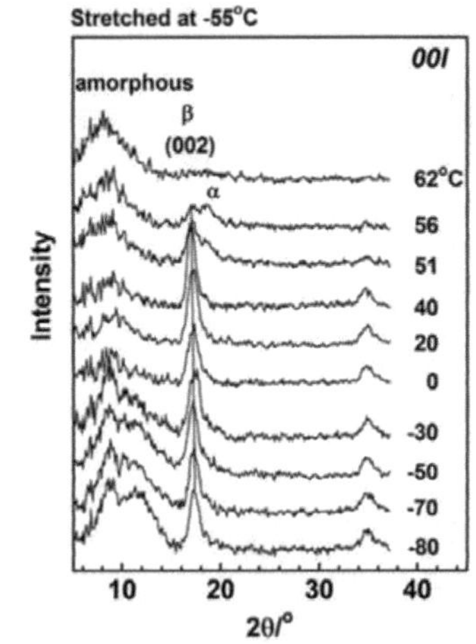

Figure 4. The *00l* profile of the oriented β sample of TPI stretched at -55°C above *Tg* in the heating process.

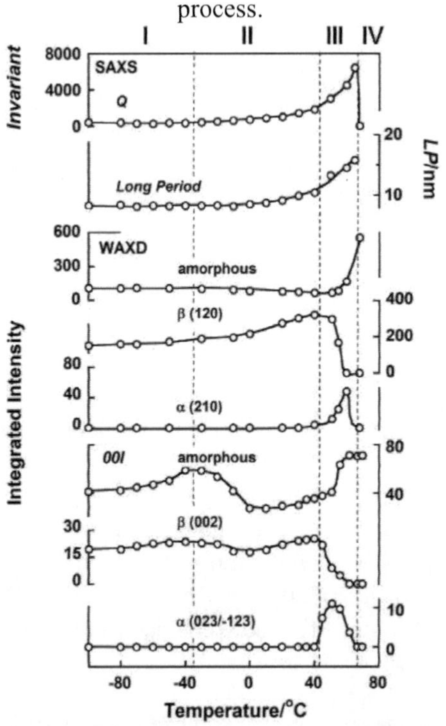

Figure 5. Various parameters plotted as function of temperature by calculating the correlation function from SAXS profiles and integrated intensity from WAXD profiles of the oriented TPI sample stretched at -55°C.

3.2. Stretching temperature at Tg (-70 ~ -60°C)

The 2D-WAXD and SAXD patterns are shown in figures 6a and 6b, respectively, for the samples stretched at -70 and -60°C. The similar WAXD patterns were observed for these two, but the SAXS patterns were different. Four scattering pattern was observed for the -70°C sample while the -60°C sample gave the meridional two scattering pattern. The SAXS profiles obtained by an azimuthal angle scan are shown in figure 7.

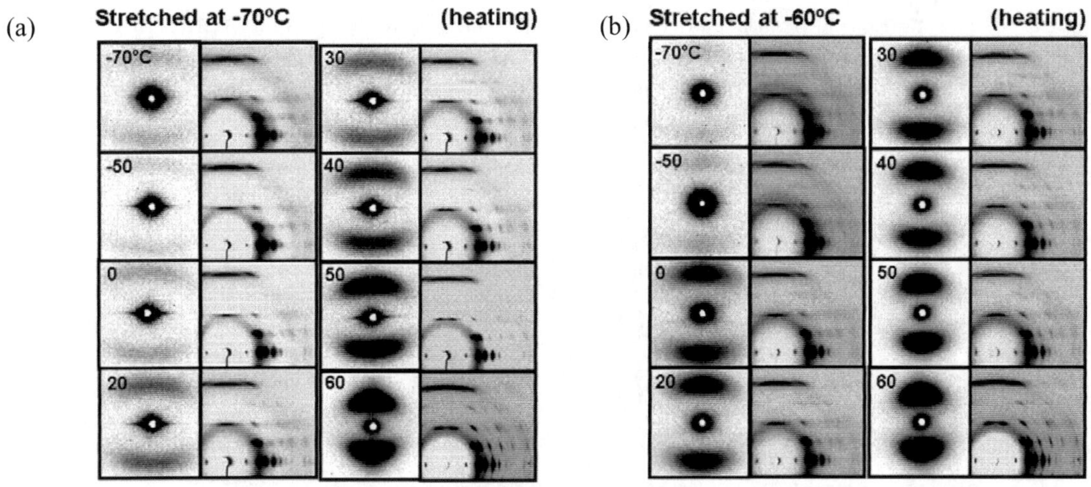

Figure 6. (a)The 2D WAXD and SAXS pattern of the oriented sample of TPI stretched at -70°C during heating. (b) The 2D WAXD and SAXS pattern of the oriented sample of TPI stretched at -60°C during heating.

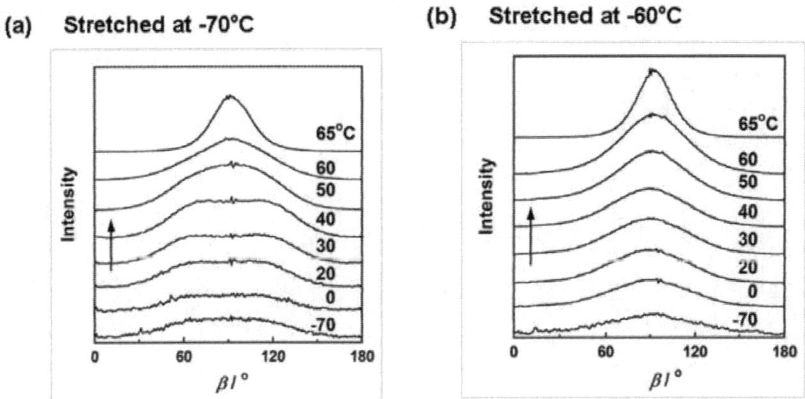

Figure 7. Azimuthally-scanned (β scan) of the oriented sample of TPI stretched at (a) -70°C and (b) -60°C during heating from -100°C to melting point.

The 2D-WAXD patterns given in figures 6a and 6b are those of the disorder α form as known from the diffuse layer lines and broader reflections. The β form content was negligible small. The 1D-profiles of WAXD pattern is shown in figure 8, where the data of the -70°C-stretched sample is given since the -60°C sample shows essentially the same behaviour about the WAXD pattern. By heating, the reflections became sharper and increased the intensity. On the first layer line, one broad reflection peak was split into two reflections (-111 and 111) at about 30°C, indicating the transformation to the regular α form. The lattice parameters were estimated for the disorder and regular α forms by assuming the similar crystal structure. The result is shown in figure 9a. In the temperature region of 40°C, the *a*- and *b*-axial lengths were detected to change, corresponding to the region where the layer line profiles became shaper.

The temperature dependence of the *00l* reflection profile is shown in figure 9b, where the result of the -70°C-stretched sample is given as an example. The meridional reflections of the disorder α form passed were sharp up to 60°C. The intensities change of SAXS, WAXD, and *00l* reflections is shown in figure 10. The meridional peak became weaker and broader in the temperature region of 65~70°C, where the equatorial line reflections were almost halo. This suggests the regular α form changes to the intermediate phase [34], before perfectly melted; that is, the phase consisting of the conformationally disordered chains but with the chain orientations kept still.

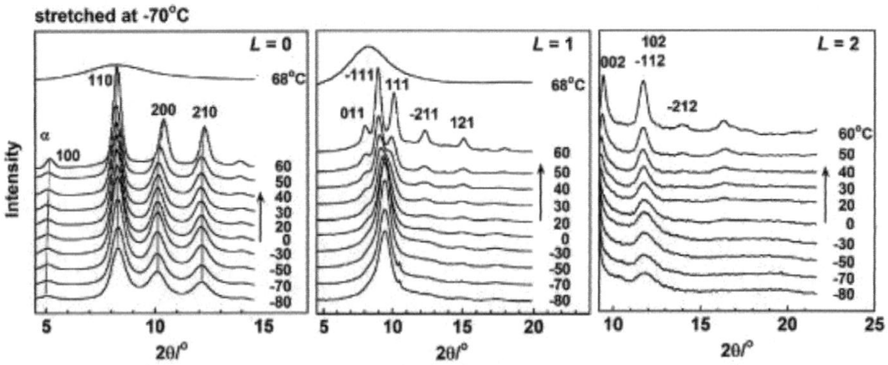

Figure 8. The 1D-WAXD diffraction profiles of the oriented sample of TPI stretched at -70°C during heating process.

(a)

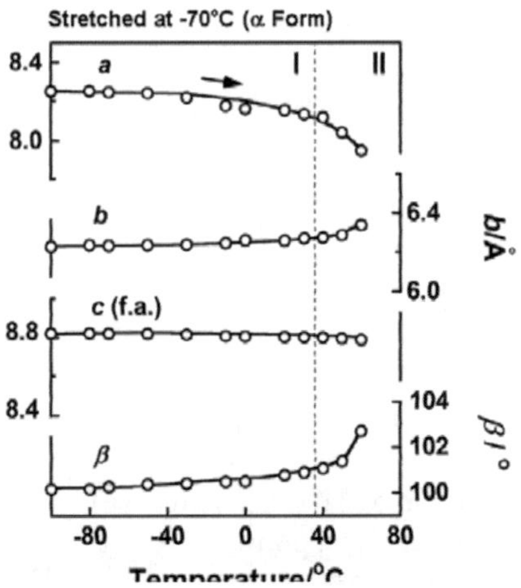

(b)

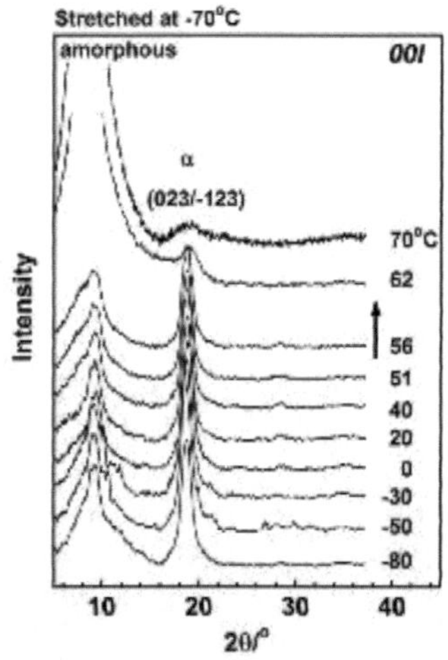

Figure 9. (a) Temperature effect to lattice cell of the oriented α sample of TPI stretched at -70°C, (b) The *00l* profile of the oriented α sample of TPI stretched at -70°C in the heating process.

Figure 11 shows the comparison of crystallite size (D_{hkl}) perpendicular to the 110 plane and lamellae thickness (<*d*>) during annealing. In the first stage, 7 monomeric units are included in the α' form lamella along the fiber axis. After the phase transition occurs, around 14 monomeric units are included in the α form lamella which is twice than the disorder form. The crystallite size of along the *ab* direction was also increased in 2 steps. When we viewed from this direction, the regular α form has three times bigger size than the disorder form. However when we compared to the crystallite size along the fiber axis toward the *ab* direction, the α form has anisotropy of crystal size.

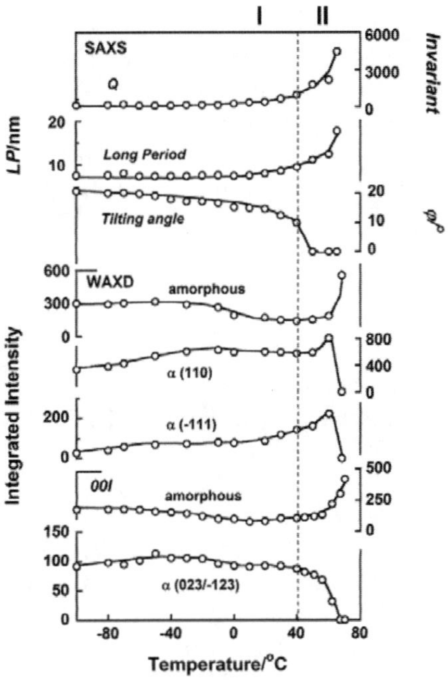

Figure 10. Various parameters plotted as function of temperature by calculating the correlation function from SAXS profiles and integrated intensity from WAXD profiles of the oriented TPI sample stretched at -70°C.

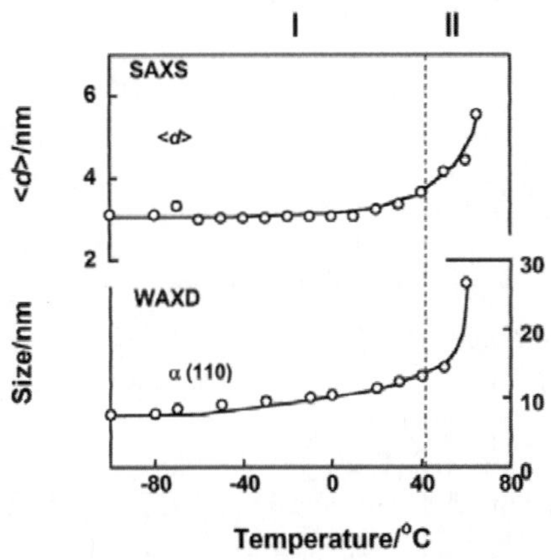

Figure 11. Crystallite size along the fiber axis and *ab* direction for the disorder-to-order transition of the α forms by stretching temperature at -70°C on the heating process.

4. Conclusion

In the present paper, the detailed information of the structural changes was investigated for the samples drawn at low temperatures by analyzing the temperature-dependent WAXD and SAXS. The stretching of the melt quenched sample at -55°C, which is below cold crystallization temperature but above *Tg*, gives the disorder β form. This disorder β form transforms to the regular β phase by heating to -30°C and to the α form at 40°C. On the contrary, when we decreased the stretching temperature of glassy sample; at -70°C and -60°C, these samples give the disordered α form which is regularized to the α form at 30°C. These two samples show the different lamellar stacking structure; tilted structure in the -70°C-stretched sample and non-tilted structure in the -60°C sample. The SAXS pattern at the high temperature is that of the non-tilted lamellar structure. The disorder α form shows the diffuse scattering along the layer lines, indicating the disorder in the relative height between the neighbouring chains. In this way the cold-drawn TPI samples show a variety of transitions attendant with the disorder-to-order structural change. The lamellar tilting is also affected by this transition: the tilted structure is created when the sample is stretched near *Tg*, while the lamellae stand vertically in the sample drawn above *Tg*.

References

[1] Fischer EW, Schmidt GF 1962 *Angew. Chem. Internat. Edit.* **1** 488-499.
[2] Tosaka M, Murakami S, Poompradub S, Kohjiya S, Ikada Y, Toki S, Sics I, Hsiao BS 2004 *Macromolecules* **37** 3299-3309.
[3] Natta G, Peraldo M, Corradini P 1959 *Rend. Fis. Accad. Lincei.* **26** 14.
[4] Wyckoff HW 1962 *J. Polym. Sci.* **62** 83.
[5] Nakaoki T, Ohira Y, Hayashi H, Horii F 1998 *Macromolecules* **31** 2705.
[6] Vittoria V, Guadagno L, Comotti A, Simounutti R, Auriemma F, De Rosa C 2000 *Macromolecules* **33** 6200.
[7] Imai M, Kaji K, Kanaya T 1994 *Macromolecules* **27** 7103.
[8] Asano T, Seto T 1973 *Polym. J.* **5** 72.
[9] Stoclet G, Seguela R, Lefebvre JM, Elkoun S, Vanmansart C 2010 *Macromolecules* **43** 1488.
[10] Wasanasuk K, Tahsiro K 2011 *Macromolecules* **44** 9650-9660.
[11] Blum J, Parahy E, Machtou P 1997 *Journal of Endodontics* **23** 307-311.
[12] Song J, Huang B, Yu D 2001 *J. Appl. Polym. Sci.* **82** 81-89.
[13] Wu Y, Casanova H, Smith WK, Edwards M, Childress DS 2003 *Prosthetics and Orthotics International* **27** 146.
[14] Tsukada G, Tanaka T, Torii M, Inoue K 2004 *Journal of Oral Rehabilitation* **31** 1139-1144.
[15] Miner MR, Berzins DW, Bahcall JK 2006 *Journal of Endodontic* **32** 683-686.
[16] Silva JBA, de Paula RCM, Feitosa JPA, Gurgel-Filho ED, Ferreira CM, Filho FJS 2006 *Journal of Applied Polymer Science* **100** 4082-4088.
[17] Zhao Y, Huang B, Yao W, Cong H, Shao H, Du A 2008 *Journal of Applied Polymer Science* **107** 2986-2993.
[18] Zhang J, Xue Z 2011 *Polymer Testing* **30** 753-759.
[19] Tangpakdee J, Tanaka Y, Shiba K, Kawahara S, Sakurai K, Suzuki Y 1997 *Phytochemistry* **45** 75-80.
[20] http://www.hitzbio.com/research%20E.html
[21] Bamba T, Fukusaki E, Nakazawa Y, Kobayashi A 2002 *Planta* **215** 934-939.
[22] Takeno S, Bamba T, Nakazawa , Fukusaki E, Okazawa A, Kobayashi A 2008 *Journal of Bioscience and Bioengineering* **106** 537-540.
[23] Bunn CW 1942 *Proc. R. Soc. Lond. A* **180** 40-66.
[24] Takahashi Y, Sato T, Tadokoro H 1973 *J. Polym. Sci.,Polym Phys. Ed.* **11** 233-248.
[26] Fischer E, Henderson JF 1967 *J. Polym. Sci. Part A2* **5** 377-390.
[27] Lovering EG, Wooden DC 1969 *J. Polym. Sci. Part A2* **7** 1639-1649.
[28] Schilder H, Goodman A, and Aldrich W 1974 *Oral Surg.* **38** 109-114.
[29] Rootare HM, Powers JM 1977 *J. Dent. Res.* **56** 1453-1462.
[30] Martuscelli E. 1972 *Die Makromolekulare Chemie* **151** 159-168.
[31] Hardin IR, Yeh GSY 1973 *J. Macromol. Sci, Part B* **7** 393-409.
[32] Yeh GSY 1973 *J. Macromol. Sci, Part B* **8** 241-257.
[33] Anandakumaran K, Kuo CC, Mukherji S, Woodward AE 1982 *J. Polym. Sci., Polym. Phys. Ed.* **20** 1669-1676.
[34] Ratri PJ, Tashiro K, Iguchi M 2012 *Polymer* **53** 3548-3558.
[35] Ratri PJ, Tashiro K 2013 *Polymer Journal* **45** 1019-1026.

IOP Conf. Series: Journal of Physics: Conf. Series **1095** (2018) 012029 doi:10.1088/1742-6596/1095/1/012029

[36] Ratri PJ, Tashiro K 2013 *Polymer Journal* **45** 1107-1114.
[37] Tashiro K, Yoshioka T, Yamamoto H, Ninh TH, Shimada S, Nakatani T, Iwamoto H, Ohta N, Masunaga H 2011 *Polym. Prepr. Jpn.* **60** 2M05.
[38] Tashiro K, Yamamoto H, Yoshioka T, Ninh TH, Shimada S, Nakatani T, Iwamoto H, Ohta N, Masunaga H. 2012 *Kobunshi Ronbunshu* **69** 213-227.
[39] Glatter O, Kratky O 1982 *Small Angle X-ray Scattering* (London: Academic Press).
[40] Hahn T 1983 *International Tables for Crystallography, Volume A, Space Group Symmetry D.* (Boston: Reidel Publishing Company)

IOP Conf. Series: Journal of Physics: Conf. Series **1095** (2018) 012030 doi:10.1088/1742-6596/1095/1/012030

Catalytic Ozonation Based Advanced Oxidation Process for Effective Treating Wastewater from Hospital and Community Health Centre Facility by FLASH WWT Catalyst System in Indonesia

R Rame[1], H Pranoto[2] , RKK Winahyu[2] , M Sofie[3] , BH Raharjo[3] , AS Utomo[3]

[1]Center of Industrial Pollution Prevention Technology, Jl. Ki Mangunsarkoro No.6, Kota Semarang, Jawa Tengah 50241,
[2]Bartec Utama Mandiri, Semarang - Indonesia

[3]Academy of Electrical Medical Engineering (ATEM) Semarang - Indonesia

email: rame@kemenperin.go.id

Abstract. Wastewater from hospital and community health centre facility can contain hazardous chemicals and harmful microbes such as viruses, multi-resistant bacteria to medical contrast agents and chemicals for treatment. Other country has placed pharmaceutical products on substances that be regulated in hospital effluent. The FLASH treatment plant involves physical and biological pre-treatment, followed by Advanced Oxidation Processes (AOPs) based on catalytic ozonation and followed by granulated activated carbon (GAC) and powdered activated carbon (PAC) filtration. Two dominant oxidants are present in ozonation applications: ozone (O_3) and hydroxyl radicals ($HO^{\cdot}$). Hydroxyl radicals are a highly reactive decomposition product of ozone and represent the main oxidant in AOPs. Conventional ozonation, both ozone and hydroxyl radicals can act as oxidants. In AOPs, hydroxyl radicals represent the primary transformation mechanism. Different ozonization types for treating wastewater from hospital and community health centre facility were considered and reviewed, and FLASH as the most common and efficient processes were discussed. In year 2015, 140 FLASH have been applied and gave satisfaction result for 100% effective meet the quality standard from regulation Ministry of Environment of Republic Indonesia for waste water quality standards Number 5/2014. FLASH treatment system could have great significance for hospitals and Community Health Centre Facility in Indonesia that face problems as removing micro pollutant. BUMA FLASH treatment is the first develop hospital wastewater treatment in Indonesia based Catalytic Ozonation Based Advanced Oxidation Process.

Keywords: catalytic ozonation, AOPs, wastewater, hospital Indonesia

1. Introduction

Wastewater from hospitals can contain hazardous chemicals and harmful microbes such as viruses, multi-resistant bacteria to medical contrast agents and chemicals for treatment [1-4]. But now a FLASH full scale plant in Indonesia has removed these residues, potentially addressing a hospital wastewater problem. The issue of hospital wastewater being sent to public sewage treatment facilities is a major problem. Municipal sewage treatment plants using conventional treatment are not designed to deal with medicinal and toxic waste [5-6]. This caused detection these substances in surface water.

Hospital wastewater can pose a health hazard to humans, especially employees at wastewater treatment plants. During heavy rains and flooding, holding tanks in the sewer system can overflow and expose the public to these hazards [8]. There is also a danger to aquatic life. Once the sewage is treated and released into the environment along with its residual content of pathogens and pharmaceuticals, the local flora and fauna are routinely at risk [9].

Even in very low concentrations, the substances in hospital wastewater can affect environment. Estrogens [10], for example, can cause hermaphroditic fish, while some painkillers [11] are poisonous to fish . Hospital wastewater and household wastewater are treated in the same way and that's not a very effective approach in Indonesia. Meanwhile, the European and other country has placed pharmaceutical products on substances that be regulated in hospital effluent [12-14]. However, there isn't Indonesia standards regulation for pharmaceuticals or pathogens in hospital wastewater.

To deal with the wastewater issue, the national company of BUMA (Bartec Utama Mandiri), established an innovation in treatment hospital wastewater that involved sending the hospital effluent at much closed through FLASH, a series of treatments. FLASH full-scale hospital and community health centre waste water Treatments in Indonesia from different province have been adopted and investigated.

The goal of FLASH Treatments was to find a solution that actually removed the problematic substances in hospital wastewater rather than simply diluting it with other wastewater streams in the public treatment system. The result was a compact, tailor-made wastewater treatment plant designed by BUMA. Local on-site treatment of wastewater is key to the success of the system. FLASH plant receives wastewater directly from the hospital. It is not mixed with the water from the public wastewater treatment system. This makes it possible to specifically target the substances in hospital wastewater.

Figure 1. FLASH Full-Scale WWT.

The FLASH treatment plant involves physical and biological pre-treatment, followed by Advanced Oxidation Processes (AOPs) based on catalytic ozonation and followed by granulated activated carbon (GAC) and powdered activated carbon (PAC) filtration units as shown on Figure 1. This treatment system is extremely flexible. Each unit can be expanded, removed or adjusted to accommodate changing needs. The physical layout of the plant is also fundamentally different from conventional wastewater facilities.

Wastewater from hospitals is typically fed into large, municipal treatment plants that take up a lot of space and require long pipelines from the hospital to the treatment plant in Indonesia. BUMA developed FLASH a compact water treatment plant that can be delivered in three or four prefabricated modules. It's no bigger than a small room and it enables hospital wastewater to be treated locally and then safely released into the local environment.

BUMA will continue to encourage hospitals and community health centre in Indonesia to apply FLASH treatment to get some benefits, namely optimal treatment efficiencies, operational cost and optimal combination of ozone and granular/powder activated carbon. During FLASH treatment, the treated water can be re-used locally as technical water at the hospital and some will be released directly into the nearby environment/ river, where it will contribute to a more stable water flow. In other words, the water that was once a risk can in future become a resource.

FLASH treatment system could have great significance for hospitals in Indonesia that face the same problems as removing micro pollutant. BUMA innovation as FLASH have the solution to a problem that exists at a great many hospitals in Indonesia. BUMA FLASH treatment is the first hospital wastewater treatment in Indonesia to invented and install this advance technology. In the years to come, FALSH technology will become extremely important especially in industrialised province/district where there is a political interest and a focus on the environment.

2. Materials and Methods

Literature were searched with original search terms were used ("Catalytic Ozonation Based AOP," "Ozone Chemistry," "Ozonation of micro pollutant," "Ozone-based AOPs, including more terms to search for ozonation ("pathogen" "antibiotic") from literature database provider such as springer, sciencedirect, and elsevier.

Data from outlet of FLASH full-scale hospital waste water treatments in year 2015 include physical parameters (temperature, dissolved solids, suspended solids), chemical parameters (pH, COD, BOD, oils and fats, MBAS, ammonia nitrogen, heavy metals), and biological parameters (total coliform) were collected from BUMA. The results of data analysis are then compared to water quality standards in accordance with their designation stipulated by regulation of the Minister of Environment Republic of Indonesia number 5:2014 about waste water quality standards.

3. Results and Discussions

3.1 Catalytic Ozonation Based AOP

Ozone has recently experienced for treatment of taste and odor compounds,[15] dissolved organic matter (DOM),[16, 17] and emerging contaminants. [18, 19]. Interest in ozone for municipal, industrial and hospital wastewater treatment is continuing to grow at a rapid pace. The spread of antibiotic resistance has become a major global challenge [20]. The emergence of antibiotic resistance has rendered the latest generation of antibiotics less effective since these new antibiotics generally rely on the same fundamental pharmacological mechanisms. Many suggest that the long-term presence of antibiotics in wastewater effluent and finished drinking water may accelerate the spread and transfer of antibiotic resistance genes [21]. To combat this threat, better removal efficiencies for antibiotics in wastewater treatment plants. The ability of catalytic ozonation to remove the threat from wastewater in FLASH is effective and efficient.

The main sources of micro pollutant to the environment are wastewater effluent, bio solids, agricultural wastewater, and manure. Depending on the wastewater composition and the process train of the wastewater treatment plant (WWTP), a range of "removal" efficiencies may be obtained for various micro pollutant. It is important to note that removal usually manifests as partitioning into the solid phase or metabolism/transformation in the aqueous phase. The presence of micro pollutant in bio solids can lead to environmental, and even food crop, contamination. Similarly, metabolites and other transformation products may retain the same pharmacological activity as parent compounds, i.e., the ability to inhibit microorganisms or cause other sub inhibitory effects. For these reasons, it is important to distinguish removal, transfer, and transformation.

One of the early reports of micro pollutant is pharmaceuticals in wastewater focused on measuring 32 drugs in German WWTPs [22]; 80% of the drugs investigated were detected in at least one treatment plant. Much of the corresponding data has been aggregated in several review articles on the presence and fate of PPCPs in wastewater [23, 24]. To date, at least 70 antibiotics have been detected in wastewater. These antibiotics include representatives from various classes, including b-lactams, cephalosporin's, FQs, lincosamides, macrolides, penicillin's, sulphonamides, and tetracycline's, among others. Commonly studied antibiotics include amoxicillin, azithromycin, chlortetracycline, ciprofloxacin, erythromycin, levofloxacin, norfloxacin, sulfamethoxazole, tetracycline, trimethoprim, and tylosin. These antibiotics are studied more often because their mass consumption tends to be higher than those of others.

As indicated above, antibiotics are not completely removed during wastewater treatment; therefore, antibiotics are present in wastewater effluent and discharged into the environment. The presence of antibiotics in surface waters raises the important question of whether antibiotics are present in drinking water supplies. In this manner, it is clear that surface water concentrations are generally an order of magnitude lower than those detected in raw wastewater. Given the expected dilution of antibiotics on discharge into surface waters, the difference in antibiotic concentrations in raw wastewater and surface waters suggests that WWTPs are the major source of antibiotics in the environment.

Inherently, the presence of antibiotics in surface waters demonstrates that conventional WWTPs do not provide sufficient removal of pharmaceuticals and results in antibiotic contamination of drinking water supplies. While the human health consequences of long-term consumption of trace concentrations of antibiotics are not yet known, eliminating antibiotics from drinking water sources seems prudent especially as antibiotics have been shown to have numerous effects at sub inhibitory levels. Antibiotics are one of the most important classes of pharmaceuticals present in wastewater because of extensive use in hospital.

The contribution of WWTPs to the spread of antibiotic resistance is an increasingly important topic. In the past few years, dozens of studies have identified the presence of antibiotic-resistant organisms and antibiotic-resistant genes in WWTPs, wastewater effluent, and surface waters [25]. studies focused on the detection and quantification of antibiotic resistance genes in various environments investigated the presence of 140 antibiotic resistance genes in activated sludge and effluent from a conventional municipal WWTP [26].

Aquatic ecologists have become increasingly concerned about impacts of trace organic contaminants on ecosystems. The continuous introduction of antibiotics into river systems can significantly change microbial communities by favouring antibiotic-resistant bacteria. As microbial communities play an important role in carbon and nutrient cycling, such changes may have drastic effects at the ecosystem level.

3.2 Ozone Chemistry

The The redox chemistry of ozone demonstrates its effectiveness as an oxidant. Reaction shows the ozone half-reaction as shown on Figure 2.

$$O_3 + 2e^- + 2H^+ \rightleftharpoons O_2(g) + H_2O \quad E_{H^o} = 2069 \text{ mV}$$

Figure 2. Ozone half-reaction.

Ozone is a stronger oxidant than most other traditional oxidants/disinfectants including hydrogen peroxide, permanganate, chlorine dioxide, chlorine gas, oxygen, and hypochlorite. Only hydroxyl radicals demonstrate a greater redox potential EHo = 2800 mV.

In solution, ozone undergoes a complex series of decomposition reactions that generate (and consume) other reactive oxygen species (ROS). ROS are exploited in advanced oxidation processes (AOPs), which focus on rapid generation of hydroxyl radicals. Regardless, an understanding of ozone decomposition is important in the context of ozone-based processes because water quality determines whether ozone decomposition kinetics are rapid or slow. Most importantly, ozone decomposition kinetics affect the ozone exposure achievable for a given applied ozone dose. A direct analogy to ozone disinfection may be helpful as ozone exposure is essentially the CT value.

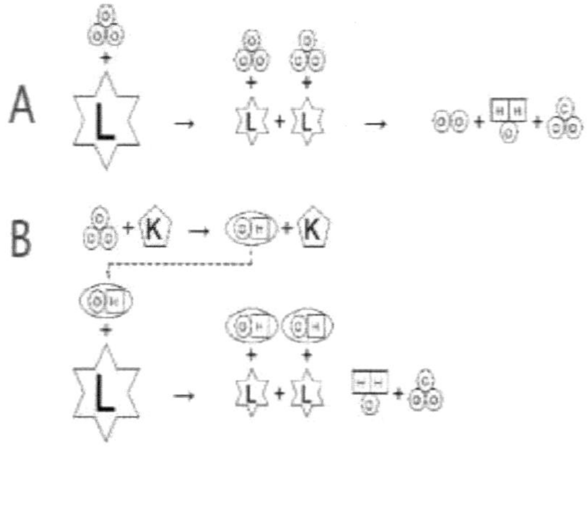

Figure 3. Mechanism of ozonation reactions: (A) directly, (B) catalytic ozonation.

Two dominant oxidants are present in ozonation applications: ozone (O_3) and hydroxyl radicals ($HO^{\cdot}$). Hydroxyl radicals are a highly reactive decomposition product of ozone and represent the main oxidant in AOPs as shown on Figure 3. Examples of AOPs include chemical (O_3–H_2O_2), photo-chemical (ultraviolet (UV)–H_2O_2), and photocatalytic (UV–TiO_2) methods, among others. In all these cases, the reaction generates hydroxyl radicals. One important difference between AOPs and conventional ozonation is that during conventional ozonation, both ozone and hydroxyl radicals can act as oxidants as shown on Figure 4. In AOPs, hydroxyl radicals represent the primary transformation mechanism.

$$O_3 + AB \xrightarrow{k''_{O_3,app,AB}} \text{products}$$
$$HO^{\cdot} + AB \xrightarrow{k''_{HO,app,AB}} \text{products}$$

Figure 4. FLASH Full-Scale WWT.

3.3 Ozonation of micro pollutant

In fact, ozone has been successfully used to treat antibiotics in these water sources, as well as specific industrial wastewaters. [27, 28]. Ozone is an excellent choice for removing antibiotics in water and wastewater supplies. Removing trimethoprim, sulfamethoxazole, clarithromycin, erythromycin, and roxithromycin, among others by ozonation of a number of PPCPs in wastewater matrices [27]. With an ozone dose of 5 mg/l found that 95% removal of a suite of seven antibiotics (carbadox, sulfachloropyradazine, sulfadimethoxine, sulfamerazine, sulfamethazine, sulfathiazole, and trimethoprim) was achieved after just 1.3 min of gaseous ozonation (2% w/w) [29]. Macrolides and sulphonamides were effectively transformed (>95%) in secondary effluent at pH 7 for ozone doses of 2 mg/l. [30]. With an ozone dose of 3 mg/l found that the total mass concentration of 26

IOP Conf. Series: Journal of Physics: Conf. Series **1095** (2018) 012030 doi:10.1088/1742-6596/1095/1/012030

pharmaceuticals, including many antibiotics, went from approximately 3800 ng/l to less than 50 ng/l [31]. All of these studies indicate that antibiotics destroy quickly with ozone.

3.4 Ozone-based AOPs

In 1987, Glaze et al. coined the term, advanced oxidation processes, to describe pro- cesses that generate hydroxyl radicals [32]. To date, a number of ozone- based AOPs have been described, including the following: O_3–$HO^{\cdot}$, O_3–H_2O_2, O_3–UV, O_3–H_2O_2–UV, O_3–catalyst, O_3–activated carbon, and O_3–ultrasound, among others [33–36]. Most ozone-based AOPs exploit the secondary reactant ($HO^{\cdot}$, H2O2, UV, etc.) to promote ozone decomposition.

$$O_3 + HO^- \xrightarrow{k_{i,1}} HO_2^- + O_2$$
$$O_3 + HO_2^- \xrightarrow{k_{i,2}} HO_2^{\bullet} + O_3^{-\bullet}$$
$$HO_2^{\bullet} \xrightarrow{k_1} H^+ + O_2^{-\bullet}$$
$$H^+ + O_2^{-\bullet} \xrightarrow{k_{-1}} HO_2^{\bullet}$$
$$O_3 + O_2^{-\bullet} \xrightarrow{k_2} O_3^{-\bullet} + O_2$$
$$O_3^{-\bullet} + H^+ \xrightarrow{k_3} HO_3^{\bullet}$$
$$HO_3^{\bullet} \xrightarrow{k_{-3}} O_3^{-\bullet} + H^+$$
$$HO_3^{\bullet} \xrightarrow{k_4} HO^{\bullet} + O_2$$
$$O_3 + HO^{\bullet} \xrightarrow{k_5} HO_4^{\bullet}$$
$$HO_4^{\bullet} \xrightarrow{k_5} HO_2^{\bullet} + O_2$$
$$HO_4^{\bullet} + HO_4^{\bullet} \xrightarrow{k_{t,1}} H_2O_2^{\bullet} + 2O_3$$
$$HO_4 + HO_3^{\bullet} \xrightarrow{k_{t,2}} H_2O_2^{\bullet} + O_2 + O_3$$

Figure 5. Mechanism of Ozone Decomposition Reactions.

Ozone decomposition result in the production of a number of reactive species such as superoxide anion radicals ($O_3^{\cdot-}$), peroxy radicals ($HO_2^{\cdot}$), ozone anion radicals ($O_2^{\cdot-}$), hydrogen trioxide radicals ($HO_3^{\cdot}$), hydrogen tetraoxide radicals ($HO_4^{\cdot}$), and hydroxyl radicals ($HO^{\cdot}$) as shown on Figure 5.

Hydroxyl radicals are more reactive than ozone, as indicated by their redox potential, 2800 mV. Hydroxyl radicals are often considered nonselective oxidants because the chemical structure and speciation of target molecules does not significantly affect reactivity. For ozone-based AOPs, the ozone exposure is typically minimal because of rapid ozone decomposition and so the majority of treatment from reaction with hydroxyl radicals.

The ideal molar ratio of ozone to hydrogen peroxide is 2 mol O_3/mol H_2O_2. If the molar ratio of ozone to hydrogen peroxide is greater than 2 mol/mol, excess ozone will be present, resulting in a dual oxidant system that first selects for hydroxyl radicals and then ozone as shown on Figure 6.

$$O_3 + H_2O_2 \longrightarrow HO^{\cdot} + O_2$$
$$H_2O_2 \xrightarrow{h\nu} 2HO^{\cdot}$$

Figure 6. Mechanism of O_3–H_2O_2 Decomposition Reactions.

Other ozone-based AOPs, like O_3–UV, O_3–H_2O_2; however, the capital costs associated with adding hydrogen peroxide and installing UV lamps are prohibitive. AOPs based O_3–catalyst are being increasingly employed [37-44].

Unlike traditional wastewater contaminants like particles, organic carbon, nutrients, and pathogens, these micro contaminants from pharmaceutical and medical industries cannot be easily treated. Therefore, increasing the efficiency of wastewater treatment of contaminants from pharmaceutical and medical industries is in the best interest of worldwide. One approach is adopt of FLASH WWT Catalyst System. By FLAST technology is hopeful that the water treatment pharmaceutical and medical industries will find a solution that enables sustainable stewardship of the environment and maintenance of effective treatment.

3.5 FLASH

Data analysis from thirty-one of parameters which were collected from BUMA such as physical parameters (pH, temperature, total dissolved solids = TDS, total suspended solids = TSS), chemical parameters (COD, BOD, oils and fats, MBAS, Ammonia, iron, Manganese, barium, copper, zinc, Hexavalent chromium, Chromium, cadmium, Mercury, lead, Arsenic, Selenium, nickel, Cobalt, cyanide, sulphide, Fluoride, Chlorine free, nitrate, Phenol, Nitrite), and biological parameters (total coliform). Data analysis showed that there were significant difference between values of water from outlet of 140 Flash Full-Scale waste water treatments and the maximum threshold value. Referring to the removal efficiency of macro-pollutants, the collected data demonstrate good removal efficiency of macro-pollutants using FLASH technologies from 140 sampling. The results of the analysis of the thirty-one at the 140 sampling from outlet FLASH could be concluded that the value of all parameters still below the maximum threshold. 140 FLASH have been applied and gave satisfaction result for 100 % effective meet the quality standard value in accordance with regulation of the Minister of Environment Republic of Indonesia number 5:2014 about waste water quality standards.

4. Conclusion

The FLASH treatment plant involves physical and biological pre-treatment, followed by Advanced Oxidation Processes (AOPs) based on catalytic ozonation and followed by granulated activated carbon (GAC) and powdered activated carbon (PAC) filtration. Two dominant oxidants are present in ozonation applications: ozone (O_3) and hydroxyl radicals (HO,). Hydroxyl radicals are a highly reactive decomposition product of ozone and represent the main oxidant in AOPs. Conventional ozonation, both ozone and hydroxyl radicals can act as oxidants. In AOPs, hydroxyl radicals represent the primary transformation mechanism. In year 2015, 140 FLASH have been applied and gave satisfaction result for 100 % effective meet regulation Ministry of Environment of Republic Indonesia for waste water quality standards Number Number 5/2014. BUMA FLASH treatment is the first develop hospital wastewater treatment in Indonesia based Catalytic Ozonation Based Advanced Oxidation Process.

Acknowledgement

The authors would also like to thank to Hospital and Community Health Centre Facility in Indonesia for using outlet wastewater treatment facilities. The authors are grateful to Sucofindo, BLKPAK, SysLab, BRSI Padang, PT Mutuagung Lestari for their assistance with the analysis.

References

[1] Arslan A, Veli S and Bingöl D 2014 *Sep. Purif. Technol.* **132** 561
[2] Lee Y, Kovalova L and Mcardell S 2014 *Water Res.* **64** 134.

IOP Conf. Series: Journal of Physics: Conf. Series **1095** (2018) 012030 doi:10.1088/1742-6596/1095/1/012030

[3] Hansen K 2016 *Chem. Eng. J.* **290** 507.
[4] Zheng J 2017 *Chem. Eng. J.*
[5] Bethi B, Sonawane S, Bhanvase B and Gumfekar S 2016 *Chem. Eng. Process.*
[6] Mahmoud W, Rastogi T and Kümmerer K 2017 *Curr. Opin. Green Sustain. Chem.*
[7] Méndez E 2016 *J. Environ. Sci. Heal.*
[8] Uhrbrand K, Schultz A, Koivisto A, Nielsen U and Madsen A 2017 *Water Res.* **112** 110.
[9] Carbajo J 2016 *Chem. Eng. J.* **283** 740.
[10] Verlicchi P and Zambello E 2014 *Sci. Total Environ.* **470**
[11] Serpone N, Artemev Y, Ryabchuk V and Emeline A 2017 *Curr. Opin. Green Sustain. Chem.*
[12] Fast S, Gude V, Truax D, Martin J and Magbanua B 2017. *Process.*
[13] Peres J, Ovejero J and García J 2017 *Chem. Eng. J.*
[14] Venditti S, Klepiszewski K and Christian K 2017 Lessons Learned from European Experiences and Presentation of Case Studies
[15] AndreadakisA, Mamais D, Gavalakis A, Noutsopoulos C, Kouris N and Nikitopoulos G 2010 *Int. J. Environ. Waste. Manag.* **5** 392.
[16] Rosal R, Rodr´ıguez A, Perdigo A, Petre A and Garc´ıa E 2009 *Chem. Eng. J.* **149** 311
[17] Matilainen A and Sillanpa M 2010 *Chemosphere* **4** 351
[18] Antoniou M, Hey G, Rodr´ıguez S, Spiliotopoulou A, Fick J and Tysklind M 2013 *Sci. Total Environ.*
[19] Hollender J, Zimmermann G, Koepke S, Krauss M, McArdell S and Ort C 2009 *Environ. Sci. Technol.* **43** 7862.
[20] Davies J and Davies D 2010 Origins and evolution of antibiotic resistance *Microbiol Mol. Biology Rev.* **74** pp 417.
[21] Auerbach E, Seyfried E and McMahon D 2007 *Water Res.* **41** 1143.
[22] Ternes T 1998 *Water Res.* **32** 3245.
[23] Ternes T, Stuber J, Herrmann N, McDowell D, Ried A and Kampmann M 2003 *Water Res.* **37** 1976.
[24] Liu J, Wong H 2013 *Environ. Int.* **59** 208.
[25] Herna´ndez-Leal L, Temmink H, Zeeman G and Buisman N 2011 *Water Res* **45** 2887.
[26] Gao P, Munir M and Xagoraraki I 2012 *Total Environ.* **4** 73.
[27] Szczepanowski R, Linke B, Krahn I, Gartemann H, Gotzkow T and Eichler W 2009 *Microbiology* **7** 2306.
[28] Acero J and Gunten U 2001 *J. Am. Water Works Assoc.* **10** 90.
[29] M. S. Elovitz M, Gunten U and Kaiser P 2000 *Ozone Sci. Eng.* **2** 123.
[30] Rice R 1996 Applications of ozone for industrial wastewater treatment - a review *Ozone Sci. Eng.* **6** pp 477.
[31] Adams C, Wang Y, Loftin K and Meyer M 2002 *J. Environ. Eng.* **3** 253.
[32] Huber M, Gabel A, Joss A, Hermann N, Laffler D and McArdell S 2005 *Environ. Sci. Technol.* **11** 4290.
[33] Okuda T, Kobayashi Y, Nagao R, Yamashita N, Tanaka H and Tanaka S 2008 *Water Sci. Technol.* **1** 65
[34] Bin A and Sobera-Madej S 2012 *Ozone Sci. Eng.* **2** 136.
[35] Wang Y, Zhang H, Chen L, Wang S and Zhang D 2012 *Sep. Purif. Technol.* **84**
[36] Rivera-Utrilla J, Sa´nchez-Polo M, Prados-Joya G, Ferro-Garc´ıa A and Bautista-Toledo I 2010 *J. Hazard Mater* **174** 880
[37] Li C, Jiang F, Sun D, Qiu B 2017 *Chem. Eng. J.* **325** 325
[38] Chen W, Li X, Pan Z, Ma S, Li L 2016 **304** 594.
[39] Sable S *Applied Catal. B, Environ.* **209** 523.
[40] Sui M, Xing L, Sheng L, Huang S and Guo H *J. Hazard. Mater.* **227** 227.
[41] Mashayekh-salehi A, Moussavi G and Yaghmaeian K 2016 *Chem. Eng. J.*
[42] Wang J and Bai Z 2016 *Chem. Eng. J.*

[43] Yaghmaeian K, Moussavi G, Mashayekh-salehi A, Mohseni-bandpei A and Satari M 2017 *Process Saf. Environ. Prot.*
[44] Gao G, Shen J, Chu W, Chen Z and Yuan L 2017 *Sep. Purif. Technol.* **173** 55–62

IOP Conf. Series: Journal of Physics: Conf. Series **1095** (2018) 012031 doi:10.1088/1742-6596/1095/1/012031

Cytotoxic activity of quinolinone Alkaloids and acylphloroglucinol from the leaves of *Melicope denhamii*

Ratih Dewi Saputri[1], Mulyadi Tanjung [1], and Tjitjik Srie Tjahjandarie[1]

[1]Natural Products Chemistry Research Group, Organic Chemistry Division, Department of Chemistry, Faculty of Science and Technology, Universitas Airlangga, Surabaya, Indonesia.

duffputri@gmail.com; tjitjiktjahjandarie@fst.unair.ac.id

Abstract. Two quinolinone alkaloids, N-methylflindersine (**1**), flindersine (**2**) and two acylphloroglucinols, evodionol (**3**), methylevodionol (**4**) were isolated from the leaves of *Melicope denhamii*. Structures of compounds were elucidated using spectroscopic methods such as UV, IR, HRESIMS, 1D and 2D NMR. Compounds **1-4** were evaluated for their cytotoxicity against P-388 cells, compound **2** showed moderate activity with IC_{50} 4.86 μg/ml.

1. Introduction

Melicope locally known as 'Ki Sampang' belongs to the Rutaceae family, consisting of about 280 species where widely distributed in Asia, Australia, Africa and Polynesia. Phytochemical studies have shown that the species produce a variety of alkaloids [1-3], flavonoids [4], coumarins [5], and acylphloroglucinols [6], which exhibit various biological activities including antioxidant, anticancer, and antiinflammatory. In continuation of our research for investigation of alkaloid and acylphloroglucinol of Indonesian *Melicope* plants, we report the isolation of N-methylflindersine (**1**), flindersine (**2**) and two acylphloroglucinols, evodionol (**3**), methylevodionol (**4**) from the methanol extract of the leaves of *Melicope denhamii*. The chemical structure of compounds **1-4** were established by UV, IR, HRESIMS, 1D and 2D NMR, as well as by comparison with those related compounds previously reported. The cytotoxic activity against murine leukemia P-388 cells of isolated compounds from this species are also briefly described.

2. Methods

2.1 General experimental

Column chromatography and planar radial chromatography were carried out using silica gel 60 and silica gel 60 PF_{254} (Merck, Darmstadt, Germany). NMR spectra were measured on a JEOL JNM-ECA 400 MHz FTNMR spectrophotometer (Tokyo, Japan) in $CDCl_3$ with TMS as the internal standard. Mass spectra were measured on an ESI-TOF Waters LCT Premier XE producing pseudo-molecular ions, $[M+H]^+$ positive ion mode (Santa Clara, CA, USA). UV spectra were recorded in MeOH on a Shimadzu series 1800 UV-VIS spectrophotometer (Kyoto, Japan). IR spectra were recorded in KBr on a One Perkin Elmer instrument (Waltham, MA, USA).

2.2 Plant Material
The leaves of *M.denhamii* was collected from the conserved forest of Gunung Salak, Bogor, West Java, Indonesia on November 2016 and was identified by Mr. Ismail Rachman from the Herbarium Bogoriense, Bogor, Center of Biological Research and Development, National Institute of Science, Bogor, Indonesia.

2.3 Extraction and isolation

The air-dried leaves of *M.denhamii* (3.5 kg) was macerated with methanol at room temperature for 24 h two times and then evaporated under reduced pressure to give a dark brown residue (250 g). The extract was redissolved in MeOH-water (9:1) and partitioned with *n*-hexane (201 g) and ethyl acetate (73 g) fractions. A part of ethyl acetate fraction (70 g) was subjected to vacum liquid chromatography over silica gel and eluted with *n*-hexane-ethyl acetate (from 9:1 to 1:9) to give fractions A-C. Fraction A (5.14 g) was added to column chromatography and eluted with *n*-hexane-ethyl acetate (from 9:1 to 7:3) to produce subfractions A_1-A_2. Subfraction A_2 was purified by planar radial chromatography using *n*-hexane-chloroform (from 9:1 to 7:3) to yielded compound **3** (68 mg) and **4** (57 mg). Fraction C (625 mg) was refractionated using column chromatography and eluted *n*-hexane-ethyl acetate (from 9:1 to 7:3) to yielded subfractions C_1-C_2. Subfraction C_1 was purified by planar radial chromatography using *n*-hexane-chloroform (from 9:1 to 1:1) to yielded compound **1** (50 mg) and **2** (10 mg).

2.4 Cytotoxic assay

Cytotoxic properties of the isolated compounds **1-4** against murine leukemia P-388 cells was evaluated according to the MTT method as previously described [10-12]. Artonin E was used as the positive control.

3. Result and Discussion

Phytochemical study on the MeOH extract from the leaves of *M. denhamii* yielded N-methylflindersine (**1**), methylflindersine (**2**), evodionol (**3**), and methyl-evodionol (**4**).

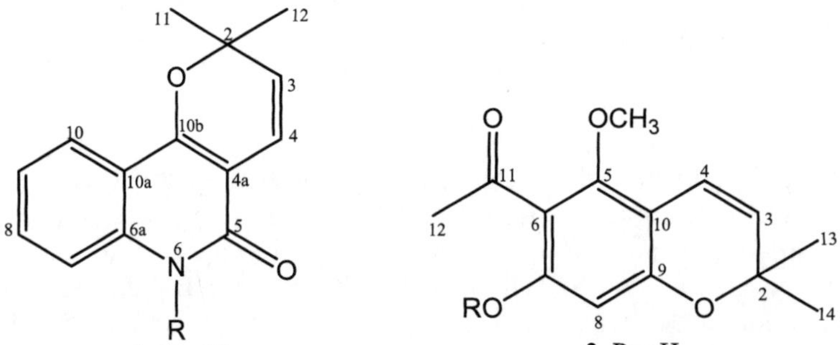

1. R = H
2. R = CH_3

3. R = H
4. R = CH_3

Figure 1. Compounds **1-4** isolated from the leaves of *M. denhamii.*

N-Methylflindersine (**1**), was isolated as a yellow solid, m.p. 84-86°C, showed a quasimolecular ion $[M+H]^+$ at *m/z* 242.1180 corresponding to the molecular formula $C_{15}H_{15}NO_2$. The UV maximum absorption at λ_{max} 226 (4.29), 285 (3.20), 333 (3.65), 348 (3.69) and 365 (3.52) nm typical for a quinolinone skeleton [1]. The IR spectrum of **1** indicated absorptions for conjugated carbonyl (1641 cm^{-1}), aromatic (1581 and 1411 cm^{-1}) and ether (1188 cm^{-1}) groups, respectively. The ^{1}H NMR spectrum of **1** (Table 1) showed the presence of four aromatic proton signals corresponding to 1,2-disubstituted benzena ring [δ_H 7.95 (1H, *dd, J* = 7.8; 1.5 Hz), 7.53 (1H, *m*), 7.30 (1H, *d, J* = 8.5 Hz), 7.21 (1H, *t,J* = 7.8 Hz)], three proton signals of 2,2-dimethylpyrano group [δ_H 6.74 (1H, *d,J* = 10.0 Hz), 5.52 (1H, *d,J* = 10.0 Hz), 1.50 (6H, *s* Hz)], and proton singlet signal of N-methyl at δ_H 3.68. The ^{13}C NMR spectrum of **1** showed 15 carbon signals and their assignments were determined by HMQC and HMBC spectra. The ^{13}C NMR spectrum of **1** corresponding to the quinolinone skeleton. The structure of **1** were confirmed by HMQC and HMBC spectra (Figure 2). The 1D and 2D NMR spectra data are consistent with published data [7].

Flindersine (**2**), was isolated as a yellow solid, m.p. 197-199°C, showed a quasimolecular ion $[M+H]^+$ at *m/z* 229.1180 corresponding to the molecular formula $C_{14}H_{13}NO_2$. The UV spectrum (λ_{max} 222, 282, 331, 347 and 363 nm) and IR spectrum (1651, 1600, 1461 and 1191 cm^{-1}) absorptions were similarity to those of **1**. The ^{1}H NMR spectrum of **2** (Table 1) showed the presence of aromatic proton [δ_H 7.89 (1H, *d, J* = 8.0 Hz), 7.48 (1H, *t,J* = 7.6 Hz), 7.30 (1H, *d,J* = 8.2 Hz), 7.19 (1H, *t,J* = 7.6 Hz)], and proton signals of 2,2-dimethylpyrano group [δ_H 6.76 (1H, *d,J* = 10.0 Hz), 5.56 (1H, *d,J* = 10.0 Hz), 1.54 (6H, *s* Hz)]. The ^{1}H NMR spectrum of **2** were very similar to those of **1** but the has demethylized at N-methyl. The ^{13}C NMR spectrum of **2** showed 14 carbon signals and the structure of **2** were confirmed by HMQC and HMBC spectra (Figure 2). The 1D and 2D NMR spectra data are consistent with published data [8].

Table-1. NMR spectroscopic data of compounds **1-2** in $CDCl_3$.

	N-Methylflindersine (1)			**Flindersine (2)**		
No.	**δ_H (mult, *J* in Hz)**	**δ_C**	**HMBC**	**δ_H (mult, *J* in Hz)**	**δ_C**	**HMBC**
2	-	78.6	-	-	79.2	-
3	5.52 (*d*, 9.8)	126.2	C-2; C-4a	5.56 (*d*, 10,0)	126.3	C-2; C-4a; C-11; C-12
4	6.74 (*d*, 9.8)	117.8	C-2; C-10b	6.76 (*d*, 10,0)	117.2	C-2; C-5; C-10b
4a	-	105.6	-	-	105.8	-
5	-	160.9	-	-	162.3	-
6a	-	139.2	-	-	137.8	-
7	7.29 (*d*, 8.5)	113.9	C-9, C-10a	7,30 (*d*, 8.2)	115.9	C-9, C-10a
8	7.52 (*t*, 7.8)	130.8	C-6a, C-10	7,48 (*t*, 7.6)	130.9	C-6a, C-10
9	7.21 (*t*, 7.8)	121.6	C-7, C-10a	7.19 (*t*, 7.6)	122.2	C-7
10	7.95 (*d*, 7.8)	123.0	C-6a, C-8	7.89 (*d*, 8.0)	122.7	C-6a, C-8, C-10b

IOP Conf. Series: Journal of Physics: Conf. Series **1095** (2018) 012031 doi:10.1088/1742-6596/1095/1/012031

10a	-	115.9	-	-	115.4	-
10b	-	155.1	-	-	157.3	-
11	1.50 (*s*)	28.1	C-2. C-3; C-12	1.54 (*s*)	28.4	C-2. C-3; C-12
12	1.50 (*s*)	28.1	C-2. C-3; C-11	1.54 (*s*)	28.4	C-2. C-3; C-11
N-CH_3	3.68 (*s*)	29.2	C-5, C-6a	-	-	-

Evodionol (**3**), was isolated as a yellow solid, the UV maximum absorption at λ_{max}; 217 (3.72), 226 (3.64) , 272 (4.21), 295 (3.76), 306 (3.70), and 349 (3.13) nm characteristic for a acylphloroglucinol structure [6]. The ^{1}H NMR spectrum of **3** (Table 2) showed the presence of a singlet signal of isolated aromatic proton at δ_H 5.88 and a singlet of methyl of acetyl at δ_H 2.59 typical of an acylphloroglucinol with five substituents [7]. The ^{1}H NMR spectrum of **3** also showed the presence of a chelated hydroxyl at δ_H 14.29, a singlet of methoxyl at δ_H 3.85, and three proton signals of 2,2-dimethylpyrano group [δ_H 6.64 (1H, *d*, *J* = 10.0 Hz), 5.44 (1H, *d*,*J* = 10.0 Hz), 1.44 (6H, *s* Hz)]. The ^{13}C NMR spectrum of **3** (Table 2) showed 14 carbon signals and confirmed by HMQC and HMBC spectra (Figure 2).

Methylevodionol (**4**), was isolated as a yellow oil, showed a quasimolecular ion $[M+H]^+$ at *m/z* 263.1283 corresponding to the molecular formula $C_{15}H_{18}O_4$. The UV maximum absorption at λ_{max}; 237 (3.80), 288 (4.30) , 297 (4.31), and 340 (3.77) nm characteristic for a acylphloroglucinol structure [6]. The IR spectrum of **1** indicated absorptions for conjugated carbonyl (1622 cm^{-1}), and aromatic (1598 and 1439 cm^{-1}) and ether (1185 cm^{-1}) groups, respectively [9].

The ^{1}H NMR spectrum of **4** were very similar to those of **3** (see Table 2). The major difference, compound **4** showed a methoxy group at C-8. The ^{13}C NMR spectrum of **2** showed 14 carbon signals and the structure of **4** were confirmed by HMQC and HMBC spectra (Figure 2). The 1D and 2D NMR spectra data are consistent with published data [6].

Table-2. NMR spectroscopic data of compounds **3-4** in $CDCl_3$.

	Evodionol (3)			Methylevodionol (4)		
No.	**δ_H (mult, *J* in Hz)**	**δ_C**	**HMBC**	**δ_H (mult, *J* in Hz)**	**δ_C**	**HMBC**
2	-	78.2	-	-	76.9	-
3	5.44 (*d*, 10.0)	125.4	C-2; C-5	5.51 (*d*, 10.0)	127.8	C-2; C-5; C-13/14
4	6.64 (*d*, 10.0)	116.1	C-2; C-10	6.46 (*d*, 10.0)	116.5	C-2; C-5; C-6; C-10
5	-	102.8	-	-	108.0	-
6	-	161.9	-	-	154.2	-
7	-	105.7	-	-	118.3	-

8	-	163.1		-	157.6	-
9	5.88 (*s*)	91.1	C-5; C-7; C-8	6.18 (*s*)	96.1	C-5; C-7; C-8, C-10
10	-	160.2	-	-	156.0	-
11	-	203.3	-	-	201.9	-
12	2.59 (*s*)	33.1	C-11	2.48 (*s*)	32.6	C-11
13	1.44 (*s*)	28.4	C-2; C-3; C-14	1.41 (*s*)	27.9	C-2. C-3; C-14
14	1.44 (*s*)	28.4	C-2; C-3; C-13	1.41 (*s*)	27.9	C-2. C-3; C-13
6-OCH_3	3.85 (*s*)	55.7	C-6	3.74 (*s*)	55.8	C-6
8-OCH_3	-	-	-	3.76 (*s*)	63.6	C-8
8-OH	14.29 (*s*)	-	C-7; C-8; C-9	-	-	-

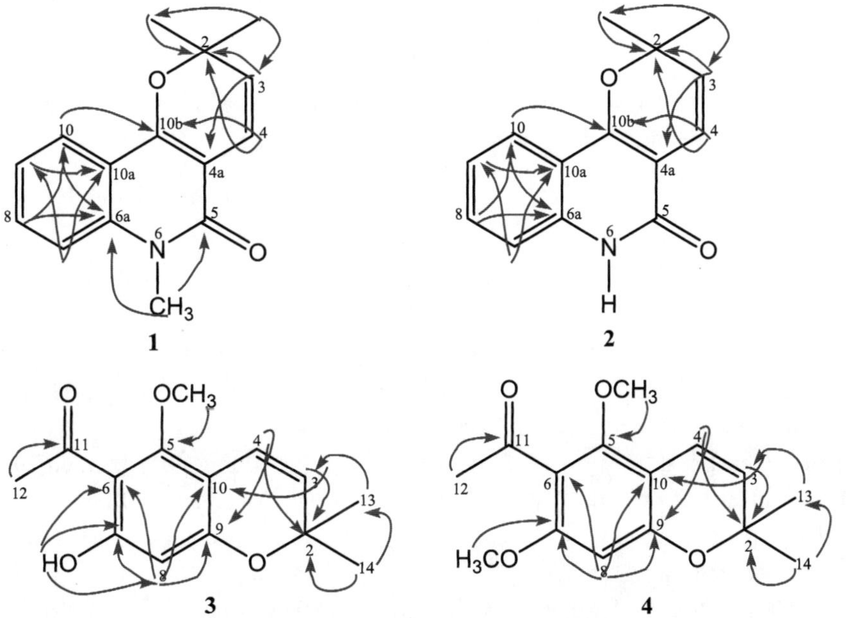

Figure 2. Selected HMBC correlations for compounds **1-4**

The cytotoxic activity (Table 3) of compounds **1-4** were evaluated for their cytotoxicity by MTT assay against murine leukemia P-388. Artonin E was used as the positive control. Those cytotoxic data for alkaloids suggested that compound **1** was inactive and compound **2** has moderate activity. Methylation at nitrogen on compound **1** can be decreased activity. On the other hand, acylphloroglucinols, compound **4** more active than **3.** In this case, methylation of compound **4** enhanced the activity.

Table-3. Cytotoxic activity data of compounds **1-4** by MTT assay.

No	Compound	IC_{50} (μg/mL)
1	N-methylflindersine (**1**)	21.06 ± 0.85
2	Flindersine (**2**)	4.86 ± 0.30
3	Evodionol (**3**)	46.46 ± 1.25
4	Methylevodionol (**4**)	11.98 ± 0.65
5	Artonin E	1.33 ± 0.18

4. Conclusions

The phytochemical investigation of the leaves of *M. denhamii* to yielded compounds, N-methylflindersine (**1**), flindersine (**2**), evodionol (**3**), methylevodionol (**4**). Flindersine (**2**) showed moderate activity against murine leukemia P-388 cells.

Acknowledgments

This research was supported by Directorate of Higher Education, Ministry of National Education, Republic of Indonesia (Penelitian Hibah Mandat, Universitas Airlangga, 2017).

References

[1] Tanjung M, Saputri R D, Wahjoedi R A, and Tjahjandarie T S 2017 *Molbank*. **M939**.
[2] Nakashima K, Oyama M, Ito T, Witono J R, Darnaedi D, Tanaka T, Murata T, and Iinuma M 2012 *Tetrahedron. Lett.* **52** 4694.
[3] Chen J-J, Chang Y-L, Teng C-M, Su C-C, and Chen I-S 2002 *Planta Med*. **68** 790.
[4] Sultana N, Hartley T G, and Waterman P G 1999 *Phytochemistry*. **50** 1249.
[5] Kassim N K, Rahmani M, Ismail A, Sukari M A, Ee G C L, Nasir N M, and Awang K 2013 *Food. Chem.* **139** 87.
[6] Kamperdick C, Van N H, Sung T V and Adam G 1997 *Phytochem.* **45**(5) 1049.
[7] Kamperdick C, Van N H, Sung T V, and Adam G 1999 *Phytochem.* **50** 177.
[8] Munoz M A, Torres R, and Cassels B K 1982 *J. Nat. Prod.* **45**(3), 367.
[9] Tjahjandarie T S, Pudjiastuti P, Saputri R D, and Tanjung M 2014 *J. Chem. Pharm. Res.* **6** 786.
[10] Tanjung M, Rachmadiarti F, Saputri R D, and Tjahjandarie T S 2018 *Nat. Prod. Res.* **32** 1062.
[11] Tanjung M, Hakim E H, and Syah Y M 2017 *Chem Nat Compd.* **53**, 215.
[12] Tanjung M, Saputri R D, and Tjahjandarie, T S 2016 *Molbank.* **3** M906.

Isolation of β-Chitosan from Squid Bone as Raw Material to Synthesize of Hybrid Photocatalysts TiO_2-Chitosan

Risfidian Mohadi [1, a)], Nurlisa Hidayati[1, a], and Aldes Lesbani [2, c)]

[1] Department of Chemistry, Faculty of Mathematics and Natural Sciences, Sriwijaya University. Jl. Raya Palembang Prabumulih Km 32 Ogan Ilir 30662, South Sumatera, Indonesia
[2] Integrated Research Laboratory, Graduate School, Sriwijaya University, Jl. Padang Selasa, Bukit Besar Palembang 30139, South Sumatera, Indonesia

[a)] Corresponding author: risfidian.mohadi@unsri.ac.id
[b)]hidayatinurlisa@yahoo.com
[c)]aldeslesbani@pps.unsri.ac.id

Abstract. Synthesis of hybrid photocatalyst of TiO_2-chitosan has been done. Chitosan as a raw material was chitosan-β obtained from chitosan isolation from squid bone. Chitosan from squid bone has a soft structure and very little to contain minerals and other impurities so easily obtained. Hybrid photocatalyst TiO_2-chitosan was synthesized by a sol-gel method in which the incorporation of chitosan as a supporting material and TiO_2 having a function as a photocatalytic agent was dispersed on the β-chitosan surface. The hybrid of TiO_2-chitosan photocatalysts was characterized by FTIR and SEM/EDX shows the functional groups and surface morphologies of hybrid photocatalyst TiO_2-chitosan covered by uniformly distributed TiO_2 nanoparticles. The FTIR spectrum showed the O-Ti-O absorption bands at the wavenumber of 678.98 cm^{-1} and the typical absorption of chitosan the -OH at the wavenumbers of 3425.58 cm^{-1}, 3834.49 cm^{-1}, and 3873.06 cm^{-1}, respectively.

Keywords: chitosan-β, photocatalyst, TiO_2, sol-gel method, nanoparticles

1. Introduction

Chitosan is a natural polymer that has a structure similar to cellulose and can be formed into thin films. Chitosan that has the properties of *nontoxic, biodegradable,* and *biocompatible* resulting from the deacetylation of chitin by using high temperature and high concentrated alkali solvents [1,2]. Chitosan has the structure of poly β-(1,4)-2-acetamide-2'-deoxy-D-glucose. Chitosan is specific because it has a reactive amine group on C-2 and primary and secondary hydroxyl groups in C-3 and C-6 which cause chitosan to have high chemical reactivity [3,4,5]. The structure of chitin and chitosan have the difference lies in the comparison of the amine group ($-NH_2$) with an acetyl group ($-CH_3CO$) called the degree of deacetylation [6]. The degrees of deacetylation depending on the raw material and the method used in the preparation of chitosan. The degree of deacetylation is a quality parameter of chitosan which shows the percentage of acetyl groups which can be removed from the chitin rendemen. The release of acetyl groups on chitin causes the positively charged chitosan, to able to bind negatively charged compounds or molecule. Methods for determining the acetyl groups released from chitin include potentiometric titration, ninhydrin test, NMR, titration of hydrogen bromide, IR, and UV-Vis spectrophotometry [7,8]. XRD analysis shows that chitosan has 3 structures, that is structure α, β, and γ. Molecule α-Chitosan has a very dense antiparallel bond because it is stabilized by its intermolecular hydrogen bond. In contrast to β and γ chitosan, where the hydrogen bond in the water increases. The high level of hydration and decreased attachment causes the β and γ chitosan structure to be more flexible and reactive. Chitosan with α structure is mostly found in animal shells of *crustaceans,* while β and γ chitosan can be found in squid bones and fungi, respectively.

Chitosan is insoluble in water but is soluble in most organic solutions, such as acetic acid, formic acid, and lactic acid, and has a finite solubility in phosphoric acid and is insoluble in sulfuric acid. The commonly used chitosan solvent is an acetic acid with a concentration of 1-2% [9]. Photocatalyst technology used for processing and degraded of wastewater contaminant, efficiently and can be done by using titanium dioxide (TiO_2) through the process photo-oxide the dye into pollutants compounds in the water. The Titanium dioxide (TiO_2) is a semiconductor material that has a relatively high photocatalytic activity. In this study, TiO_2 will be dispersed into chitosan which acts as a *supporting material,* wherein the chitosan-TiO_2 formed to combine the functions of chitosan as an adsorbent and TiO_2 that has high photocatalytic activity [10,11].

2. Experiment

Materials used in this research were squid bone, hydrochloric acid, sodium hydroxide, nitrogen gas, and distilled water, while instrumentation applied for analysis were spectrophotometer FTIR Shimadzu Prestige-21, and SEM-EDX JEOL JED-2300.

2.1. Preparation of Chitosan from Squid Bone

Fresh squid bone washed and dried in the sunlight for about four days. Once the squid bone dry milled using a mortal and sieved to pass the 80 *mesh* size. Amount of 50 g of 80 mesh squid bone put into a 500 mL beaker glass, then adding 1 M HCl solution with a ratio of 1:10 (w/v) for demineralization process. The mixture is stirred with a magnetic stirrer at room temperature for 3 hours and then filtered through filter paper while continuously washed with distilled water until no residual chloride ion remains in the samples. The washing process is stopped if no turbid solution formed when the filtrate drops with an $AgNO_3$ solution. Residue from demineralization put into a 500 mL beaker glass and added with 1 M NaOH solution at a ratio of 1:10 (w/v). The mixture stirred and heated at 60 °C on a hotplate stirrer for 1 hour and then filtered with Whatman 41µm filter paper. The residue was found on the filter paper was washed with distilled water until the residue namely chitosan in neutral pH. Chitosan was dried in an oven at a temperature of 70 °C until dry to constant weight. Chitosan has been obtained then characterized by FT-IR spectrophotometer to identify the functional groups of chitosan, and determine the degree of deacetylation product by base line method.

The FT-IR spectrum analysis for chitosan does area of functional groups and the fingerprint region with a frequency of 4000 cm^{-1} - 400 cm^{-1}. Deacetylation degree of chitosan determined by *base line* by FT-IR spectra, with the formula:

$$DD = 100 - \left[\left(\frac{A_{1655}}{A_{3450}}\right) x \ \frac{100}{1.33}\right] \quad (1)$$

Where, A_{1655} showed absorption at amide band, A_{3450} showed absorption at hydroxyl and a factor of 1.33 shows the value of the ratio A_{1655}/A_{3450} to completely deacetylation degrees of chitosan.

2.2. Synthesis of Hybrid Photocatalysts TiO_2-Chitosan from Squid Bone

Amount of 10 g of chitosan from squid bone added into 100 mL of 0.1 M HCl solution in 250 mL beaker glass, while continuously stirring until it dissolved evenly and the TiO2 particle was added with a ratio of 1:1 (w/w). The solution stirred to homogenize for 20 minutes then filtered through Whatman 41µm filter paper, then the precipitate obtained is dried in an oven at a temperature of 50°C about 4 h in the form of TiO_2-chitosan gel. The product of TiO_2-chitosan gel washed with distilled water until the pH of water residue is neutral, then dried in the oven at a temperature of 70 °C for 12 h to obtain a hybrid of TiO_2-chitosan. The hybrid TiO_2-Chitosan characterized by a FT-IR spectrophotometer to see functional groups and the elemental contained by SEM-EDX analysis.

3. Results and Discussion

Squid bone powder (80 mesh) that has been prepared was added 1 M HCl to remove mineral content, and continue to the second stage is deproteinize, which in this process aims to break the protein bonds and other crosslinked intramolecular in squid bone chitin using 1M NaOH. The yield produced at this stage is about 65.6% which estimated that the protein content in squid bone around 30-40% [12],13. The deacetylation process of chitin to chitosan by removed of the acetyl group (CH_3CO^-) by using a concentrated solution of NaOH 50% (v/v). The process of deacetylation of chitin takes place under alkaline conditions because the N-acetyl group is not removed with an acid reagent without polysaccharides [14]. Thus, in this process the yield obtained from squid bone about 80.15% (w/w) with the texture of chitosan in shaped hydrogel [15].

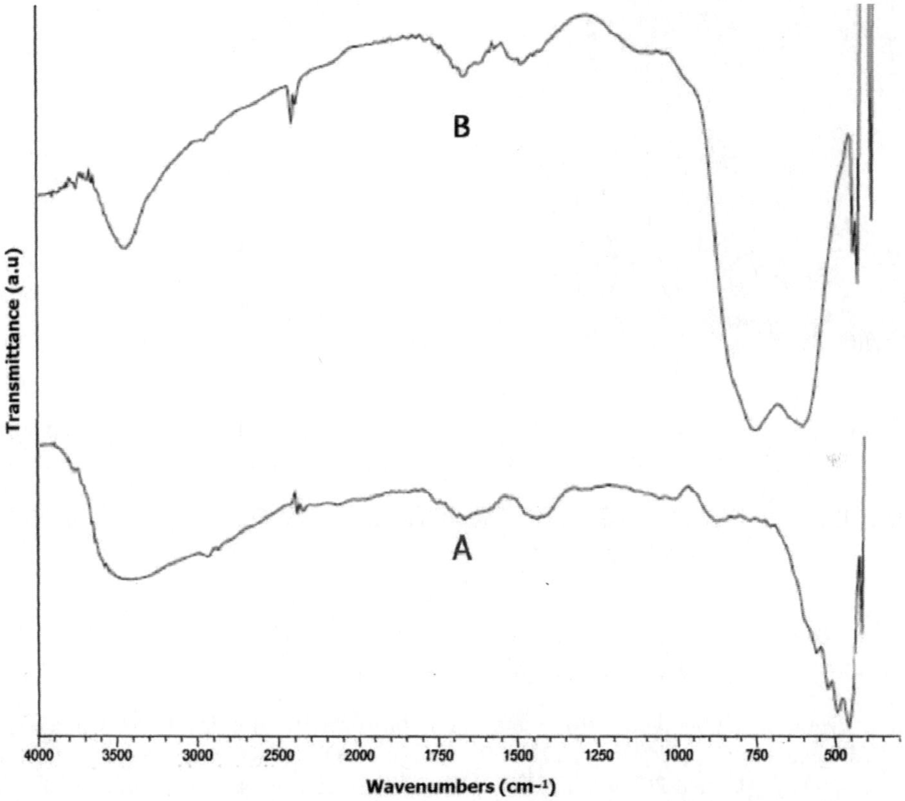

Figure 1. FT-IR spectrum of chitosan (A), and TiO_2-Chitosan (B)

Based on the FT-IR spectra on **Figure 1**. (A) and the calculation using equation (1) the degree of deacetylation by methods *base line* on a FT-IR spectrum obtained the degree of deacetylation of chitosan from squid bone ranged around ± 93.6%. The degree of deacetylation levels off at relatively high values while normal commercial chitosan has of ± 80% deacetylation. **Figure 1**. was confirmed the characterizations of chitosan (A), and a hybrid of TiO_2-chitosan (B), the presence of carboxyl and

IOP Conf. Series: Journal of Physics: Conf. Series **1095** (2018) 012032 doi:10.1088/1742-6596/1095/1/012032

hydroxyl groups [16]. The peak at (3380 to 3480) cm^{-1} is ν(O-H) and $\nu(NH_2)$; the peak at (2920 to 2940) cm^{-1} is from $\nu(CH_3$, CH_2, CH, and OH); the peak at (1630 to 1650) cm^{-1} is from ν(C=O) in low absorbances indicate that the C=O group of chitin have had reduced in the form of chitosan [17]. The adsorption band of O-Ti-O for hybrid of TiO_2-chitosan from squid bone appear at (524 to 680) cm^{-1} showed in **Figure 1**. (B) with presence of ν(O-H) and $\nu(NH_2)$ stretching vibration overlap at 3425 cm^{-1} indicate that that functional group is still active as adsorbent in the form of hybrid of TiO_2-chitosan.
The hybrid of TiO_2-chitosan surface morphology characterized by SEM and the elemental analyses performed by EDX. According to the SEM photograph in **Figure 2.** it is seen that the TiO_2 particles have had evenly distributed over the surface of chitosan in the hybrid of TiO_2-chitosan with particle size almost in nano size. The surface of hybrid of TiO_2-chitosan mass percentage had dominated by TiO_2 particle while the N atom from the aminne group (NH_2) of chitosan confirmed as the active site of the hybrid agrees to the FTIR spectra in **Figure 1**. and the detail of elementals percentage on the hybrid of TiO_2-chitosan given in **Table 1.**

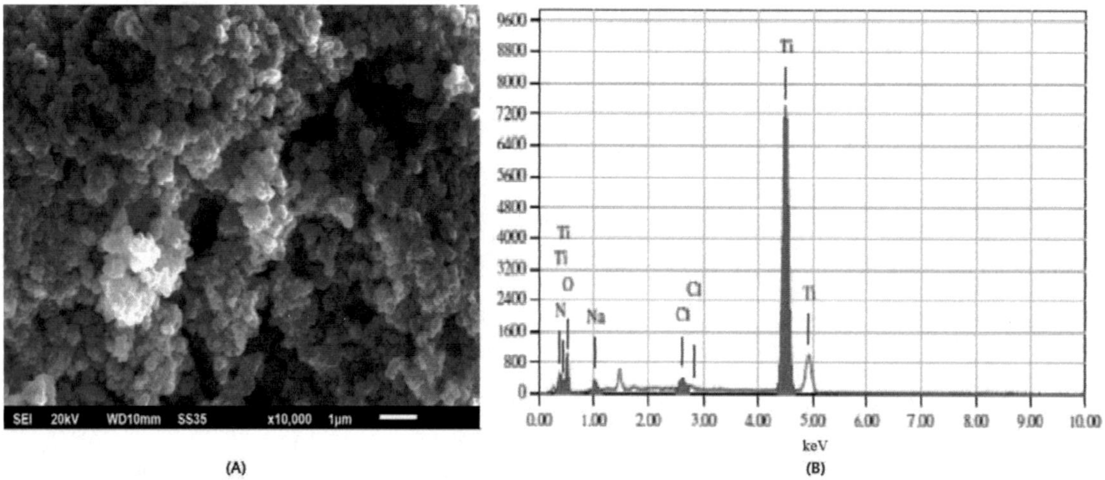

Figure 2. *Scanning Electron Microscope* (SEM) photograph of hybrid of TiO_2-chitosan from squid bone with (A) 10.000x and (B) spectrum of *Energy Dispersive X-ray* (EDX)

Table 1. The percentage of elements and oxides on hybrid of TiO_2-chitosan from squid bone

Elements	% Mass	% Atom	Oxides
O	44.20	62.10	TiO_2
Ti	43.66	20.49	NO
N	9.36	15.03	

The TiO_2 on a hybrid of TiO_2-chitosan from squid bone has 62.10% in oxide form, while the N atom contributes 15.03%, and 20.49% of NO, respectively. That mean 15.03% of nitrogen atom from chitosan exist as aminne group (NH_2) and 20.49% also in the form of oxide. The in oxide form (NO) of chitosan has been by oxidation process when amine direct contacted to the TiO_2 particle on chitosan surface to form of hybrid of TiO_2-chitosan.

4.Conclusions

The product of β chitosan isolated from squid bone gain about 80.15% (w/w) yield. The degree of deacetylation of the β chitosan determines by *base line* methods on a chitosan FT-IR spectra results about ± 93.6% degree of deacetylation. TiO_2-chitosan from squid bone was used as raw material to synthesis of hybrid TiO_2-chitosan. Base on FT-IR spectra and SEM-EDX analysis the absorbances appear at (524 to 680) cm^{-1} and ν(O-H) and $\nu(NH_2)$ stretching vibration overlap at 3425 cm^{-1} indicate that those functional group of NH_2 from β chitosan is still active as adsorbent in the form of hybrid of TiO_2-chitosan even though TiO_2 particle were distribute evenly on the surface of chitosan.

5. Acknowledgments

This research supported from Hibah Kompetitif of Sriwijaya University 2017 through PNBP founding with contract number 1012/UN9.3.1/PP/2017. The authors wish to acknowledge gratefully to all research group members of Inorganic Laboratory Dept. of Chemistry, Faculty of Mathematics and Natural Sciences, Sriwijaya University.

6. References

[1] Chandumpai, A., Narongsak S., Damrongsak F., and Prasait S., *Carbohydr. Polym.* 58, 467-474 (2004).
[2] Modrzejewska Z., Maniukiewicz W., Wojtasz-Pająk A., *Polish Chitin Soc.*, Monograph XI, 113–121 (2006).
[3] Jolanta K., Małgorzata C., Zbigniew K., Anna B., Krzysztof B., Jorg T., and Piotr S., *Mar Drugs.* 8(5):1567–1636 (2010).
[4] Cuong H., Minh N., Van Hoa N., Trung T., *Int J Biol Macromol.* 93:442-447 (2016).
[5] Alessandro I., Matteo Di G., Simona F., Chiara S., Marianna B., Francesco V., Graziella P., Fabio B., Francesco Z., Matteo C., and Giuseppe F., *Mar Drugs.* 12: 5979–5992 (2014).
[6] Islem Y., and Marguerite R. *Mar. Drugs*, 13, 1133-1174 (2015)
[7] Muzzarelli R.A.A. *Mar. Drugs.* 9:1510–1533 (2011).
[8] Kurita K., Kaji Y., Mori T., Nishiymama Y. *Carbohydr. Polym.* 42:19–21 (2000).
[9] Youn D.K., No H.K., Prinyawiwatkul W. *Int. J. Food Sci. Technol.* 48:571–577 (2013).
[10] Cardenas G., Cabrera G., Taboada E., Miranda S.P. *J. Polymer Sci.* 93:1876–1885 (2004).
[11] Tolaimate, A.; Desbrieres, J.; Rhazi, M. *Polymer* 44, 7939–7952 (2003).
[12] Rinaudo M. *Prog Polym Sci.* 31:603–632 (2006).
[13] Clasen C, Wilhelms T, Kulicke W-M. 7:3210–3222 (2006).
[14] Synowiecki J, Al-Khateeb NA. *Crit Rev Food Sci Nutr.* 43:145–171 (2003).
[15] Zhang Y, Xue C, Xue Y, Gao R, Zhang X. *Carbohydr Res.* 340:1914–1917 (2005).
[16] Mohadi R., Kurniawan Ch., Yuliasari N., Hadayati N., Lesbani A., *Prog. on Chem. and App. of Chitin and its Deriv.* 20:213-221 (2015)
[17] Kasaai MR. *Carbohydr Polym.* 71:497–508 (2008).

IOP Conf. Series: Journal of Physics: Conf. Series **1095** (2018) 012033 doi:10.1088/1742-6596/1095/1/012033

Rare Earth Elements on the A-type Unggan Granite and Its Comparison to the A-type Section of Sibolga Granite

Ronaldo Irzon[1,2], Ildrem Syafri[1], Kurnia[2], Purnama Sendjadja[2], Verry Edi Setiawan[2], Johannes Hutabarat[1]

[1]School of Geological Technique Padjadjaran University, Jl. Raya Bandung-Sumedang Km. 21. Jatinangor, Sumedang, Indonesia
[2]Ceenter for Geological Survey, Jl. Diponegoro 57, Bandung, Indonesia

ronaldoirzon18@gmail.com

Abstract. Rare earth elements are often correlated to A-type granite. Unggan Granite is A-type granite which is located in Sijunjung Regency, West Sumatra. This study discusses the rare earth elements composition on A-type Unggan Granite and compares the element abundance to the nearest A-type granite from Sibolga. Total REE of the studied location in the part of Solok Quadrangle is averagely 860 ppm. XRF and ICP-MS at Center for Geology Survey were applied to measure the composition of major oxides, trace elements, and rare earth elements in the selected granite. Megascopically, the samples are dark red and coarse grain granites with quartz and feldspar as major minerals. Major oxides content of the samples affirms the samples as granite based on geochemistry composition. Ga, Nb, and Y amounts in the granites reveal the A-type affinity. The average REE abundance of Unggan Granite is higher than the A-type section of Sibolga Granite. The selected rock are classified in shoshonitic series with lanthanum and neodymium as the major REE. Negative Ce and negative Eu anomalies in normalized spider plot are detected in the samples. The granite characters are worth to be explored more for national income from rare earth elements resource.

1. Introduction

Rare Earth Elements is a set of chemical elements, especially the lanthanides, and were considered rare and isolated in the 19th century. The development of earth science uncovered that some parts of the REE are more abundant than silver (Ag), lead (Pb), copper (Cu), and mercury (Hg) in the Earth's crust (Castor and Hedrick, 2006). Scandium (Sc, Z = 21) and yttrium (Y, Z = 39) are sometimes considered rare earth elements because they tend to occur in the same ore deposits as the lanthanides and exhibit similar chemical properties (i.e. Zepf, 2013; Migaszewski and Galuszka, 2015). This suite of elements is required in a variety of high-tech industries, such as permanent magnets, superconductors, rechargeable batteries, mobile phones, and security system devices. In Indonesia, REE has not become a major focus for both the government and industry although it is proven to be a strategic resource (Irzon et al., 2014).

Advanced analytical instruments are required along with careful preparation procedures to measure the REE content in any samples precisely. The integration of Instrumental Neutron Activation Analysis (INAA) with Inductively Coupled Plasma - Mass Spectrometry (ICP-MS) was used in the study of soil samples (Salvini et al.. 2006) while the Multi Collector (MC)-ICP-MS isotope dilution device was applied on analysing the REE abundance in various certified reference material (CRM) (Kent et al., 2004). ICP-MS at the Geological Laboratory of the Center for Geological Surveys has been frequently utilized to identify, characterized, and predict the REE mobility in Indonesia. REE in the parent granite, weathered horizons, and tin mining tailings in Lingga Regency was studied using the Thermo Elemental X-Series ICP-MS device (Irzon et al., 2014; Irzon, 2015; Irzon et al., 2016). Although ultramafic rocks are not the primary domain of the REE, the differences between lherzolite and olivin-hornblende-pyroxenite rocks in North Konawe were portrayed on the spider diagram patterns (Irzon and Baharuddin, 2016).

The most abundant intrusive rock in upper continental crust, phaneritic, granular, and consist of three major minerals (feldspar, quartz, and mica) is the character of granitoid. The rock range from diorite to granite (i.e. Kurniawan, 2014; Hutabarat et al., 2016). Granitoids are interested to be explored because of their REE content, especially Light-REEs (i.e. Castor and Hedrick, 2006; Sahoo et al., 2011). Sumatra-Peninsular Malaysia is a complex tectonic region and was developed by several microplates (Pulunggono, 1985). Various granitoid locations were outcropped in the West Sumatra Province and tend to be emplaced in 122 ± 24 Ma (Katili, 1968). Granitoids in Sumatra and Peninsular Malaysia are generally grouped on four provinces: Eastern, Main Range, Western, and Volcanic Arc (Cobbing, 2005). The plutons in this study are situated on Volcanic Arc Province and are adjoining the Bukit Barisan Mountains as shown in figure 1.

Geochemistry uses the tools and principles of chemistry to explain the mechanisms behind geological systems. The combination of chemical and geological concepts have been applied in exploring the genesis of rocks and/or formations, trace mineral deposits, and even environmental pollutions. Geochemistry including REE, the characterization of granitoids from Unggan, Sijunjung Regency - West Sumatra Province, is the aim of this study. The REE abundance on Unggan Granite is then compared to the nearest A-type granite from Sibolga (Setiawan et al., 2017). The geochemistry composition is then utilized in chemistry based diagrams to understand about the granitoids more broadly. REE identification might encourage further exploration activities as a strategic resource.

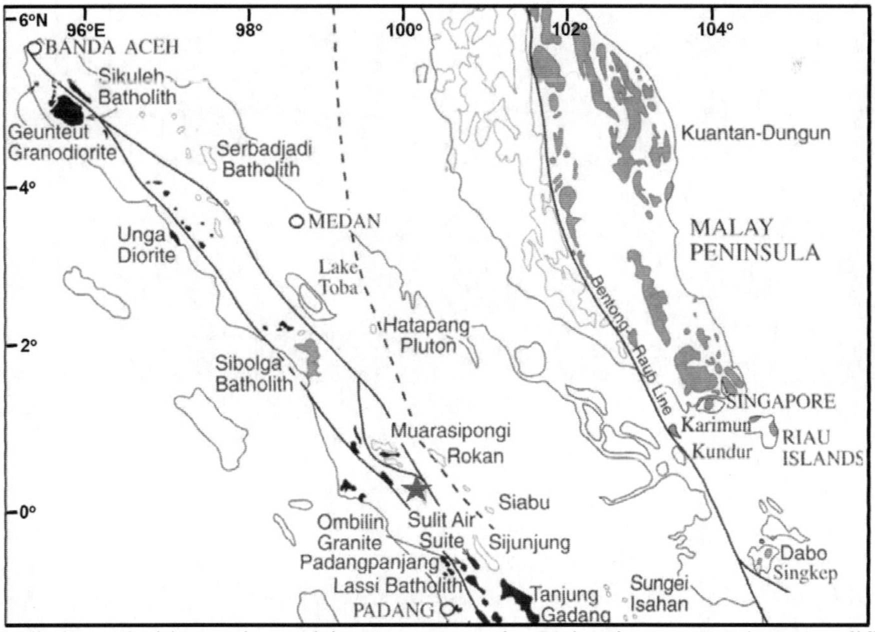

Figure 1. Studied granitoids are located in Unggan on the Volcanic Arc Province (modified from Cobbing, 2005). ★ = studied location

2. Material and Methods

Two granitoid samples were taken during fieldwork at Unggan, Sumpur Kudus – West Sumatra. The first rock, P-30, was obtained near the roadway and was described as reddish brown, granular, very coarse grain, and slightly weathered granitoid. P-31 is brighter but smaller grain pluton just near a small river in Unggan. Megascopically, the rocks are majorly composed of quartz and K-feldspar. Some field conditions are illustrated in figure 2.

The selected samples were analyzed in Geological Laboratory – Center for Geological Survey. Prior to chemical preparation, the samples were dried under direct sunlight and then were crushed to 200 mesh grain size to simplify the destruction process because the area of the reaction is directly proportional to the speed of reaction. One gram of sample that was transformed into pressed pellets then was analyzed using XRF to determine the major oxides contents. The Loss on Ignition (LOI) was measured to determine the amount of volatile material in samples such as hydrate compounds, hydroxy compounds, and carbon dioxide. Only 0.1 gram of rock was needed for subsequent destruction with three types of acids prior to ICP-MS analysis: nitric acid (HNO3, ultra-pure grade), formic acid (HCOOH, ultra-pure grade), and perchloric acid ($HClO_4$, pro analysis grade). The blank solution and five levels of the calibration solutions (0.1 ppb, 1 ppb, 5 ppb, 10 ppb, and 50 ppb) were prepared to be measured together with the selected samples. Geochemistry components of the samples were obtained by comparing the count per second (CPS) of the calibration solution series against the sample solutions.

ICP-MS should be tuned up before analyzing element concentration in the samples to gain its good stability and the detection capabilities which affect the measurement result. Sample introduction to ICP-MS device begins with the pumping process and ends in a mass spectrometer which detects the element concentration of the sample (Irzon, 2010). The sample solution is nebulized toward sample injector and then is transformed into plasma with the argon gas in the plasma torch. The computer program converts CPS of the samples from the mass spectrometer to become element abundance based on a comparison of the calibration solution curve.

Figure 2. Field conditions in Unggan Area, West Sumatra: a) P- 30 is situated on the hillside near roadway; and b) P- 31 is relatively experienced lower weathering process than P-30.

3. Result and Discussion

The composition of major oxides in the two Unggan Granite samples is not much different. The rocks show felsic character on the high SiO_2 content (> 68%). The abundance of Fe_2O_{3T} is lower than K_2O and Al_2O_3 of 2.61%, 5.18%, and 16.7% respectively. The significant difference between the two examples is that the composition of CaO at P-31 (1.6%) is almost twice that of P- 30 (0.88%). Although the amount of Na_2O in P- 31 was detected lower than P- 30, the first sample contains more TiO_2, MgO, and P_2O_5 than the later one. Higher LOI in P- 30 is parallel to the megascopic description that the rock tends to be more weathered. The major oxide, rare elements, and REE compositions in the examples of the study are summarized in Table 1.

Based on the geochemical composition, the two samples belong to the granite group in De la Roche et al. (1980) diagram as shown in figure 3a. P-30 and P-31 are derived from the same magma series (shoshonite) with reference to the SiO_2 versus K_2O diagram (Peccerillo and Taylor (1976)) (figure 3b).) On the other hand, the A-type section of Sibolga Granit depicts high-K calc-alkaline alkali feldspar granite characters. The peralkaline rock is indicated on smaller aluminum oxide content than sodium oxide plus potassium oxide. High Al rocks were further classified into two groups: metaluminous with Al_2O_3 <(CaO + Na_2O + K_2O) and peraluminous with opposite conditions (Shand, 1943). Both of Unggan Granite and the A-type section of Sibolga Granite are peraluminous based on A/CNK - A/NK diagrams (Shand, 1943) as shown in figure 3c.

The high contents of SiO_2, Na_2O+K_2O, Nb, Ta, Y, Cs, Ga, U, Th, and REE are the A-type granite characters. However, the abundances of MgO, CaO, Ba, Sr, P, Ti, Ni, Cr, Co, V in this type of pluton tend to be low (i.e. Singh and Vallinayagam, 2012; Irzon, 2017; Setiawan et al., 2017). Grouping the two samples as A-type granite are supported by the high sodium oxide plus potassium oxide contents (Table 1 and Figure 4b) of 8.15% - 9.04%. The A-type granite is attractive to be explored in correlation to the economic REE content as previous studies stated that REE is more concentrated in this type of pluton (i.e. Irzon, 2017). The number of REE in the Unggan Granite samples is 1,369 ppm (P-30) and 355 ppm (P-31). This REE composition is much higher than the average of the B facies of Muncung Granite Muncung (163 ppm) which was classified as S-type granite (Irzon, 2015). Lanthanum and neodymium were detected as the major REE of 270 ppm and 245 ppm respectively. High Ga, Ce, Nb, and Y contents support the typical of A-type granite on Whalen et al. (1987) diagrams as shown in figure 4. However, Ga and Nb compositions of Sibolga Granite is not provided in the previous study (Setiawan et al., 2017).

REE numbers should be normalized to a certain parameter to eliminate the Oddo-Harkins effect, and make it easier to be studied (i.e. Irzon and Baharuddin, 2016). The abundance of REE in this study was normalized using the primitive mantle value (McDonough, 1995). All REE in both samples is very high, 10-800 times than the primitive mantle as shown in figure 5. In the spider diagram, similar patterns are detected: a fairly sharp subduction of the Light-REE and the relatively flat pattern flat on Heavy-REE to conclude that the two samples are originated from the similar magma source. Subsequent uniformity is the negative Ce and negative Eu anomalies to explain that the rocks formed on the reductive condition and the magma source experienced plagioclase reduction (i.e. Irzon, 2017). REE composition in Unggan Granite is relatively higher than the Sibolga Granite. The next difference between the two plutons is the occurrence of positive Tm anomaly which might reflect the Tm-bearing mineral in the A-type section of Sibolga Granite.

Tabel 1. Major oxides, trace elements, and REE composition of the two Unggan Granite samples.

Major Oxides (%)	P- 30	P- 31	Trace Elements (ppm)	P- 30	P- 31	REE (ppm)	P- 30	P- 31
SiO_2	68.50	68.80	Sc	3.86	4.42	La	435	105
TiO_2	0.26	0.34	Ga	37	25.3	Ce	94.6	80.7
Al_2O_3	17.10	16.30	Sr	125	184	Pr	106	22.9
Fe_2O_{3T}	2.32	2.90	Y	233	34.8	Nd	401	89
MnO	0.04	0.04	Nb	25.4	24.9	Sm	88.9	17
CaO	0.88	1.60	Cs	5.16	4.05	Eu	6.37	1.88
MgO	0.46	0.59	Ba	676	615	Gd	32.5	6.62
Na_2O	3.59	3.24	Ta	2.07	1.33	Tb	10.7	1.74
K_2O	5.45	4.91	Tl	6.57	4.19	Dy	82.5	12.8
P_2O_5	0.10	0.12				Ho	15	2.36
SO_3	0.02	0.02				Er	42.8	6.84
LOI	1.50	0.98				Tm	5.8	0.92
						Yb	44	6.72
						Lu	4.26	0.67

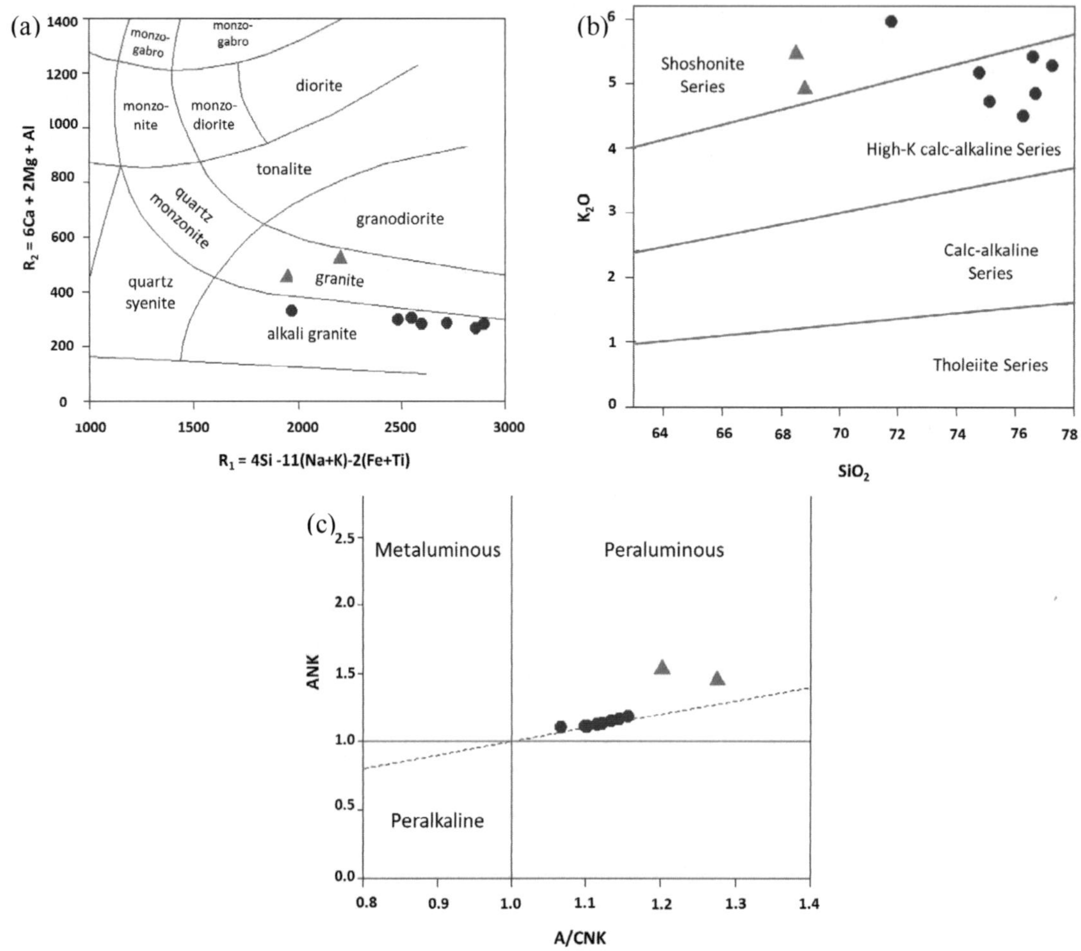

Figure 3. Geochemistry based diagrams: a) R1-R2 classification (De la Roche et al., 1980); b) Tectonic diversification of Peccerillo and Taylor (1976); and c) A/CNK-A/NK diagram (Shand, 1943). ▲ = Unggan Granite and ● = A-type section of Sibolga Granite.

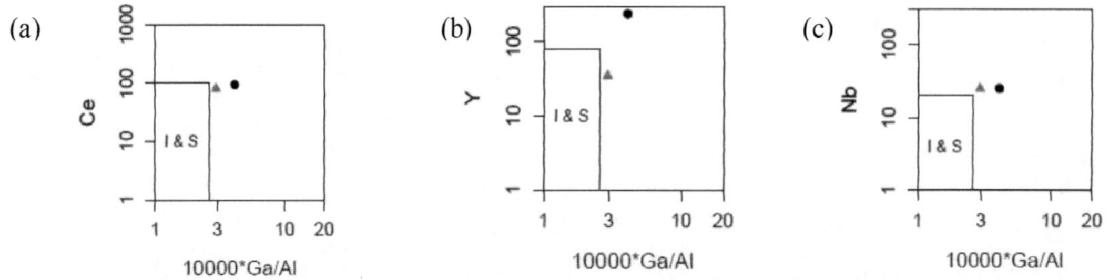

Figure 4. Samples plot on Whalen et al. (1987) discriminations that show A-type granite characters: a) 1000*Ga/Al vs. Ce; b) 1000*Ga/Al vs Y; and c) 1000*Ga/Al vs Nb. ● = P- 30 and ▲ = P- 31.

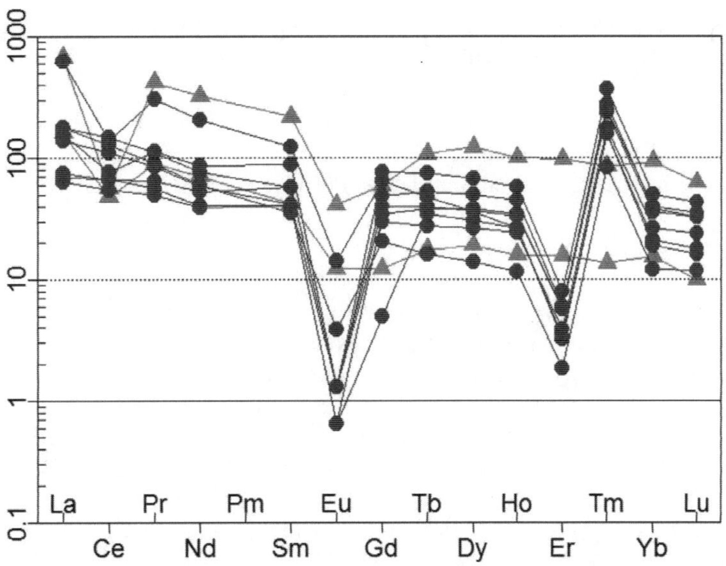

Figure 5. REE spider diagram which is normalized to primitive mantle value of Sun and McDonough (1995). Symbols are the same as figure 3. Positive Tm anomaly is detected on Sibolga Granite.

4. Conclusion

The combination of chemical and geological based analysis is very useful in understanding the natural process. The composition of major oxides, trace elements, and REE of the two pinkish granitoid rocks from the Unggan region was measured using XRF and ICP-MS devices. The rocks are categorized as granite, formed on the shosonite series, and are peraluminous based on their geochemical composition. The samples are relatively fresh refer to LOI of 0.98% -1.5%. REE in the samples is 355 ppm and 1,369 ppm with lanthanum and neodymium as the most abundant elements. The high contents of SiO_2, Na_2O + K_2O, Ga, Y, and REE amplify A-type granitoid character. Rocks were mostly formed in the oxidative state of a parental magma which experienced plagioclase reduction during the differentiation process. The REE content in Unggan Granite is relatively higher than the A-type section of Sibolga Granite which indicates positive Tm anomaly.

References

[1] Castor S B and Hedrick L B 2006 *Rare earth elements*; in Kogel J E, Trivedi N C, Barker J M, and Krukowski S T, ed., Industrial Minerals volume, 7th edition: Society for Mining, Metallurgy, and Exploration, Littleton, Colorado, pp. 769-792.

[2] Cobbing E J 2005 *Granites*. In: Barber A J, Crow M J, Milburn J S (Eds.), Sumatra: Geology, Resources and Tectonic Evolution. Geological Society of London, London, pp. 54–62.

[3] De La Roche H, Leterrier J, Grandclaude P and Marchal M 1980 A classification of volcanic and plutonic rocks using R1R2-diagram and major element analyses – its relationships with current nomenclature *Chemical Geology* **29** 183–210.

[4] Hutabarat J, Syafri I, Sulaksana N, Sukiyah E, Fauzielly L and Sudrajat A 2016 Fenomena Pelapukan Granit "Corestones" Di Daerah Bagian Timur dan Utara Pulau Bintan (Untuk Menunjang Pengembangan Kawasan Taman Bumi di Kepulauan Riau) *Seminar Nasional Ke – III Fakultas Teknik Geologi Universitas Padjadjaran.*

[5] Irzon R 2010 Pengujian Trace-Rare Earth Elements Terhadap SRM AGV2 dan GBW 07113 dengan ICP-MS *Kumpulan Makalah Sarana Teknik Pusat Survei Geologi* **39** 51-64.

[6] Irzon R, Sendjaja P, Kurnia, Imtihanah, and Subandrio J 2014 Kandungan Rare Earth Elements dalam Tailng Tambang Timah di Pulau Singkep *Jurnal Geologi dan Sumberdaya Mineral,* **15(3)** 143-151

[7] Irzon R 2015 Contrasting Two Facies of Muncung Granite in Lingga Regency Using Major, Trace, and Rare Earth Element Geochemistry *Indonesian Journal on Geoscience* **2(1)** pp. 23-33

[8] Irzon R and Baharuddin 2016 Geochemistry of Ophiolite Complex in North Konawe, Southeast Sulawesi *Eksplorium* **37(2)** pp.101-114.

[9] Irzon R, Syafri I, Hutabarat J and Sendjaja P 2016 REE Comparison Between Muncung Granite Samples and their Weathering Products, Lingga Regency, Riau Islands *Indonesian Journal on Geoscience* **3(3)** pp.149-161.

[10] Irzon R 2017 Geochemistry of Late Triassic weak Peraluminous A-Type Karimun Granite, Karimun Regency, Riau Islands Province *Indonesian Journal on Geoscience* **4(1)** pp. 21-37

[11] Katili J A 1968 Permian Volcanism and Its Relation to the Tectonic Development of Sumatra *Bulletin of National Institute of Geology and Mining* **1(1)** pp. 3-13.

[12] Kent A J R, Jacobsen B, Peate D W, Waight T E and Baker J A 2004 Isotope Dilution MC-ICP-MS Rare Earth Element Analysis of Geochemical Reference Materials NIST SRM 610, NIST SRM 612, NIST SRM 614, BHVO-2G, BHVO-2, BCR-2G, JB-2, WS-E, AGV-1 and AGV-2 *Geostandards and Geoanalytical Research* **28(3)** pp.417–429.

[13] Kurniawan A 2014 Geologi Batuan Granitoid di Indonesia dan Distribusinya *Masyarakat Ilmu Bumi Indonesia* **1** (E-3).

[14] McDonough W F and Sun S 1995. Composition of the Earth *Chemical Geology* **120** pp.223-253.

[15] Migaszewski Z M and Galuszka A 2015 The Characteristics, Occurrence, and Geochemical Behavior of Rare Earth Elements in the Environment: A Review *Critical Reviews in Environmental Science and Technology* **45(5)**, pp. 429-471.

[16] Peccerillo A and Taylor S R 1976 Geochemistry of Eocene calc-alkaline volcanic rocks from the Kastamonu area, Northern Turkey *Contributions to Mineralogy and Petrology* **58**, pp. 63–81.

[17] Pulunggono A 1985 The changing pattern of ideas on Sundaland within the last hundred years and its implication to oil exploration *Indonesia Petroleum Association, Proceeding of the 15th Annual Convention, Jakarta, 1986* pp.187-214.

[18] Salvini A, Cattadori C, Broggini C, Cagnazzo M, Ori G G, Nisi S, Borio A, and Manera S 2006 Ir and Rare Earth's Elements Determination by Neutron Activation analysis and ICP-MS In soil samples *Journal of Physics Conference Series* **41** pp.551–554.

[19] Sahoo S K, Hosoda M, Kamagata S, Sorimachi A, Ishikawa T, Tokonami S and Uchida S 2011 Thorium, Uranium and Rare Earth Elements Concentration in Weathered Japanese Soil Samples *Progress in Nucleas Science and Technology* **1** pp. 416-419

[20] Setiawan I, Takahashi R, and Imai A 2017. Petrochemistry of Granitoids in Sibolga and its Surrounding Areas, North Sumatra, Indonesia. *Resource Geology* **67(3)** pp.254-278.

[21] Shand S J 1943 *Eruptive Rocks. Their Genesis, Composition, Classification, and Their Relation to Ore-Deposits with a Chapter on Meteorite* New York: John Wiley & Sons

[22] Singh I G and Vallinayagam G 2012 Petrological and Geochemical Constraints in the Origin and Associated Mineralization of A-Type Granite Suite of the Dhiran Area, Northwestern Peninsular India *Geosciences* **2(4)** pp.66-80.

[23] Whalen J B, Currie K L and Chappell B W 1987 A-type granites: geochemical characteristics, discrimination and petrogenesis *Contributions to Mineralogy and Petrology* **95** pp. 407-419.

[24] Zepf V 2013 *Rare earth elements. A new approach to the nexus of supply, demand and use: exemplified along the use of neodymium in permanent magnets* Ph.D. thesis of the University of Augsburg, Germany. Springer Publisher, 157 p.

Antihypertensive Bioactive Peptides From Hydrolysates of Soy milk Yoghurt (Soygurt)

Sandra Hermanto[1], F Hatiningsih[1] , D K Putera[1]

Chemistry Department, Faculty of Science and Technology
Syarif Hidayatullah State Islamic University Jakarta
Jl. Ir. H. Juanda No. 95 Ciputat Jakarta Indonesia 15412

Email : hermantokimia@uinjkt.ac.id

Abstract. Soyghurt is a product of fermented soy milk which beneficial nutrients and functional food. This study was conducted to isolate and characterize bioactive peptides from soyghurt hydrolysates which potential as antihypertensive drugs. Soyghurt precipitated by the addition of HCl 1M and centrifuged. Pepsin as a proteolytic enzyme was used to hydrolyzed soygurt with enzyme/substrate ratio (1/20) at 37 °C and pH 4.5 by interval of hydrolysis time 2, 4, 8 and 16 hours. The quantity of protein hydrolysates determined by Lowry method and the peptide pattern analyzed by SDS-PAGE (*sodium dodecyl sulphate polyacrylamide gel electrophoresis*). Antihypertensive activity of soyghurt hydrolysates determined by angiotensin converting enzyme (ACE) inhibitory assay. Protein hydrolysates with the highest inhibition capacity purified by gel filtration column G-15. The results of this study show that soygurt hydrolysates with the highest of ACE inhibition obtained by 16 hours of hydrolysis with the highest degree of hydrolysis (51.6%) and antihypertensive activity 90.2%. This soygurt hydrolysates could potentially be used as natural antihypertensive agent.

Keywords: Antihypertensive, bioactive peptides, soybean yoghurt

IOP Conf. Series: Journal of Physics: Conf. Series **1095** (2018) 012034 doi:10.1088/1742-6596/1095/1/012034

1. Introduction

One of disease that affects many modern societies is high blood pressure. Recently, high blood pressure (hypertension) was a major risk factor in cardiovascular disease. High blood pressure is one of the main factors causing stroke, heart attack, and kidney failure. Many factors triggers hypertension such as the consumption of salt, cholesterol, excessive alcohol and caffeine, obesity and genetic predispositions that can not be controlled [1].

Angiotensin converting enzyme (ACE) is a key enzyme which plays an important role in regulating blood pressure in humans. ACE catalyzes the hydrolysis of inactive angiotensin decapeptide to vasoconstrictor angiotensin II, an octapeptide that plays a role in regulating blood pressure. In addition, ACE will also cut a vasodilator bradykinin into inactive fragments to raised blood pressure [2].

Some of synthetic drugs such as ACE inhibitor agent has been widely used in the treatment of hypertension [3]. The use of synthetic antihypertensive drugs contain some risks and may cause side effects, such as hypotension, decreased renal function, a dry cough, and fetal abnormalities. To solved this limitations many researchers considered to find out the alternative natural ACE inhibitor which not risk and not addictive at all. Some of ACE inhibitors which has extensively studied among others derived from animal protein and vegetable protein [4].

Milk protein is one of bioactive peptides source that can serve as an antihypertensive [5]. These peptides can be produced by enzymatic hydrolysis with proteolytic enzymes such as pepsin or papain or through by lactic acid fermentation of bacteria [6]. Some of milk bioactive peptides capable in inhibiting ACE activity *in vitro* and *in vivo* [7], [8], [9]. These peptide fragments have also been found to have antihypertensive properties in animals and humans [10].

Over the past decade, there has been increasing use of food sources producing bioactive peptides in addition to dairy products, for example in cereal products. Fermented cereal products containing peptides that have the effect of lowering blood pressure [11]. Several other studies shown that foods containing animal protein or vegetable protein is an important source of bioactive peptides. The plant sources include cereals (wheat, barley, maize, rice) pseudo cereals (buckwheat and spinach), legumes (soy, lentils, beans), brassica species and others (sunflower) is also sources of bioactive peptides. The presence of bioactive peptides in cereals and legumes that can contribute to an increase in the value of quality protein and add functionality as a functional food that can be consumed daily [12].

Soygurt or soybean yoghurt is a functional food product with highly protein content and soy ingredients processing are continuously improved. For example in fermented soymilk, like the traditional food (soy sauce and tempeh), it has been known a significant role of antihypertensive peptides by acts as an ACE inhibitor [13]. [14]. Another antihypertensive peptides from soybean product are "*chunggugjang*" from Korean food derived from fermented soybeans with *Bacillus subtilis* CH-1023 [15] and soybean paste [16], soy sauce [17], [18], natto and tempeh [19}, and another soy fermentation products [20].

Study on soygurt hydrolysate by proteolytic cutoff as antihypertensive bioactive peptides has not been done. Therefore this study conducted to investigate soygurt bioactive peptides from local soygurt produced on industrial scale as a source of antihypertensive agents. This objective of research is to isolate and characterize an antihypertensive bioactive peptides from soygurt hydrolysate by enzymatic treatment with pepsin. We are expected that the ACE inhibitory activity from this hydrolysates are more better than commercial ACE inhibitor so its could be applied as an alternative antihypertensive drug.

2. Materials and Method

2.1 Sample and reagents

Fresh soygurt purchased from Local Supplier in Depok Regions West Java Indonesia. Other reagents consist of Pepsin (Porcine gastric mucosa, activity 0.8-2.5 units g-1 protein) was purchased from Sigma Chemical Co. (USA), the enzyme ACE (Angiotensin Converting Enzyme, rabbit lung) were obtained from Sigma Chemical Co. (USA). N-Hippuryl-His-Leu (Hydrate Powder) ≥ 98% (HPLC grade), sodium dodecyl sulfate (SDS) 10% (Sigma Sigma Chemical Co., USA); Lowry Lowry solution I and II, Gel Acrylamide solution (30% T; 2,67C) Bio-Rad; Bis-Acrylamide Sigma, Resolving Buffer (1.5M Tris-HCl pH 8.8 Bio-Rad); Stacking Buffer (0.5M Tris-HCl pH 6.8 Bio-Rad); Ammonium peroxide disulfate (APS) 10%; N, N,N'N'-Tetramethylethylenediamine (TEMED); Sephadex G-15 (medium), 0.01 M phosphate buffer solution pH 7.2; Running buffer Sigma; Staining solution Coomasie blue R-250 Bio-Rad; 0.25M phosphate buffer pH 7.4. hydrochloric acid (HCl, Merck), sodium acetate, acetic acid (Merck), trochloroacetic acid (TCA, Merck).

2.2 Preparation of soygurt hydrolysates

Soygurt hydrolysate prepared by dissolved in distilled water and the pH set to 2 with the addition of HCl 1M. After coagulation, the fraction separated from the whey protein precipitates by centrifugation (5,000g, 10 min). Protein precipitates were dissolved again in distilled water to obtain a protein concentrate 10% (w/v). To the precipitate was added pepsin with ratio of enzyme: substrate (1:5) and incubation carried out at 2, 4, 8 and 16 hours at 37°C in 0.05 M acetate buffer pH 4.5 [21]. The optimum conditions of pepsin hydrolysis of soygurt analyzed by SDS-PAGE electrophoresis. During the incubation process, approximately 5 mL of hydrolysate mixture is taken at 2, 4, 8 and 16 hours and the degree of hydrolysis measured by Alder-Nissen method [22]. After incubation was completed, each of the hydrolysate mixture was heated at 98°C for 5-10 minutes to inactivate the enzyme. The mixture was centrifuged at 12000 rpm and the supernatant measured an antihypertensive activity by using ACE inhibitory assay. Samples dried in a freeze dry for 24 hours and then stored at -20°C before analysis.

2.3 Detemination of protein molecular weight of hydrolysate

Protein degradation and molecular weight of hydrolysates was identified by SDS-PAGE (Sodium Dodecyl Sulphate Acrylamaide Gel Electrophoresis) standard methods [23], with a Mini-Protean II Slab Electrophoresis Cell (Bio Rad). Soygurt hydrolysates from enzymatic hydrolysis denatured in sample buffer (Tris-Cl 150 mM pH 6.8 SDS 25%, β-mercaptoethanol, 25% glycerol and 2.5 mM Bromophenol blue) and boiled for 10 min. Electrophoresis SDS-PAGE performed at 16% separating gels by the voltage 200 volts for 60 minutes. Visualization of protein bands staining by Coomasie brilliant blue 0.1% (w/v).

2.4 ACE-inhibitory activity of Soygurt Hydrolysates

The ACE-inhibitory activity was measured by monitoring the release of hippuric acid from hippuryl-Lhistidyl-L-leucine (HHL) using ACE solution [24]. The ACE solution contained 5 mM HHL, 200 mM potassium phosphate buffer (pH 8.3), 25μL ACE from rabbit lung. After 30 min of incubation at 37°C the reaction was stopped by addition of 200 μL of 1N HCl. The hippuric acid formed was extracted with ethyl acetate and removal of ethyl acetate by heat evaporation. The hippuric acid formed was measured by uv-vis spectrophotometer at 228 nm. The percentage of ACE inhibition was calculated based on the absorbance value of the samples **:**

$$\text{ACE inhibition } (\%) = (A-B)/(A-C)$$

where A = absorbance of solution without sample, B = absorbance of solution with sample, C = Absorbance of blank solution (without ACE solution and sample).

3. Results and Discussions

3.1 Hydrolysis of soygurt

The enzymatic hydrolysis using pepsin enzyme on soygurt hydrolysates was monitored by the SDS-PAGE method and the degree of hydrolysis (DH) was calculated by the Adler-Nissen methodx This has been shown that hydrolysis reaction using pepsin enzyme can achieve high DH value with various peptides sizes. SDS-PAGE analysis of soygurt hydrolysates revealed that smaller peptides (<10 kDa) were successfully produced and has a great potential to be used as an antihypertensive. Based on the SDS-PAGE analysis (Figure 1) shows that the optimum hydrolysis of soygurt precipitates by pepsin digested obtained at 16 hours with lost of intensity of the higher bands on molecular weight (MW) of protein hydrolysate (25-37 kDa) and increase in the lower molecular weight of protein bands (<10 kDa).

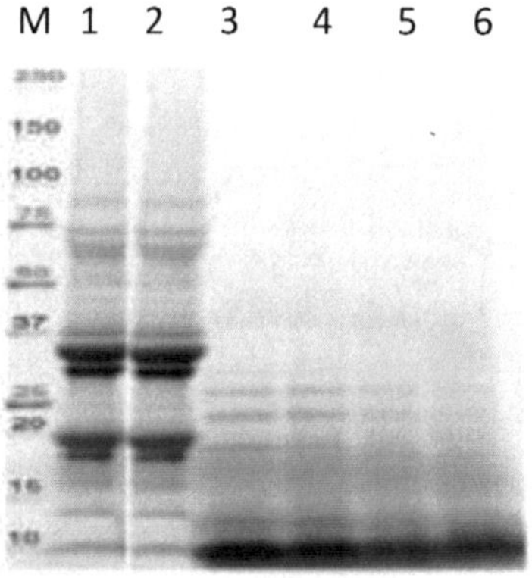

Figure 1. SDS-PAGE pattern of soyghurt hydrolysates. M : Protein markers. 1 : negative control (not hydrolysed soygurt), 2 : hydrolysis 0 hours, 3 : hydrolysis of 2 hours, 4 : hydrolysis of 4 hours, 5 : hydrolysis 8 hours, 6 : hydrolysis 16 hours.

Loss of higher MW of protein fractions indicated that the molecular digest from enzymatic hydrolysis would be eliminate protein bands to produce biopeptide fractions with a lower molecular weight. No bands were visible for the MW>20 kDa after 16 hours incubation indicated that the hydrolysis of protein had the optimum time at 16 hours. According to the results of the quantitatively analysis of the soluble proteins level of hydrolysate, it was found that soluble protein with lower molecular weight increased after 16 hours hydrolysis as follows in Figure 2.

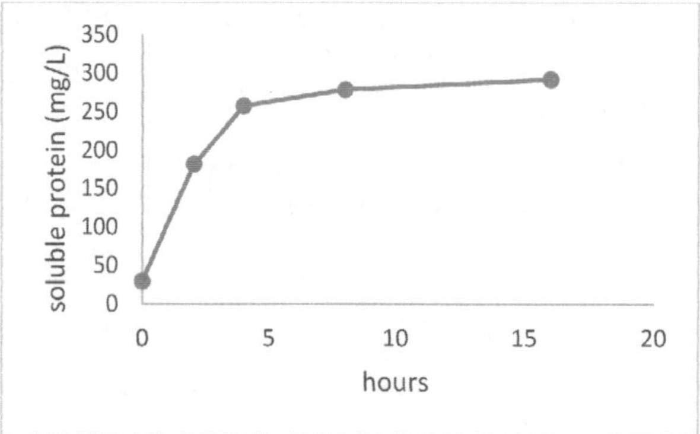

Figure 2. The soluble protein of hydrolyzed soyghurt 0-16 hours (treatment by TCA).

The results of enzymatic hydrolysis of soygurt indicate that the protease preparation is efficient to form smaller peptides by pepsin hydrolysis. In comparison to the chemical hydrolysis, enzymatic hydrolysis is preferable due to several advantages, such as mild reaction conditions, low undesirable products, and high product quality and yield [25].

3.2 Degree of hydrolysis (DH)

The DH for soygurt protein hydrolysates was determined according to the method of Adler-Nissen [22]. The DH gives an initial indication for the change in the molecular integrity. During protein hydrolysis, the large complex structured protein molecules are broken down into smaller sized peptides and specific amino acids [25]. The DH was continuously monitored during enzymatic treatment of soygurt hydrolysis. As shown in Figure 3, the unhydrolyzed soygurt showed an average DH value of 5.2%. In the case of enzymatic hydrolysis, the DH increased significantly ($P < 0.05$). The highest DH value of 51.6% was achieved after enzymatic treatment of soygurt for 16 h with pepsin enzyme.

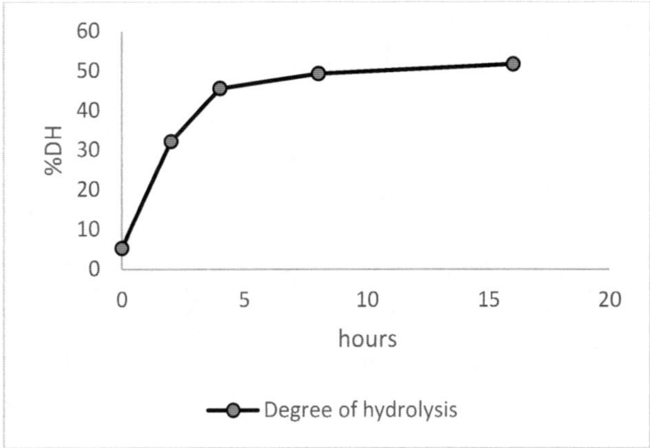

Figure 3. The degree of hydrolysis of soygurt (0-16 hours)

Degree of hydrolysis is affected by the length of incubation and ratio of enzyme used. The enzyme concentration is one of the factors that affect the rate of degradation of proteolytic enzymes. In this process, soygurt hydrolysates will be digested by the proteolytic enzyme in the active site of pepsin to produce an oligopeptides with the small molecular weight. Schematic reaction of proteolytic hydrolysis can be seen at Figure 4.

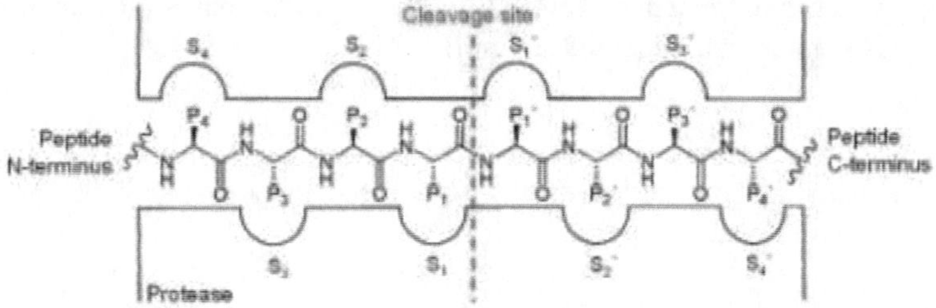

Figure 4. Schematic representation of complex enzyme-substrate with 5 binding site. The position of the peptide bond to the substrate is calculated from left to right where the disconnection occurred in bonding P1-P1' in the pepsin active site' [26].

Pepsin is a proteolytic enzyme that will break the peptide bonds in several amino acid residues such as leucine, phenylalanine, and tyrosine triptopan. The mechanism of peptide bond disconnection occurs at position P1-P1' which is occupied by the amino acid residues polar through the mechanism of acid-base reactions with active sites occupied by the carboxyl group of Asp-32 and Asp-215 [27].

3.3 ACE-Inhibitory Activity
Determination of antihypertensive activity was performed by ACE-inhibitory assay refers to method developed by Cushman and Cheung. Based on investigation it was found that the highest antihypertensive bioactive peptide was obtained at 16 hours of hydrolysis with % ACE inhibition 90.9% (Figure 5). Increasing of antihypertensive activity in fraction hydrolysates probably caused by a large part of the protein fraction has been degraded into more active oligopeptides. According to Lee *et al.* {21], the proteolytic hydrolysis of goat milk casein fraction gave the highest inhibitory activity (87.8%) on a casein hydrolysates after pepsin hydrolysis at 48 hours.

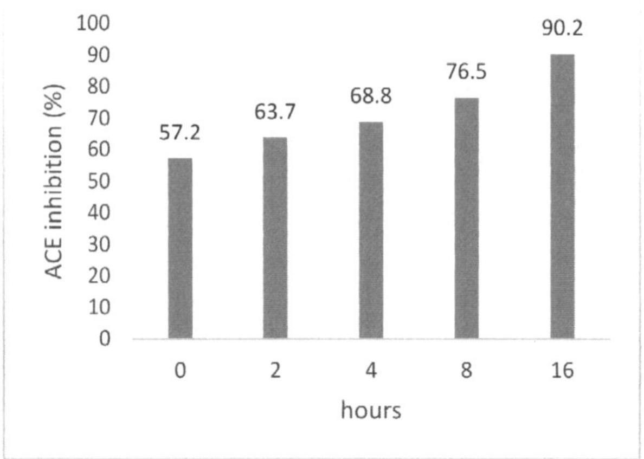

Figure 5. Antihypertensive activity of soygurt hydrolysates (ACE assay).

Soy foods such as soy sauce, miso, and natto show antihypertensive effect by inhibiting ACE, the key enzyme of renin-angiotensin system [18], [19], [20], [28]. Shimakage *et al.* [29] had identified 5 novel ACE inhibitory peptides containing Ile-Ile and Ile-Asp from protease-treated hikiwari-natto, whose ACE inhibitory activity was about 1.4 times higher than that of protease-untreated hikiwari-natto. In addition, they identified 8 novel ACE inhibitory peptides containing Phe-Phe-Tyr-Tyr and Trp-His-Pro derived from protease-treated soymilk, whose ACE inhibitory activity was about 36 times higher than that of protease-untreated soymilk. The results show that ability in inhibiting ACE of soygurt hydrolysates is greater than untreated soygurt.

4. Conclusions
In this study, we conclude that enzymatic hydrolysis of soygurt by pepsin enzyme has been effective to provide an antihypertensive bioactive peptide with the percent of ACE inhibition up to 90%, so it could be potentially used as an antihypertensive agents for hypertension treatment.

Acknowledgements
This work was supported by Institute for Research and Community Service of SyarifHidayatullah Jakarta State Islamic University 2016 (SK Number. Un.01/KPA/371/2016).

References

[1] Japanese Society of Hypertension Committee.2009 Tokyo.
[2] Brown N J and Vaughan D E 1998 *Circulation* **97** 1411
[3] Parmley W 1998 *American Journal of Medicine* **105**, 27S
[4] Ariyoshi Y 1993 *Trends Food Sci. Technol.* **4** 139.
[5] Yamamoto N and Takano T 1999 *Nahrung* **43**(3) 159.
[6] FitzGerald R J Murray B A and Walsh D J 2004 *J. Nutr.* **134** 980S.
[7] Maeno M, Yamamoto N and Takano T 1996 *J. Dairy Sci.* **79** 1316.
[8] Abubakar A, Saito T, Kitazawa H, Kawai Y and Itoh T 1998 *J. Dairy Sci.* **81** 3131.
[9] Pihlanto-Leppala A, Koskinen P, Piilola K, Tupasela T and Korhonen H 2000 *J. Dairy Res.* **67** 53.
[10] Nakamura Y, Yamamoto N, Sakai K, Okubo A, Yamazaki S and Takano T 1995 *Journal of Dairy Science,* **78** 777.
[11] Malaguti M, Dinelli G, Leoncini E, Bregola V, Bosi S 2014 *Int J MolSci* **15(11)** 21120.
[12] Martinez-Maqueda D, Miralles B, Recio I, Hernandez-Ledesma B 2012 *Food Funct* **3(4):** 350
[13] Cha M and Park J R 2005 *J Med Food* **8**(3) 305.
[14] Vallabha V S and Tiku P K 2014 *J Pept Res Ther* **20**(2) 161.
[15] Korhonen H and Pihlanto A 2003 *Curr Pharm* **9(16)** 1297.
[16] Shin Z I, Yu R, Park S A, Chung D K and Ahn C W 2001 *J Agric Food Chem* **49(6)** 3004.
[17] Okamoto A, Hanagata H, Matsumoto E, Kawamura Y and Koizumi Y 1995 *Biosci Biotechnol Biochem* : **59(6)** 1147.
[18] Nakahara T, Sano A, Yamaguchi H, Sugimoto K R I and Chikata H 2010 *J Agric Food Chem* **58**(2) 821.
[19] Gibbs B F, Zoygman A, Masse R and Mulligan C 2004 *Food Res Int* **37**(2) 123.
[20] Rho S J, Lee J S, Chung I Y, Kim Y W and Lee H G 2009 *Process Biochem* **44**(**4**) 490.
[21] Lee K J, Kim S B, Ryu J S, Shin H S and Lim J W 2004 *Asian-Aust. J. Anim. Sci* **18**(5) 741.
[22] Adler-Nissen, J 1986 *Enzymic hydrolysis of food proteins* (New York - Elservier)
[23] Laemmli U K 1970 *Nature* **227 (5259)** 680.
[24] Cushman D W and Cheung H S 1971 *Biochem.Pharmacol.*20 1637.
[25] Meinlschmidt P, Sussmann D, Schweiggert-Weisz U and Eisner P 2016 *Food Science & Nutrition* **4(1)** 11.
[26] Karstad R, Isaksen G, Wynendaele E, Guttormsen Y, De Spiegeleer B and Brandsdal B O, *J Med Chem.* 55 6294.
[27] Fujinaga M, Chernaia M M, Tarasova N I, Mosimann S C and James M.N G 1995 *Protein Science* **4** 960.
[28] Ibe S, Yoshida K, Kumada K, Tsurushiin S and Furusho T 2009 *Food SciTechnol Res:* **15**(2) 199.
[29] Shimakage A, Shinbo M and Yamada S 2012 *J. Biol. Macromol,* **12**(3) 72.

The Characterization of Mango (*Mangifera indica L*) Powder of Various Drying Temperature

Sri Agustini[1]

[1] Palembang Institute For Industrial Research and Standardization , Ministry of Industry, Indonesia

Corresponding author: sragustini@yahoo.com

Abstract. The research was intended to characterize mango powder which produced by small industries. Research using randomized complete design with drying temperature (50°C, 60°C,70°C) as the treatment with 3 replications. Mango used in this research were varieties mangga harum manis and fillers used were sucrose at concentration of 20%. The physicochemical properties of powders produced observed were water content, ascorbic acid, color, hygroscopicity, flowability, index solubility and total plate count investigated by using the gravimetric method, titrimetric, and plate count respectively. While for color were investigated by using Munsell color system consisting of lightness, hue angle, and chroma. The results showed that oven drying successfully produced mango powder which has water content ranged from 2.366 ± 0.576% up to 2.938±0.691%, and total plate count 1.667±0.039 log cfu up to 1.748±0.038 log cfu. The highest ascorbic acid content was obtained from treatment at 50°C namely 0.146±0.016%. Test on the color showed that mango powder which was dried at 50°C have the highest lightness of 75.6±1.356, chroma of 41.00±1.414 and hue of 82.4±1.02°. The flowability of all powder indicated fair flow properties (angle of repose 37.77-38.356°), the hygroscopicity of 8.32-9.43%, the degree of caking was 9.54-10.00% and index solubility were 97.73%-98.90%.
Keywords: mango powder, drying temperature, physicochemical properties

1. Introduction

Mango (*Mangifera indica* L.) is one of the important tropical fruits in Indonesia due to it has good taste and bioactive compound. Mango is a source of sucrose, fibers, fat, protein, beta carotene, kalium, vitamin A, B6, and C [45]. Mango is a seasonal fruit and classified as a perishable fruit. Damage and quality deterioration can occur at various levels of handling, such as harvesting, distribution and transportation, and insufficient storage during peak times. High moisture and high sugar content cause mango quickly deteriorated so that it has a short shelf life. As a perishable fruit mango stored at room temperature will cause quality deterioration [4, 35, 38, 20]. Processing the discarded and unutilized fruit into juice or powder can provide an additional advantage [43].

Processing of mango into mango juice powder is a good choice in preserving of mango. Mango juice powder can be used as a flavor enhancer in various types of foods and beverages such as a mixture in baby and child food, bread and cakes, jelly, pudding, and flavor enhancers for ice cream and yogurt [29]. Fruit juice powder has potential economic advantages compared to fruit juice and fruit in general. Powdered juice had a smaller volume and weights thus require a smaller packaging and easier to transport. In addition, fruit juice powder has a longer shelf life [16].

The principle of processing the fruit into a fruit powder is to lower the water content in the juice to some extent so that the microbial growth and enzyme activity can be prevented [21, 27]. Fruit juice can be converted into powder particles using a variety of methods of drying, such as sun drying, oven drying, freeze drying, vacuum drying, spray drying and microwave drying [5, 6, 8, 14, 23]. Drying method will affect the physical and biochemical properties of fruit juice powder produced, for example, vacuum microwave drying can improve the texture, color, and flavor of dried fruit [14].

Sun drying and oven drying are the most common drying method used by the home industry because of its simplest technology. The oven drying method requires a shorter drying time when compared to sun drying because the drying temperature can be adjusted according to the need so that the powder produced has the lower water content. Unfortunately decreasing water content by using heat will affect the nutrient content [1]. This research was conducted to know the effect of drying temperature on the physicochemical properties of mango powder and to obtain the most suitable drying temperature to get mango powder which has physical properties similar to fresh mango juice for home industry.

2. Materials and Methods

The research was conducted at Baristand Industri Palembang Laboratory in February till July 2017. Mango used in this study was ripe mango (variety of indramayu) and sucrose as fillers obtained from the traditional market in the city of Palembang.

2.1. ***Preparation***. The preparation of mango powder by drying temperature treatment consisting of 50°C, 60°C, and 70°C. Mango fruit was cleaned, peeled and deseeded using a stainless steel knife. The flesh cuts into pieces. They were then made into juice through juicer at high speed using a commercial juicer (Model HR 1811, Philips). The Mango juice obtained then filtered by using a muslin cloth. The obtained mango juice was then added with 20% (w/w) sucrose. The mixture was stirred until the sucrose dissolves completely. Finally, the mixture was then poured into stainless steel tray and dried in a drying oven for 8 hours at three different temperature of 50°C, 60°C, and 70°C. Semi dried mat was peeled and reversed at around 20% moisture for better mass transfer and faster drying. The dried mats were pulverized to be powder and packed. All the powders produced were then subjected to physicochemical analysis including moisture content, ascorbic acid, color, solubility index, hygroscopicity, the degree of caking, flowability, and total plate count. All experiments were conducted in triplicates and an analysis of variance (ANOVA) and LSD tests were performed using SAS software. The data were expressed as mean ± standard deviations. The confidence interval at 95% ($p < 0.05$) was used for statistical significance.

2.2. ***Physicochemical Analysis.*** The moisture content of the powders was determined based on AOAC [3]. Color measurement using Konica Minolta color reader, using Munsell color system including lightness (L), c (chroma), and h (hue). The pH value was measured using a digital pH meter (Jenway, model 3505, UK).

Hygroscopicity was determined according to the methods of Cai and Corke [8], with slight changes. 2 g of mango juice powder was placed in a desiccator containing a Na_2SO_4 solution (75% RH) at 25 ± 1.0°C. After one week the sample was weighed. Hygroscopicity is shown as g moisture per 100 g dry solids (g/100g).

Solubility index determined according to Cano-Chauca *et al.* [9] with slight modification. One gram sample of mango powder was dissolved in 100 ml aquadest. 25 ml of aliquote was taken and put into a glass cup weighed and dried in an oven at 105°C to obtain a constant weight. The solubility index was calculated as a percentage difference in the weight of the cup. Total plate count of the powders was determined based on AOAC [3].

Flowability was determined using an angle of repose method described by Chauhan and Patil [11]. A funnel is set as high as 2 cm on paper. Furthermore, the mango powder was poured into the funnel and allowed to flow through the funnel and form a cone-shaped heap until the top of the cone touches the funnel handle. Mark the base of the heap using a pencil, and remove the mango juice powder. The angle of repose was calculated using the following formula:

Tan θ = (2 x Height of heap) / (diameter of the heap-diameter of the funnel)
where θ = angle of repose

3. Results and Discussion

Table 1 shows the effect of temperature on the moisture, ascorbic acids, TPC and color of mango powder.

Table 1. Effect of Temperature on Moisture, Ascorbic acid, TPC and Color

Temp	Moisture (%)	Ascorbic Acid (%)	TPC (log cfu)	Color		
				L	c	h
50°C	2.938±0.691a	0.146±0.016b	1.748±0.038c	75.6 ± 1.354e	41.0± 1.414g	82.40± 1.02i
60°C	2.760±0.598a	0.140±0.006b	1.712±0.022c	72.6 ± 1.625e	42.4± 1.019g	81.20± 0.75i
70°C	2.366±0.576a	0.130±0.009b	1.667±0.039d	70.4 ± 1.823f	45.6± 1.600h	78.00± 0.63j

Data were expressed as the mean ± standard deviation (n=3).
[a-b] Values in each column with different superscripts are significantly different (P<0.05).

3.1. Moisture Content. Table 1 showed that moisture content of mango powder ranged from 2.366±0.576% up to 2.938±0.691% which considered as stable and microbiologically safe. According to Noor Aziah and Komathi [28], moisture contents higher than 14% will affect storage quality of powders and flour because mold growth, insect infestation, and agglomeration could occur in this condition. Wong *et al.* [43] and Sangamitrha *et al.* [37] stated that fruit powder containing less than 5% moisture content can prevent enzymatic reactions and damage caused by microbial activity. Fruit juice powder containing less than 10 % moisture content is classified as microbiologically safe. This result similar to pulp mango powder which is foam mat dried at 65°C,75°C, and 85°C [42].

The results showed that drying at 70°C gave the lowest moisture content when compared with other treatments. This is because, at that temperature, evaporation of water occurs faster than evaporation at 50°C and 60°C. According to Chegini and Ghobadian [12] and Phisut [32], the increase in drying temperature caused the rate of heat transfer between materials with air faster, resulting in faster water evaporation of the material. However, ANOVA showed that at the temperature ranged 50°C to 70°C, the drying temperature had no significant effect on the moisture content of mango powder.

3.2. Vitamin C (Ascorbic acid). Vitamin C (Ascorbic acid) is classified as heat sensitive [17,33]. Fresh mango containing vitamin C about 6-80 mg/100g [30]. The highest vitamin C content obtained was at 50°C (0.146±0.016%) and the lowest at 70°C (0.13±0.009%). However, the results showed there were no significant differences (p< 0.05). Drying at different temperature did not have a significant effect on the vitamin C content. This was due to the drying time for each treatment in the experiment was the same. This result was in accordance with Qadri and Srivastava [33] who reported that nutrients more sensitive to the drying time than to temperature. This result differs to the work of Kandasamy *et al.* [25] who reported that drying temperature had a significant effect on vitamin C.

The decreased of vitamin C in mango powder can be caused due to the nature of vitamin C which is very sensitive to the heat. This result is consistent with several other studies showing decreased levels of vitamin C for example, Kadam *et al.* [24] and Kandasamy *et al.* [25].

Vitamin C is an essential nutrient that has reducing properties, antioxidative properties and also has functional properties for health as a component of an antioxidative defense system. It is able to reduce free radical scavenging and oxygen scavenging so as to protect oxidizing compounds easily, inhibit the formation of nitrosamines in processed meats and reduce metal ions. In food processing, vitamin C is very effective in inhibiting enzymatic browning by reducing orthoquinone [17].

3.3. Total Plate Count. The total plate count (TPC) is the enumeration of aerobic, mesophilic organisms that grow in aerobic conditions under moderate temperatures of 20-45°C. This includes all aerobic bacteria, yeast, molds, and fungi that grow in the specific agar. This count includes all pathogens and nonpathogens and is used to determine the hygienic status of the food product. The presence of TPC in the food in an amount exceeding of the certain limit can be a sign that the food has been exposed to conditions that allow the reproduction of pathogenic microbes. Foods containing TPC load more than specified limits indicated the problem of cleanliness, stagnation or biofilm formation [7].

Test results showed that all treatment gave mango powder with TPC less than 2 log cfu (1.667-1.748 log cfu). This suggests that the oven drying treatment was able to suppress the amount of TPC in mango powder produced. This was closely related to moisture content, where the moisture content of mango powder was <3% (ranging from 2.366 ± 0.576% up to 2.938 ± 0.691%). Wong *et al.* [43] stated that the microbes would be difficult to grow in foods with moisture content less than 10%. In addition, the use of sucrose as the filler in the processing of fruit powder also provides its own advantages, because sucrose is a microbiologically safe, so it is often used as a food preservative.

3.4. Color. Color is an important parameter in food. In addition, color may also indicate the occurrence of chemical changes in food [2]. Hidalgo and Zamora [19] suggest that color is often used as an indicator of sensory quality in foods.

The lightness indicated by the value 0 reflects black and 100 represents white. The mango juice powder which was dried at 50°C has the highest lightness (75.6 ± 1.356) compared to which dried at 60°C (72.6 ± 1.625) and at 70°C (70.4 ± 1.825). This means that an increase in drying temperature causes the color of the resulting mango juice powder to be darker. Anova test showed the increase in drying temperature caused significant difference on lightness. A similar finding was reported by Ekpong *et al.*[15] who conducted research on the processing of instant Javanese fruit powder using the method of foam mat drying at a temperature of 55°C to 70°C.

The chroma of mango juice powder ranged between 41 ± 1.414 (dried at 50°C), 42.4 ± 1.019 (dried at 60°C) and 45.6 ± 1.60 (dried at 70°C). Similar to the lightness, ANOVA test showed that drying temperature had a significant effect on chroma of mango juice powder. These results are consistent with Roongruangsri and Bronlund [36] who examined the effect of drying temperature on pumpkin powders.

The hue angle of the mango juice powder produced at 70°C was 78.27 ± 0.03 representing the yellow-red color. This value was lower when compared to the hue angle of mango juice produced at 60 °C of 81.71 ± 0.15 and 82.27 ± 0.08 for the dried mango juice powder at 50°C. The color of the mango comes from the carotenoid pigment. Light, oxidation, and heat treatment can cause degradation of carotenoids. Decreased levels of carotenoids may lead to discoloration [34]. Von-Elbe and Schwartz [41] suggest that carotenoids can be easily isomerized by heat, acid and light. Oxidation in carotenoids is very easy to occur because it has many conjugated double bonds. Oxidation causes carotenoids to lose color which is a color degradation mechanism. ANOVA test, however, showed that the drying temperature had a significant effect on hue angle of mango juice powder. The hue angle of mango juice dried at 50°C was significantly different from that dried at 70° C.

3.5. Hygroscopicity. Hygroscopicity is the ability of a powder to absorb moisture in environments containing high relative humidity [22]. Hygroscopicity measurements were performed to determine the rate of change in water content so as to indicate the tendency of sticky properties [13]. The materials have a high hygroscopicity has a tendency to absorb much more water so it is easy to agglomerate [39].

Hygroscopicity mango juice powder ranged from 12.00 ± 1.081% up to 12.80 ± 1.481%. This result was lower than hygroscopicity level of papaya extract powder by spray drying method that was equal to 24.57 ± 0.53% [43]. However, hygroscopicity powder mango juice in this study was higher than the hygroscopicity of mango pulp powder which is dried by vacuum drying method that was 7.2% [22].

Table 2. Effect of drying temperature on physical properties

Temp	Hygroscopicity (%)	The degree of caking (%)	The Angle of repose (°)	Solubility index (%)
50°C	12.80±1.481a	9.62±1.483b	38.06±1.093c	98.90 ± 0.47d
60°C	12.51±0.577a	9.54±0.476b	37.77±0.652c	98.30 ± 0.53d
70°C	12.00±1.081a	10.00±0.265b	38.06±1.313c	97.73 ± 0.24d

Data were expressed as the mean ± standard deviation (n=3).
[a-b] Values in each column with different superscripts are significantly different ($P<0.05$).

3.6. Degree of caking. The degree of caking is expressed as the percentage of powder left in the sieve at a certain size after sieving of re-drying the agglomerates formed in hygroscopicity testing [22]. The tendency of the powder to agglomerate is expressed as the percentage of powder left on a 500-micron sieve after sieving of re-drying the agglomerates formed on hygroscopicity testing.

The degree of caking of the mango juice powder ranged from 9.54±0.476% to 10.00± 0.265%. According to Chiou and Langrish [13], powder that has a high hygroscopicity will absorb the surrounding water vapor easier so that the powder becomes sticky and cause caking. This result was lower than that of Jaya and Das [22] which obtained the degree of caking of mango pulp powder by the method of vacuum drying of 18.38%.

3.7. Flowability. Flowability represents the ability of the powder to flow freely which is measured as the angle at which the material stops moving (Angle of Repose). According to Xinde *et al.* [44] the flowability of a material can be measured using the flow index, Hausner ratio and angle of repose methods. Prescott and Barnum [31] defined flowability as the ability of the powder to flow in the desired way on certain equipment. Flowability is the result of a combination physical properties of the material that affect the flow of materials, equipment used for handling, storage, or processing of materials. This means that the flowability of the mango juice is necessary for determining the type of storage and packaging equipment.

Test results showed that the angle of repose of the mango juice powder ranged from 37.77 ± 0.652° up to 38.356 ± 0.93°. Xinde *et al.* [44], state that materials having an angle of repose between 45°-60° are freely flowing and materials with an angle of repose more than 60° can not flow and are cohesive. US Pharmacopeia defines that the material having an angle of repose of 25°-30° has excellent flow properties, 31°-35° good flow properties, 36°-40° fair flow properties and does not require materials aids and materials that have an angle of repose more than 40° has flow properties that quite tolerable and can clog. This means that the mango juice powder has fair flow properties. ANOVA test showed that drying temperature did not have a signicant effect on the flowability of mango powder.

3.8. Solubility Index. The solubility index indicates the capacity of mango powder to continue to form a homogeneous mixture in water [40]. The ideal fruit powder will dissolve completely in water without forming a clot [43].

The solubility index of mango juice powder ranged from 97.73 ± 0.24 % to 98.90 ± 0.47 %. These results indicate that the mango juice powder produced from all treatments had a pretty good solubility.

This was because the filler used was a sucrose which can dissolve 100% in water. This result was better than the solubility index of dried mango pulp powder by spray drying method of 86.7% to 91.5% [26].

Anova showed that the drying temperature had no significant effect on the solubility index of mango juice. This was because the drying temperature used was low, far below the melting temperature and glass transition temperature of sucrose, so there was no change on the sucrose molecule.

4. Conclusions

Choosing a suitable processing parameter that can produce good quality of food product was very important for food industries. It can be summarized that processing mango juice powder by using sucrose as a filler which was dried in an oven drying at 50°C, 60°C, and 70°C could produce mango powder with moisture content less than 3%, ascorbic acid of 0.146 %, TPC less than 2 log cfu. All mango juice powder produced have good physical properties. This study suggested that drying at 50°C was better suited for producing mango juice powder as it preserves most of the physicochemical properties and microbiological stable.

Acknowledgments

The authors would like to thank you to Palembang Institute for Industrial Research and Standardization, Ministry of Industry, in providing the laboratory facilities for this study.

References

[1] Agoreyo, B.O., Akpiroroh, O., Orukpe, O.A., Osaweren, O.R., and Owabor, C.N. (2011). The Effects of Various Drying Methods on the Nutritional Composition of Musa paradisiaca, Dioscorea rotundata, and Colocasia esculenta. Asian Journal of Biochemistry, 6: 458-464.

[2] Agustini, S., Priyanto, G., Hamzah, B., Santoso, B. and Pambayun, R. (2014). Changes on the physical-chemical properties of kue delapan jam on various steaming time. International Journal of Science and Engineering 7(2): 161-165.

[3] AOAC. (1984). Official methods of analysis. Association of Official Analytical Chemists. p. 1010-1011. USA: Arlington, VA.

[4] APO.(2006). Postharvest Management of Fruit and Vegetables in the Asia-Pacific Region. Tokyo and Rome:Asian Productivity Organization and Food and Agriculture Organization of the United Nations.

[5] Bala, B.K., Mondol, M.R.A., Bsiwas, B.K., Das Chowdury, B.L., and Janjai, S. (2003). Solar Drying of Pineapple Using Solar Tunnel Dryer. Renewable Energy, 28:183-190.

[6] Botha,G.E., Oliveira, J.C., and Ahrne, L. (2012). Microwave assisted air drying of osmotically treated pineapple with variable power programmes. Journal of Food Engineering 108:304-311.

[7] Codex Alimentarius Comission (2011). CAC/RCP 33-1985. Code of Hygine and Practice for Collecting, Processing and Marketing of natural mineral water. Rome, Italy.

[8] Cai,Y.Z. and Corke, H. (2000). Production and Properties of Spray-dried Amaranthus Betacyanin Pigments. Journal of Food Science: Sensory and Nutritive Qualities of Food 65 (6): 1248-1252.

[9] Cano-Chauca, M., Stringheta, P.C., Ramos, A.M., and Cal-vidal, J. (2005). Effect of the carriers on the microstructure of mango powder obtained by spray drying and its functional characterisation. Innovative food science and engineering technologies 6:420-428.

[10] Caparino, O.A. (2012). Mango (Philipine Carabao var.) Powder Made From Different Drying Systems.Washington State University (Dissertation).

[11] Chauhan, A. K. and Patil, V. (2013). Effect of packaging material on storage stability of mango milk powder and the quality of reconstituted mango milk drink. Power Technology 239: 86-

93.

[12] Chegini, G.R. and Ghobadian , B. (2005). Effect of spray-drying condition on physical properties of orange juice powder. Drying Technology 2(2): 108-113.

[13] Chiou, D. and Langrish, T. A. G. (2007). Development and characterization of novel nutraceuticals with spray drying technology. Journal of Food Engineering 82: 84-91.

[14] Chong, C.H., and Law, C.L. (2011). Drying of exotic fruits in Jangam, S.V., Law, C.L. and Mujumdar, A.S. Drying of foods,vegetables and fruits.Volume 2, 2010, ISBN-978-981-08-6759-1, Published in Singapore.

[15] Ekpong, A., Phomkong, W., and Onsaard, E. (2016). The Effects of Maltodextrin as a Drying Aid and Drying Temperature on Production Tamarind of Powder and Consumer acceptance of the Powder. International Food Research Journal 23(1):300-308.

[16] Fazaeli, M., Tahmasebi,M., Emam.Djomeh, Z. (2012). Characterization of Food Texture: Application of Microscopic Technology in Mendez-Vilaz, A. Current Microscopy Contribution to Advances in Science and Technology.Formatex.

[17] Fennema O.R, Foods Chemistry. University of Wisconsin Madison, Marcel Dekker Inc, New York, Basel, Hongkong.

[18] Gabas, A.L., Telis, V.R.N, Sobral, P.J.A and Telis-romero, J. (2007). Effect of Maltodextrin and Arabic Gum in Water Vapor Sorption Thermodynamic Properties of Vacuum Dried Pineapple Pulp Powder. Journal of Food Engineering, 82(2):246-252.

[19] Hidalgo, F.J. and Zamora, R. (2000). The role of lipids in non enzymatic browning. Grasas y Aceites 51:35-49.

[20] Hii, C.L., Ong, S.P., and Law, C.L. (2011). Drying studies of tropical fruits cultivated in Malaysia: a review, Journal of Applied Sciences, 11(24):3815-3820.

[21] Jangam, S.V., Joshi, V.S, Mujumdar, A.S, and Thorat, B.N.(2008). Studies on Dehydration of Sapota (Achras zapota). Drying Technology 26(3):369-377.

[22] Jaya, S. and Das, H. (2005). Accelerated storage, shelf life and color of mango powder. Journal of Food Processing and Preservation 29 (1): 45-62.

[23] Jittanit, W., Nitti-Att, S., and Techanuntachaikul, O. (2010). Study of Spray Drying of Peanapple Juice Using Maltodextrin as an Adjunct. Chiang Mai J.Sci. 37(3):498-506.

[24] Kadam D.M., Wilson R.A., and Kaur S. (2010). Determination of Biochemical Properties of Foam-Mat Dried Mango Powder.Int. J. Food Sci. Technol., 45(8), 1626-1632.

[25] Kandasamy, P., Varadharaju, N., and Kalemullah, S. (2012). Foam-mat Drying of Papaya (Carica Papaya L.) Using Glycerol Monostearate as Foaming Agent. Food Science and Quality Management, 9: 17-27.

[26] Mahendran, T. (2008). Effects of Drying Methods on the Quality Characteristics of Mango Powder. JSc-EUSL, 5(1):54-64.

[27] Mishra, P., Mishra, S. and Mahanta,C. (2014). Effect of Maltodextrin Concentration and Inlet Temperature During Spray Drying on Physicochemical and Antioxidant Properties of Amla (Emblica officinalis) Juice Powder. Food and Bioproducts Processing 92 (3): 252-258.

[28] Noor Aziah, A.A and Komathi, C.A (2009). Physicochemical and Functional Properties of Peeled and Unpeeled Pumpkin Flour. J Food Sci 74(7): S328–S333.

[29] Paramita, O. (2012). Kajian Proses Pembuatan Tepung Buah Mangga (Mangifera Indica,L) Varietas Arumanis Dengan Suhu Perendaman yang Berbeda. Jurnal Bahan Alam Terbarukan, 1(1):32-41.

[30] Pasaribu, H.U., Ali, A., and Hamzah F.(2015). Utilization of Arum manis Mango In Dodol Making With Different Concentrations of Glutinous Flour and Seaweed. Jom Faperta 2(2).

[31] Prescott, J.K., Barnum, R.A. (2000). On Powder Flowability. Pharmaceutical Thecnology: 60-84.

[32] Phisut, N. (2012). Spray Drying Technique of Fruit Juice Powder: Some Factors Influencing the Properties of Product. International Food Research Journal 19(4): 1297-1306.

[33] Qadri, O.S, and Srivastava, A.K. (2014). Effect of Microwave Power on Foam-mat Drying of

Tomato Pulp. Agric Eng Int: CIGR Journal, 16(3):238-244.

[34] Que, F., Mao, L., Fang, X., Wu, T. (2008). Comparison of hot air drying and freeze drying on the physicochemical properties and antioxidant activities of pumpkin (Cucurbita moschata Duch) flours. Int. J. Sci. Technol, 43:1195-1201.

[35] Rajkumar, P., Kailappan, R., Viswanathan, R., and Raghavan, G.S.V. (2007). Drying Characteristics of Foam Alphonso Mango Pulp in a Continous Type Foam Mat Dryer. Journal of Food Engineering, 79:1452-1459.

[36] Roongruangsri, W and Bronlund, J.E (2016). Effect of Air-drying Temperature on Physico-chemical, Powder Properties and Sorption Characteristics of Pumpkin Powders. International Food Research Journal 23(3):962-972.

[37] Sangamithra, A., Venkatachalam, S., John, S.G., and Kuppuswamy, K. (2015). Foam Mat drying of Material : A Review. Journal of Food Processing and preservation 39:3165-3174.

[38] Syafutri, M.I, Pratama, F., and Saputra, D. (2006). Sifat Fisik dan Kimia Buah Mangga (Mangifera). Industri Pangan xvii (1):1-11.

[39] Tonon, R. V., Brabet, C. and Hubinger, M. D. (2008). Influence of Process Conditions on the Physicochemical Properties of Acai (Euterpe olearaceae Mart.) Powder Produced by Spray Drying. Journal of Food Engineering 88: 411-418.

[40] Vissotto, F. Z., Montenegro, F. M., Santos, J. M. And Oliveira, S. J. R. (2006). Avaliacáo da influencia dos processos de lecitinacao e de aglomeracao nas propriedades fisicas de achocolatado em pó. Ciéncia e Tecnologia de Alimentos,Campinas 26 (3): 66-671.

[41] Von Elbe, J.H., Schwartz, S.J. (1996). Colorant in Fennema O.R. Foods Chemistry. University of Wisconsin Madison, Marcel Dekker Inc, New York, Basel, Hongkong.

[42] Wilson, R.A., Kadam, D.M., Chadha, S. and Sharma, M. (2012). Foam Mat Drying Characteristics of Mango Pulp. International Journal of Food Science and Nutrition Engineering, 2(4): 63-69

[43] Wong, C.W., Pui, L.P., and Ng, J.M.L. (2015). Production of Spray-dried Sarawak Pienapple (Ananas comosus) Powder from Enzyme Liquefied Puree. International Food Research Journal, 22(4);1631-1636.

[44] Xinde, X., Shanjing, X., Ning, H., Bin, S. (2007). Measurement and Influence Factor of the Flowability of Microcapsules with High-Content β Carotene. Chin.J.Chem.Eng., 15(4):579-585.

[45] Zaini, S.N.(2009). Production of Mangifera Indica Powder Using Spray Dryer and The Effects of Drying on Its Physical Properties. Universiti of Malaysia Pahang (Thesis).

[46] http://www.pharmacopeia.cn/v29240/usp29nf24s0_c1174.html acceced on 21 March 2017

Determination of yeast co-culture ratio and stirring for optimization of bioethanol content of garlic (*Allium sativum*) peels and corn (*Zea mays* L.) cob

Sri Hartini and A Ign Kristijanto

Chemistry Study Program, Faculty of Science and Mathematics
Universitas Kristen Satya Wacana, Salatiga
Jln. Diponegoro no 52-60 Salatiga50711 Jawa Tengah – Indonesia

Email : sri.hartini@staff.uksw.edu

Abstract. Bioethanol as a source of energy from biomass of garlic peels and corn (*Zea mays* L.) cobs waste come into one alternative to reduce the dependence of fossil fuels and the increasingly severe environmental damage. The aims at the research were bioethanol producing with optimum level of garlic (*Allium sativum*) peels and corn cob revealed from yeast co culture and stirring ratio and interaction between both . Fermentation performed during 3 days with various yeast co culture of yeast tape and bread yeast with stirring ratio and control (without stirring). Data analyzed on each type of waste source by design of factorial treatment (4x2) and Randomized Block Design (RBD) with 4 repetitions. As the first factor is the ratio of yeast tape co-culture and bread yeast consisting of 4 levels, that is 7.5%:7.5%; 10%:5%; 15%:5% and 20%:5%. A second factor is done stirring and without stirring. Comparison of the ethanol production percentage_performed between the raw materials source. The research can be concluded that the most optimal of ethanol content of garlic peels are 7.428 ±0.072% on interaction ratio of tape yeast 15% and bread yeast 5% with stirring. The most optimum ethanol product of corn cobs is 8.19% produced at ratio of 10% yeast tape and 5% yeast bread with stirring.

Keywords: *bioethanol, garlic peels, corn (Zea mays L.) cobs, yeast, stirring*

1. Introduction

Fuel energy requirements derived from fossil exploration continue to increase in line with increasing industrial and economic growth. This can be a major problem, when the state has not been able to reduce dependence on fossil fuels or fuel oil while the reserves of these energy sources are increasingly limited [1]. It is time for the use of renewable energy sources of the form of biofuels or increased biofuels, replacing the fossil fuels that are dwindling as in Indonesia Presidential Instruction No . 1/2006 and Indonesia Presidential Regulation No . 5/2006 on national energy policy. One of the liquid vegetable fuel is a substitute for gasoline named bioethanol.

From the last several years, the worldwide economic and environmental pollution issues there has been increasing research interest in the value of bio-sourced lignocellulosic biomass. Agro-industrial biomass comprised of lignocellulosic waste is an inexpensive, renewable, abundant and provides a unique natural resource for large-scale and cost-effective bio-energy collection. In this background green biotechnology present a promising approach to convert most of the solid agricultural wastes particularly lignocellulosic materials into liquid bio based energy-fuels [2]. The use of much cheaper

IOP Conf. Series: Journal of Physics: Conf. Series **1095** (2018) 012036 doi:10.1088/1742-6596/1095/1/012036

and widely available feedstock, lignocellulosic materials from agricultural and forestry industry wastes needs to be further developed.

One of the most widely found agricultural waste is garlic peels and corn cobs. Garlic peels are an industrial waste and agriculture that has not been utilized maximally, especially in its potential as an energy source alternative. An investigation into the composition of garlic peels showed the presence of proteins, lipids, lignin, mannitol, pectin and polysaccharides. Garlic peels are rich in pectin (27%), combined rhamnose (11.42%) and galactose (5.6%) [3]. Production garlic Indonesia in 2016 according to data sourced from Directorate General of Horticulture is amounted to 21.15 thousand tons [4]. According to the data also only 71% of garlic can be consumed. This means that about 29% is the waste of garlic products. Based on the percentage, the waste of garlic product would be 6.1335 thousand tons, garlic peels included. Research on garlic peels content that could be used as raw material for bioethanol performed by [5], where there was 26.58% of total carbohydrate, 18.62% cellulose and 0.4% protein.

The content of carbohydrates from garlic peels could be hydrolyzed into glucose, when it is fermented, becoming ethanol. The corn cobs are preferred over other agricultural by-products due to its composition which is easily converted to bioethanol. A viable alternative to bio-fuel can be pursued by utilization of corn cobs due to suitable lignocellulosic waste for bioethanol production of chemical pre-treatments. The lignocellulosic bio mass comprises of cellulose, hemicellulose and lignin. Now-a-day the demand of bioethanol has increased considerably because of its use as a gasohol in addition to the other application for industries which need production of alcohols on large scale. According to [6], the advantageous utilization of co-cultures instead of single cultivations included the production of bulk chemicals, enzymes, food additives, antimicrobial substances and microbial fuel cells. Co-culture fermentations may result in increased yields, improved control of product qualities and the possibility of utilizing cheaper substrates. The use of co-culture techniques in the fermentation process to produce bioethanol proved to provide better results than the use of pure culture of *S. cerevisiae* and single yeast tape culture [7].

Based on the above scientific explanation, the research aims were bioethanol producing with optimum level of garlic peels and corn (*Zea mays* L.) cobs revealed from yeast co-culture and stirring ratio and interaction between both.

2. Materials and method

2.1. Materials

The garlic peels is obtained from a nut factory in Pati, corn (*Zea mays* L.) cobss, molasses, tape yeast bought from the market in Salatiga, and yeast bread bought from a bakery in Salatiga. Chemicals used include NaOH p.a (Merck), HCl p.a (Merck), H_2SO_4, Nutrient Broth, glucose standard, DNS reagents, and KNa Tartrate. The tools used include drying cabinet, TomY SS-240 autoclave, Autonics TC45 incubator, magnetic stirrer, Optizen 2120 UV-Vis spectrophotometer, 1 set of distillation equipment, alcohol meter

.

2.2. Methods

2.2.1. Delignification ([8] modified)

The garlic peels powder was delignification with 15% NaOH (1:10) (b/v) in 121°C autoclave for 15 minutes, while the corn cobs powder delignification using 0.25 M NaOH (1: 8) (w / v), then rinsed until the neutral pH is then dried in a 50°C drying cabinet for 24 hours.

2.2.2. Cellulose hydrolysis ([9] modified)

Delignified garlic peels powder was hydrolyzed with 4N HCl (1:10) (b/v) and corn cobs powder with 15% H_2SO_4 (1:10) (w / v) in reflux at a temperature of 100°C for 120 minutes.

2.2.3. Bioethanol fermentation

Hydrolyzed substrate was neutralized to pH 4.6 then added molasses and aquadest with ratio of substrate: molasses: aquadest (6: 2: 2), then solution fermented with ratio of yeast tape: bread yeast (% v/v) 7.5: 7.5; 10: 5; 15: 5; and 20: 5. Yeast tape is given first then after 24 hours bread yeast added and fermentation continued up to 72 hours. All ratios are treated with stirring and without stirring.

2.2.4. Bioethanol distillation

The solution was distilled at 78.5°C for ± 4 hours until no drips and ethanol content were measured with alcohol meter.

2.3 Data Analyse

Data analyzed on each type of waste source by design of factorial treatment (4x2) and Randomized Block Design (RBD) with 4 repetitions. As the first factor is the ratio of yeast tape co-culture and bread yeast consisting of 4 levels, that is 7.5%:7.5%; 10%:5%; 15%:5% and 20%:5%. A second factor is done stirring and without stirring.

3. Result and discussion

3.1. The effect of various yeast co-culture ratios on ethanol content

The mean ethanol content (in % ± SE) in the various yeast ratios of tape and bread yeasts ranged from 4.690 ± 0.545% to 6.894 ± 0.272% (**Table 1**).

Table 1. Average garlic content among various yeast ratios

	Tape yeast : Bread yeast (%)			
	7.5 : 7.5	10 : 5	15 : 5	20 : 5
Average ± SE	4.768 ± 0.145	5.715 ± 0.222	6.894 ± 0.272	4.690 ± 0.545
W = 0.143	(a)	(b)	(c)	(a)

Note : *W = HSD 5%

* The numbers followed by the same letter indicate that the treatments were not significantly different, while the numbers followed by the unequal letters showed different treatments. This description also applies to the next Table.

Table 1 shows that tape yeast up to 15% in the co-culture ratio increases in the ethanol content. The optimal level is produced by 15% yeast tape ratio and 5% yeast bread is 6.894 ± 0.272%. This value is lower when compared with research conducted by [7] with cassava base, 5% yeast tape ratio and 5% yeast bread produce 11% ethanol. The decrease in ethanol contents after a 15% tape yeast ratio due to glucose on the fermented substrate is no longer sufficient to convert to ethanol.

Average of ethanol content (% ± SE) from corn cobs ranged from 5.113 ± 0.221 to 7.510 ± 0.345 (Table 2).

Table 2. Average ethanol levels of corn (*Zea mays* L.) cobs among various yeast ratio

	Tape yeast : Bread yeast (%)			
	7.5 : 7.5	10 : 5	15 : 5	20 : 5
($\bar{X} \pm SE$)	6.199 ± 0.234	7.510 ± 0.345	6.255 ± 0.320	5.113 ± 0.221
W = 0.097	(b)	(c)	(b)	(a)

IOP Conf. Series: Journal of Physics: Conf. Series **1095** (2018) 012036 doi:10.1088/1742-6596/1095/1/012036

Table 2 shows the same phenomenon, showing that the addition of tape yeast increases to the ethanol content and the addition of bread yeast does not affect the ethanol content of the corn at 15% yeast tape ratio and 5% yeast yeast.

3.2. The influence of stirring on bioethanol product

The average ethanol content of Garlic peels with stirring ranged from 5.157 ± 0.383% to 5.950 ± 0.431% (Table 3). Purify levels of bioethanol (in % ± SE) in various yeast ratios of tape and bread yeasts ranged from 4.690 ± 0.545% to 6.894 ± 0.272%.

Table 3. The average of bioetanol content of garlic peels among various yeast ratio

	Tape Yeast : Bread Yeast (% ± SE)			
	7.5 : 7.5	10 : 5	15 : 5	20 : 5
Average ± SE	4.768 ± 0.145	5.715 ± 0.222	6.894 ± 0.272	4.690 ± 0.545
W = 0.143	(a)	(b)	(c)	(a)

Average ethanol content of corn cobs fermentation ranges of 5.113 ± 0.221 up to 7.510 ± 0.345 (Table 4).

Table 4. Average ethanol levels corn cobs among various yeast ratio

	Tape yeast : Bread yeast (% ± SE)			
	7.5 : 7.5	10 : 5	15 : 5	20 : 5
Average ± SE	6.199 ± 0.234	7.510 ± 0.345	6.255 ± 0.320	5.113 ± 0.221
W = 0.097	(b)	(c)	(b)	(a)

Table 4 shows that the addition of tape yeast increases the ethanol content and the addition of bread yeast does not affect the ethanol content. Ethanol content increased at 10% yeast tape ratio and 5% yeast bread then decreased at 15% yeast tape ratio and 5% yeast bread.

3.3. Interaction between various yeast co-culture ratios and stirring on bioethanol product

The average garlic peels content of the interaction of the various yeast co-culture and stirring ratios ranged from 4.188 ± 0.157% to 7.428 ± 0.072% (Table 5 and Table 6). Table 5 shows that the ethanol content between ratios will increase from the addition of tape yeast to 15%, then decrease to the addition of tape yeast 20%. Furthermore, the highest ethanol content was obtained with 15% yeast tape ratio and 5% yeast yeast in stirring of 7.428 ± 0.072% Decrease in ethanol levels over 15% tape yeast is likely due to the availability of fermented glucose that is no longer sufficient to convert to ethanol. Stirring serves to help cell and substrate contact, keep the microorganisms from settling down and flatten the temperature so that the yeast can work more optimally [9].

Table 5. Content of ethanol of garlic peels based on interaction between yeast ratio and stirring

Tape yeast : Bread yeast	Stirred				Without stirring			
W = 0.151	Ratio (% ± SE) Tape yeast : Bread yeast				Ratio (% ± SE) Tape yeast : Bread yeast			
	7.5:7.5	7.5:7.5	15:05	20:05	7,5:7,5	10:05	15:05	20:05
	5.045 ± 0.072 (a)	6.135 ± 0.168 (b)	7.428 ± 0.072 (c)	5.193 ± 0.046 (a)	4.490 ± 0.118 (b)	5.045 ± 0.072 (a)	6.360 ± 0.087 (d)	4.188 ± 0.157 (a)

Ethanol content average of corn cobs of interaction ratio of yeast co-culture and stirring ranged from 4.683 ± 0.121 until 8.190 ± 0.019 (Table 6). Ethanol content increased in the ratio of yeast tape 10%: yeast bread 5% with stirring of 8.19% and without stirring that is 6.830%. Ethanol contents at ratio of 10% yeast tape: 5% yeast bread increased due to glucose used for fermentation more than other ratios. However, the ethanol content of the stirring treatment was wholly lower than that without stirring. The stirring causes the yeast cell to be homogeneous and non-settling so that the sugar conversion to ethanol is more optimal [10].

Table 6. Ethanol content of corn cobs based on interaction between yeast ratio and stirring

Tape yeast : Bread yeast	Stirred				Without stirring			
	Ratio (% ± SE) Tape Yeast : Bread Yeast				Ratio (% ± SE) Tape Yeast : Bread Yeast			
	7.5:7.5	7.5:7.5	15:05	20:05	7.5:7.5	10:05	15:05	20:05
W = 0.103	6.658 ± 0.087 (b)	8.190 ± 0.019 (d)	6.878 ± 0.200 (c)	5.543 ± 0.081 (a)	5.740 ± 0.062 (a)	6.830 ± 0.073 (c)	5.633 ± 0.0657 (b)	4.683 ± 0.121 (a)

Figure 1 illustrates the linkage of added yeast ratio and the effect of stirring on the production of ethanol. The highest production of ethanol was obtained from garlic peels fermentation (KBP) by stirring and ratio of the yeast tape (15): bread yeast (5). Different results are shown by the fermentation of the stirred corn cobs, the highest ethanol product actually produced by the ratio of tape yeast (10): bread yeast (5).

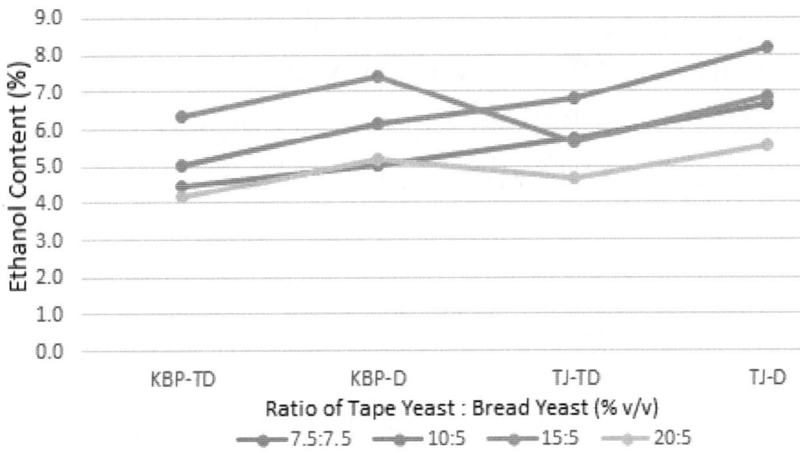

Figure 1. Ethanol content viewed from interaction among the yeast ratio and stirring
(KBP = Garlic Peels, TJ = Corn (*Zea mays* L.) cobs, TD = Without Stirrer, D = Sturred)

4. Conclusion

The research can be concluded that the most optimal of ethanol content of garlic peels are 7.428 ±0.072% on interaction ratio of tape yeast 15% and bread yeast 5% with stirring. The most optimum ethanol product of corn cobs is 8.19% produced at ratio of 10% yeast tape and 5% yeast bread with stirring.

Acknowledgments
Thank you to Valeri Stefania and Tiara Osa Meutia who have helped a lot in the laboratory during the research.

References
[1] Zou C, Qun Z., Guosheng Z and Bo X. 2016 Natural Gas Industry B. 3(1) 1.
[2] Anwar Z., Gulfraz M and Muhammad I 2015 Journal of Radiation Research and Applied Sciences. 7(2) 163.
[3] Abdel-Fattah A F and Edrees M 1972 *Journal of the Science of Food and Agriculture*. **23**(7) 871.
[4] Pusat Data dan Sistem Informasi Pertanian, 2017. Buletin Konsumsi Pangan Semester 2.
[5] Sugave D, 2014. Characterization Of Garlic Peels And Its Evaluation As Biomaterial. Rourkela.
[6] Bader J, Mast-Gerlach E, Popović M K, Bajpai R and Stahl U 2010 *J. Appl. Microbiol.* **109**(2) 371.
[7] Arnata I W and Anggraeni A A.M D 2013 *J. Agrointek* **7** (1).
[8] Sukumaran R.K, Singhania RR, Mathew G.M and Pandey A 2009*Renewable Energy*. **34** (2) 21.
[9] Kurniawan S, Juhanda, S. Syamsudin, R and Lukman M.A 2011 *J. STU*, ISSN: 1693.
[10] Hossain A B M S, Boyce A N, Salleh A and Chandran S 2010 *African Journal of Agricultural Research* **5(14**) 1851

Phytoremediation Of Coal Mining Acid Water In PT Bukit Asam Tanjung Enim South Sumatera

Sri Pertiwi Estuningsih[1], Bambang Yudono[2], Yonanda[2]

[1,3]Department of Biology,University of Sriwijaya
[2]Department of Chemistry, University of Sriwijaya

Email: yudonob@hotmail.com Phone:+628153821994

Abstract. The using of some water plants to improve the quality of waters which is polluted by acid mine drainage (AMD) containing sulphate(SO_4^{2-}) and heavy metal(Fe and Mn) can be used as one of the ways of biological controls which is environmental friendly, accurate, effective and efficient. Research about "The Ability Test Several Types of Water Plants as the Agents Phytoremediation Acid Mine Drainage Coal PT Bukit Asam, TanjungEnim, South Sumatera" aims to know the ability of *Eichhorniacrassipes*, *Limnocharisflava*, and *Neptuniaoleraceain* phytoremediation acid mine drainage coal at many concentrations. This research use complete randomly plans (CRP), the pattern factorial with two factors, they were water plants (*Eichhorniacrassipes, Limnocharisflava*, and *Neptuniaoleracea*) and concentration of acid mine drainage coal (0%, 25%, 50%, 75% and 100%), every combination was repeated twice. Based on the data from the research, *Eichhorniacrassipes, Limnocharisflava*, and *Neptuniaoleracea* are potentially as agent of phytoremediation AMD. The high percentage of decreasing sulphate (SO_4^{2-}) at 75% concentration of AMD as media for water plant but the ability of water plant to decreased sulphate is same level. Interaction between kinds of water plants with concentration of AMD effect the percentage of decreasing manganese (Mn), *L. flavain* 75% concentration of AMD has the percentage of decreasing manganese which is high rather than *E. crassipesand N. oleraceain* different concentration of AMD. By considering its ability of growth in AMD, *N. oleraceahas* a better capability than *E. crassipes* and *L. flava*.

Keywords : Phytoremediation, Water Plants, Acid Mine Drainage Coal

1. Introduction

Mining in Indonesia uses an open mining system, due to the release of high coal pyrite (FeS) concentration and the further oxidation of FeS into $FeSO_4$, these compounds in water decompose produce potential H_2SO_4 in lowering the pH of liquid waste to form acid mine water (AMW). The acidity condition of mine water will dissolve other heavy metals such as Mn, Co, Hg, which are detrimental to the environment. The impact of AMW can be felt directly, if AMW flows into the waters then aquatic life will be poisoned. People in the mining area feel the effects of AMW that pollute the river, they have itchy diseases and river water can not be utilized. If the AMW pollutes the soil, the soil may become acidic and increase the solubility of heavy metals, so the soil becomes an unfavorable medium for plant growth [1].

Acid water resource management activities have been conducted in various ways, both physically and chemically, but they are relatively expensive, inhospitable, ineffective and incomplete. Biological management such as phytoremediation uses a variety of plant species, the implementation is environmentally friendly, thorough, effective, and efficient control of pollutants in the waters [2,3,4].

Research on AMW waste phytoremediation has not been widely discussed. Hyacinth is known as a weed plant, but its ability to live in extreme conditions and multiply rapidly then this plant can be used as a remediator agent. *Eichhorniacrassipes* has the ability to adapt and be tolerant of acid mine acid wastes with concentrations of 25%, 50%, 75% and 100%, the root canopy ratio in the treatment shows a proportional comparison respectively [5].

Limnocharis flava plant effectively reduces Fe, Mn, BOD, COD, TSS, sulphate, and phosphate levels in Cisande River waters that have contaminated iron (Fe) and manganese (Mn) content higher than the standard applicable water quality. *Limnocharis flava* can absorb Fe as much as 63.99% at Fe 3 ppm concentration and absorb Mn equal to 63.21% at concentration Mn 3 ppm. *Neptunia oleraceae* was able to absorb heavy metals Cd, Pb, and Hg in liquor polluted water in Thailand at a concentration of 20% of 89.3% -96.3% within 30 days and also increase the pH [6,7].

The type of plant that suits the waste characteristics as phytoremediation agents is one of the essential elements for successful phytoremediation. The determination of the concentration of the waste also has a role in the success of phytoremediation, since the response of plants to different levels of waste concentration varies, depending on the genetic characteristics and characteristics of the plant [3,8].

This research needs to be done to see the ability of several types of aquatic plants as phytoremediation agents in various concentrations of coal acid mine waste water by measuring the percentage of sulphate and manganese reduction and to see the characteristics of the aquatic plants at the end of the study. It is assumed that in this study *Eichhorniacrassipes, Limnocharisflava* and *Neptuniaoleracea* have different capabilities in the phytoremediation process of acid mine coal water at various concentrations [5,9].

2. Materials and Methods

This study used a complete randomized design (RAL) factorial pattern consisting of two factors: water plants (*Eichhorniacrassipes, Neptuniaoleracea* and *Limnocharisflava*) and concentrations of AMW coal (0%, 25%, 50%, 75% and 100%). Each treatment combination was repeated 3 times. This research begins with AMW sampling, then preparation and acclimatization of water plants for 7 days, making bioreactors, measuring the initial content of AMW (sulphate (SO_4^{2-}) and manganese (Mn), planting water plants into bioreactors in accordance with experimental design, maintenance and observation of water plants for 30 days of treatment. At the end of the treatment, we measured the dry weight of each plant, measurement of final sulphate content (SO_4^{2-}) and manganese (Mn) and observation of physical characteristics of aquatic plants [10].

3. Results and Discussion

Based on the research that has been done about the ability test of some water plants as phytoremediation agent of acid mine coal water got the following result:

3.1 Percentage of Sulphate Reduction (SO_4^{2-})

Based on the Variance Analysis test, it is found that the concentration of acid mine water has a significant effect on the percentage of the decrease of sulphate content, while the plant species, the interaction between plant species and the concentration have no significant effect because plants (*Eichhorniacrassipes, Neptunia oleracea* and *Limnocharis flava*) ability that is not different in absorbing and adapting in sulphate stress. Sulphate will be reduced to sulfihidril in order to be utilized by plants to perform its metabolism [7,11].

Further test results of DMRT α 5% AMW concentration on the percentage decrease in sulphate content is presented as follows:

Table 1. Percentage of Sulphate Reduction in Phytoremediation Process at Various AMW Concentrations

AMW Conc. (%)	Sulphate (SO_4^{2-})		Procent AMW reduction (%)
	Initial (mg/L)	Final (mg/L)	
K_0 (0)	0	0	0 [a]
K_1 (25)	637,50	526	17,49 [a]
K_2 (50)	1275	762,50	40,20 [b]
K_4 (100)	2550	1769,83	69,41 [b]
K_3 (75)	1912,50	1414,66	71,76 [b]

Description: The numbers followed by the same letters show no significant difference in the DNMRT test

Based on Table 1. The high percentage of sulphate reduction in concentrations ie 75%, 100% and 50% is significantly different from the percentage of sulphate decrease at 0% and 25% concentrations due to concentrations of 75%, 100% and 50%, respectively, the sulphate content in the media is thought to be utilized by plants for its metabolism such as photosynthesis. The sulphate absorbed by the plant as an important part of ferodoxin of Fe and S complexes in chloroplasts, is involved in the formation of chlorophyll associated with photosynthesis and some of the carbohydrate, fat and protein formation reactions. Sulphates present in the acid mine water medium at concentrations of 50%, 75% and 100% are thought to be described with the help of microbes present in the rhizosphere so they can be absorbed by plants. Sulphate will be absorbed by plants through roots with the help of sulphate reducing bacteria (SRB) [7].

The low percentage of sulphate reduction at 25% concentration is not significant with 0% concentration, this is because the agent at 25% concentration can adapt to slightly sulphate medium so that its ability is the same on unsullied media. Low sulphate causes the phytoremediation process to be ineffective and efficient. Sulphate deficiency will inhibit protein synthesis and decrease chlorophyll levels.

3.2 Percentage of Manganese Reduction (Mn)

Based on the Variance Analysis test, it is found that the interaction between acid concentration of coal mine and water plants has significant effect on the percentage of manganese degradation as shown in Table 2.

Table 2. Percentage of Manganese Decrease after 30 Days of AMW Phytoremediation Process

Treatment combinations	Mn (mg/L)		Av. Procentage Mn Reductions (%)
	Initial	Final	
J_1K_0	0	0	0 [a]
J_2K_0	0	0	0 [a]
J_3K_0	0	0	0 [a]
J_2K_4	8,20	8,18	0,30 [ab]
J_1K_4	8,20	8,10	1,22 [ab]
J_1K_1	2,05	2,02	1,46 [ab]
J_2K_3	6,15	5,40	12,20 [bc]
J_2K_1	2,05	1,68	18,05 [bc]
J_1K_3	6,15	4,84	21,30 [bc]
J_1K_2	4,10	3,15	23,17 [bc]
J_2K_2	4,10	3,15	23,17 [bc]
J_3K_4	8,20	3,60	56,10 [cd]
J_3K_1	2,05	0,60	70,73 [de]
J_3K_2	4,10	0,35	91,46 [de]
J_3K_3	6,15	0,35	94,31 [e]

Description: The number followed by the same letters show no significant difference in the DNMRT test

The plant is able to perform rhizofiltrat ion and rhizodegradation in the absorption of heavy metal manganese [1,4]. DMRT test results α 5% interaction between species of aquatic plants and acid concentration of mine water to the percentage of manganese decrease is presented Table 2. *E. crassipes* and *N. oleracea* at 100% concentration, low percentage of Mn decrease is due to the concentration of 100% manganese is too high and the plant can not tolerate it anymore, so the ability of plants to absorb manganese and other nutrients decreases. Nutrient uptake by roots is selective, but some nutrient ions uptake can be competitively inhibited by other ions with different valencies such as K^+ ions whose uptake can be inhibited by Cl^-, Mg^{2+}, SO_4^{2-}, Ca^{2+} ions or Mn^{2+}. At too high a concentration, essential elements can cause poisoning and inhibition of the absorption of other nutrients will increase [12].

The percentage of manganese *E. crassipes* decrease at 50% and 75% concentrations was not significant with percentage of manganese decrease by *N. oleracea* at concentrations of 50% and 75%. At this concentration plants can still utilize the manganese contained in the growing medium for metabolic processes, especially in the light phase of photosynthesis. Manganese serves as activators of various enzymes, breaking H_2O molecules (water photolysis) in the light phase of photosynthesis and is a structural component of the chloroplast membrane system used for chlorophyll synthesis [13].

Based on Table 2. it is expected that *Limnocharisflava* has better ability to decrease manganese than *E. crassipes* and *N. oleracea* because *L. flava* has long and thick fibrous roots which helps the process of absorption of pollutant through roots. *L. flava* can decrease manganese in B3 contaminated waste, the highest accumulation of manganese is found in root organ, because the root is directly related to contaminated manganese planting media. Manganese that have been absorbed, will be distributed to the leaves [14,15].

L. flava is also thought to work with microoganism for manganese degradation process in its root zone. The plant will form chelate (phytochelatin) if it is in heavy metal stress, the formation of this chelate is the response of plants to be able to adapt in a poisonous environment. Manganese absorbed in the form of valence two (Mn^{2+}) will chelate with organic compounds produced by microbes in the area around the roots. Substances that chelated manganese is the type fitosiderofor, then manganese will be brought to the root cell through active transport [15,16].

3.3 Measurement of Dry Weight

Variant Analysis Test (ANAVA) got result that interaction between acid concentration of acid water and type of aquatic plants significantly affect dry weight *E.crassipes, N.oleracea* and *L.flava*. The nutrient content contained in the acid mine water medium is an additional nutrient for the plant. The plant can utilize well the nutrients in the form of sulphate (SO_4^{2-}) contained in acid mine water media to form proteins as the constituent parts of the body cell [17].

Further test results of DMRT α 5% interaction between species of aquatic plants and acid concentration of mine water to average dry weight of aquatic plants are presented in Table 3.

Table 3. Average Dry Weight of Water Plant after 30 Days of Phytoremediation Process

Treatment combinations	Av. dry weight (gram)
J_3K_4	8,04 ^{a}
J_3K_0	8,70 ^{a}
J_3K_3	8,46 ^{a}
J_1K_0	13,29 ^{ab}
J_1K_1	16,75 ^{bc}
J_3K_1	16,84 ^{bc}
J_1K_3	17,88 ^{bc}
J_1K_2	18,04 ^{bc}
J_1K_4	21,13 ^{bc}
J_2K_4	22,97 ^{c}
J_2K_3	32,20 ^{d}
J_2K_1	36,06 ^{d}
J_2K_2	37,11 ^{d}
J_2K_0	38,43 ^{d}

Description: The numbers followed by the same letters show no significant difference in the DMRT test

The average dry weight of low consecutive on treatment of *L. flava* concentration of 50%, 100%, 0% and 75% acidic water was not significant with *E. crassipes* concentration 0%, respectively. The low average dry weight of *L. flava* due to *L. flava* is not resistant to acid acid water stress so many organs are dead. *L. flava* is a type of herbaceous plant in which the organ is relatively susceptible to pollutant stress and can not be used as a phytoremediation agent for long periods (30 days). *L. lava* usage is used as phytoremediation agent of river water contaminated by hazardous materials in the form of Fe and Mn content only within 6 days. The average dry weight of low consecutive on treatment of *L. flava* concentration of 50%, 100%, 0% and 75% acidic water was not significant with *E. crassipes* concentration 0%, respectively. The low average dry weight of *L. flava* due to *L. flava* is not resistant to acid acid water stress so many organs are dead. *L. flava* is a type of herbaceous plant in which the

organ is relatively susceptible to pollutant stress and can not be used as a phytoremediation agent for long periods (30 days). *L. lava* usage was used as phytoremediation agent of river water contaminated by hazardous materials in the form of Fe and Mn content only within 6 days.
Physical Characteristics of Water Plants Changes in root color at AMW concentrations (0%, 25%, 50%, 75% and 100%) in Root's*E.crassipes* as in **Figure 1**

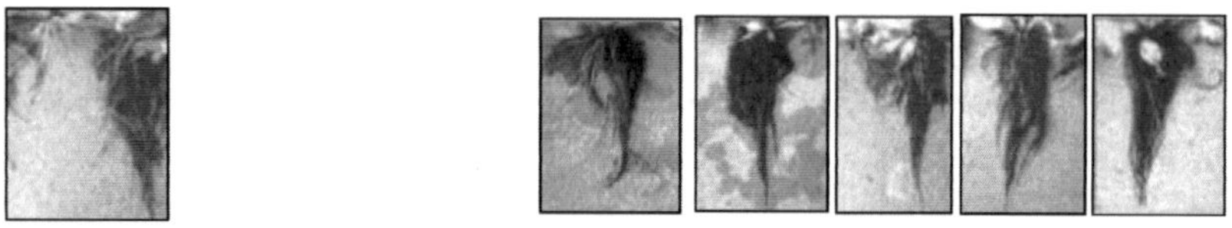

0% 25% 50% 75% 100%

Before treatment After treatment

Figure 1.Root's *E.crassipes*

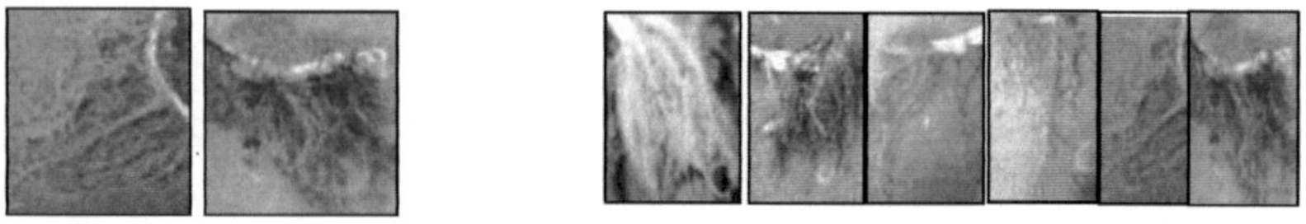

0% 25% 50% 75% 100%

Before treatment After treatment

Figure 2. Root's *N. oleracea*

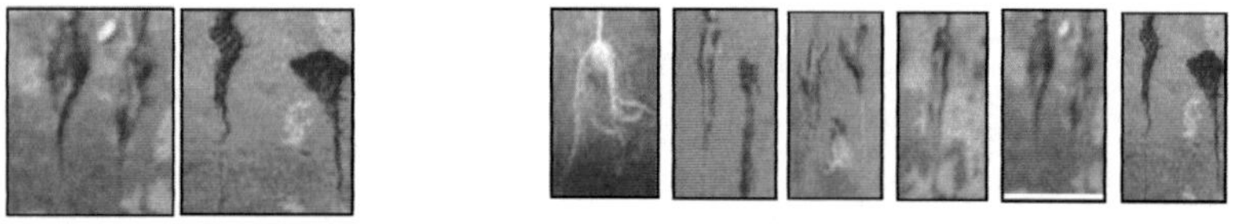

0% 25% 50% 75% 100%

Before treatment After treatment

Figure 3. *Root's L Flava*

The higher the acid water concentration, the root color on *E. crassipes* and *N. oleracea* are fade, hair loss of roots also increases whereas for *L. flava* at 100% concentration blackened this is because the agent die and cause root rot as shown in Fig. 1, 2, and 3. High concentration sulphate and heavy metals contained in the media allegedly affect the physical condition of the roots causing the falling of new roots or root hair because the roots are the organs that directly intersect with the media. Root is the organ that will experience the first damage due to the roots that intersect directly with contaminated planting media [18].

Acid water contains micro nutrients used by plants for the metabolism of one of the micro nutrients ie copper (Cu). The presence of Cu in the media affects the growth and root characteristics. *E. crassipes* that are in heavy Cu metal stress will give a fading color response with increasing acid concentration of acidic water. Heavy metal stress causes intracellular and extracellular damage due to the presence of Cu triggering the seizure of other binding protein minerals so that the absorption of other elements decreases. Excess Cu also inhibits the level of permeability plasmalema and resulted in the destruction of the cell wall of the roots. Damage to the root cell wall can cause a fall of new roots and root hairs. The more heavy metal absorbed, the color of the roots will fade [1, 4].

3.4. Leaf color changes at AMW concentration (0%, 25%, 50%, 75% , and 100%) in leaf's E.crassipes

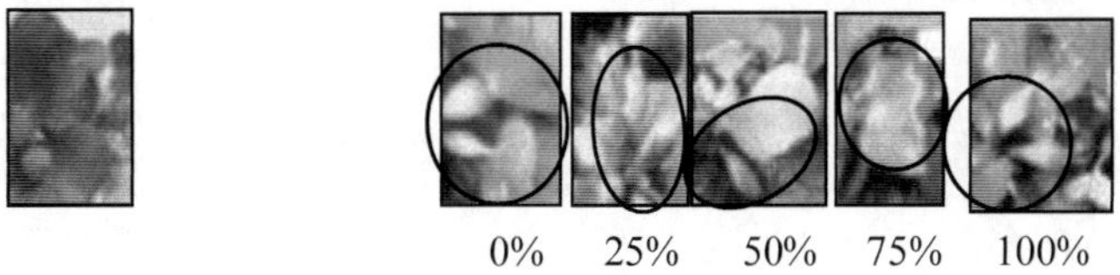

Before treatment After treatment

Figure 4. Leaf's *E. crassipes*

Figure 4 shown that the morphology of *E. crassipes* leaves at 0% concentration (control) and 100% of severe necrosis and chlorosis, whereas at concentrations of 25% to 75% of leaf states is better this shows that *E. crassipes* is less able to adapt, can not grow well on non-pollutant media (0%) and with pollutants at high concentrations (100%) [19].

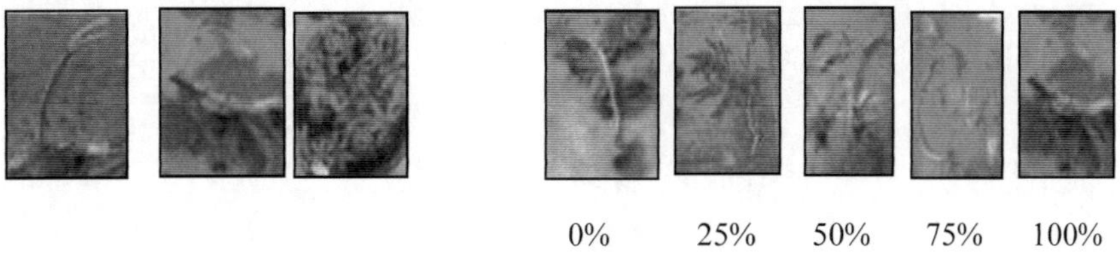

Before treatment After treatment

Figure 5.Leaf's *N. oleracea*

Leaf color of *N. oleracea* before treatment at each concentration is almost the same that is dark green. Leaf color changes in *N. oleracea* is indicated by the occurrence of chlorosis until the leaf falled. At 0% concentration, the leaves of N. oleracea still look healthy while at concentrations of 25% to 100% of *N. oleracea* leaves have chlorosis and leaf loss as shown in Figure 5 [4].

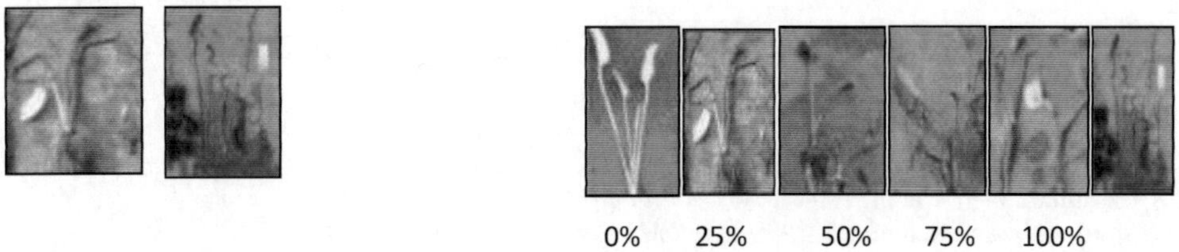

Before treatment After treatment

Figure 6. Leaf's *L. flava*

In L.flava chlorosis occurs in the leaves. The chlorosis in *L. flava* begins with a green leaf color that changes into a reddish and then turns yellow and eventually dies. The increasing concentration causes growth to be further hampered as shown in Figure 6. According to Salisbury & Rossa (1995) chlorosis that begins with a leaf turns red is a symptom of phosphorus deficiency. Acid water is suspected to contain no phosphorus. *E. crassipes*, *N. oleracea* and *L. flava* at a concentration of 100% damage and leaf fall in a high degree, suspected leaf tissue and chlorosis due to the high accumulation of heavy metals [20].

4. Conclusions

Based on research that has been done about the ability test of several types of aquatic plants as phytoremediation agent of acid mine coal water of PT. Bukit Asam, TanjungEnim, South Sumatera can be summarized as follows:

- *Eichhornia crassipes*, *Neptunia oleracea* and *Limnocharis flava* have potential as phytoremediation agents of acid mine coal water.
- High percentage of sulphate reduction (SO4) is concentration of 75% acidity of mine water asa medium for growing aquatic plants whereas for the species of water plants (*Eichhornia crassipes*, *Neptunia oleracea* and *Limnocharis flava*) have the same ability to lower sulphate.
- *Limnocharis flava* at a concentration of 75% mine acid water is more likely to decrease manganese (Mn) in phytoremediation of acid mine coal water compared to *Eichhornia crassipes* and *Neptunia oleracea* in various concentrations of acidic acid water.
- *Neptunia oleracea* at various concentrations has a high dry weight value than *Eichhornia crassipes* and *Limnocharis flava* which show that *Neptunia oleracea* is able to grow well in acid mine water media than *Eichhornia crassipes* and *Limnocharis flava*.

Prospective

Based on the results of research that has been obtained that can be done to study the environmental aspects of AMW phytoremediation activities on the microcosm scale and can then be applied to the AMW processing in the coal mining industry.

Acknowledgment

Thank you to DP2M DIKTI for Research funding through Grant Research Program competing in 2015/2016

References

[1] Renault S, Sailerova E, Fedikow MAF 2002 Progress Report on Seed Germination, Survival and Metal Uptake of Seedlings Planted at Central Manitoba (Au) Minesite (NTS 52L13) 138
[2] Schwab P, Banks MK, Kyle WA 2006 *Water. Air. Soil Pollut.* **177(1-4)** 239
[3] Wenzel WW 2009 *Plant Soil* **321(1-2)** 385
[4] Monitoring F and Luan Y 2009 *Application of hyperaccumulators in the phytoremediation of heavy metals-contaminated soils*
[5] Bauder JW *Final Technical Report Evaluation of Phytoremediation of Coal Bed Methane Product Water and Waters of Quality Similar to that Associated with Coal Bed Methane Reserves of the Powder River Basin , Montana and Wyoming Evaluation of Phytoremediation of Coal* 1
[6] Bhadbhade BJ, Sarnaik SS, and Kanekar PP 2002 *Curr. Microbiol.* **45(5)** 346
[7] De Almeida DG *et al.* 2016 *Frontiers in Microbiology* **7**6.
[8] Zakharova EA *et al.* 2000 *Environ. Sci. Pollut. Res.* **7(4)** 191
[9] Singh DN and Tripathi AK 2013 *Bioresour. Technol.* **128** 215
[10] Hasanuzzaman M *et al.* 2004 *Curr. Microbiol.* 49(2) 108
[11] Margesin R and Schinner F 1998 *Appl. Microbiol. Biotechnol.* 49 (4) 482
[12] Young CC, Lin TC, Yeh MS, Shen FT, and Chang JS 2005 *World J. Microbiol. Biotechnol.* 21(8-9) 1409
[13] Chaîneau CH, Yepremian C, Vidalie JF, Ducreux J, Ballerini D 2003 *Water. Air. Soil Pollut.* **144(1-4)** 419
[14] Kirk JL, Moutoglis P, Klironomos J, Lee H, Trevors JT 2005 *Plant Soil* 270 (1) 23
[15] Qadir M, Oster J 2002 *Irrig. Sci.* **21(3)** 91
[16] Banaszak JE, Rittmann BE, Reed DT 1999 *J. Radioanal. Nucl. Chem.* **241(2)** 385
[17] Shah K and Nongkynrih JM 2007 *Biol. Plant.* **51(4)** 618
[18] Child R 2007 *Appl. Microbiol. Biotechnol.* 75(3) 655
[19] Charlson J 2006 Phytoremediation of Arsenic Joe Charlson

[20] Prince RC and Suflita JM *Biodegradation* 18(4) 515

IOP Conf. Series: Journal of Physics: Conf. Series **1095** (2018) 012038 doi:10.1088/1742-6596/1095/1/012038

Bio-concentration factors of copper (Cu) and lead (Pb) in seagrass and some fish from coast Batam, Riau Islands, Indonesia

S Suheryanto[1], I Ismarti[2,3]

[1]Department of Chemistry, Faculty Math and Natural Sciences, Sriwijaya University, 30662, Indralaya, South Sumatera, Indonesia
[2]Department of Mathematics, Faculty Teaching Training and Education, Riau Kepulauan University, Batuaji Baru Street, 29438, Batam, Indonesia
[3]International Institute for Halal Research and Training, International Islamic University Malaysia, Gombak, Kuala Lumpur, Malaysia

Email :ismarti78@gmail.com

Abstract. Heavy metals pollution in aquatic environment has a significant adverse effect for biota especially invertebrates, fishes and also humans. Lead is toxic even at low concentration, whereas copper toxicity requires high concentration. The bioconcentration factors (BCFs) of heavy metals in seagrass and some fishes from coast Batam, Riau Islands, Indonesia were studied. Tissues from *Enhallus accoroides, Siganus sp, Sphyraena sp, Caesionidae sp* and *Chaetodontidae sp* were collected from six locations and the bioconcentration factors separately determined. The concentration of Cu in samples still meets with the permissible limit. However the concentration of Pb in four species of fishes were exceed the permissible limit. The BCFs study revealed the ability of fishes and seagrass to accumulate Cu and Pb. Level of Cu in seagrass tissues obtained higher than Pb. As demonstrated in BCFs study, the accumulation of Cu in fish was decreasing as follows: *Siganus sp>Sphyreaena sp> Chaetodontidae sp>Chaesionide sp*. The accumulation of Pb in fish was decreasing respectively as follow: *Sphyreaena sp>Chaetodontidae sp>Siganus sp> Chaesionidae sp.*

1. Introduction

The coasts Batam, Riau Islands as one of the busiest area in South East Asia have been received waste discharges from two sources, land and sea as well as natural and anthropogenic sources. Various human activities, shipping activities, and industries potentially contain hazardous substances which are harmful to both human health and marine ecosystem. Heavy metals are the most important pollutants which cause considerable harm to environment when they exceed certain concentrations [1]. The contamination of the aquatic system with heavy metals from natural and anthropogenic sources has become a global problem which poses threats to ecosystems and natural communities [2,3].

Heavy metals are deemed serious pollutants due to their toxicity, persistence [1], and non-biodegradable in the environment [4,5]. In another side, heavy metals taken up by marine organisms can move to the higher trophic levels and enter the human food chain through the contamination seafood [6]. According to [2], the increasing pollution by heavy metals has a significant health effects to invertebrates, fishes, and humans. Fishes can accumulate heavy metal in their body through absorption while the humans can be exposed to heavy metals through their meal [7]. Exposure of heavy metals has liked to several human diseases such as development retardation or malformation, kidney damage, cancer, abortion, effect on intelligence and behavior, and event death [5].

Heavy metals such as lead (Pb), copper (Cu) and cadmium (Cd) are of interest because of their toxicity. Cadmium and lead are toxic at low level, whereas copper is toxic at high concentration [6]. The increasing of activity around coast Batam by shipyard industries and transportations has an impact to level of cadmium, copper and lead in water, sediment, angel fish [7] and seagrass [8].

Metal accumulation is a tool to identify the impact of heavy metals in aquatic ecosystems because it shows the adverse of effect on the organism [7]. The metals are separated into two categories: essential and non-essential. The essential elements such as Fe, Cu, Zn and Mn have defined biological functions in organism whereas non-essential element such as Pb, Cd, Cr, As, and Hg are not involved in any metabolic mechanism [9].

Levels of heavy metals in biota are fluctuating, influenced by age, size, level of trophic, species, and feeding behavior [10]. Seagrass and fish are good accumulators of heavy metals. In another side, fish have good of nutritive and economical values [11] so that the consumers may be exposed to heavy metal toxicity if bioaccumulation results due to regular consumption [12]. Despite the many studies carried out in heavy metals, scanty information is available on the bioconcentration factors of heavy metals in biota from coast Batam. Consequently, the current work was carried out to determine the BCF values of seagrass (*E. accoroides*) and some fishes species from coast Batam in order to know whether or not the metals are bio-accumulated in the tissues of the seagrass and fishes so as to protect the consumers from exposure to the heavy metal toxicity as a result of regular consumption.

2. Materials and Methods

2.1 Sampling method

The fresh fish samples were collected from 6 sites (Tanjung Pinggir, Sekupang, Tanjung Riau, Marina Beach, Tanjung Uncang and Sagulung) distributed along the west side of coast Batam (See Table 1). The fish samples (Table 1) were collected by using traps that were planted in the sea bottom. The fishes were kept in an ice box and then transported to the laboratory until dissection took place. The water samples were collected from the bottom water (depth 2-2.5m). Samples were collected from the marine used Van Dorn horizontal water sampler into polyethylene bottles and transported to laboratory. 20-25 specimens of seagrass were collected from each location. The seagrass removed from the seawater, washed with tap water, kept in a plastic bag, and transported to the laboratory.

Table 1. Seagrass and fish species collected from the sampling sites

Sampling sites	*E.accoroides*	*Siganus sp*	*Sphyraena sp*	*Chaesionidae sp*	*Chaetodontidae sp*
Tanj. Pinggir	+	+	+	-	+
Sekupang port	+	+	+	+	-
Tanj. Riau	+	+	+	-	+
Marina Beach	+	+	+	+	+
Tanj. Uncang	+	+	+	+	+
Sagulung	+	+	+	+	-

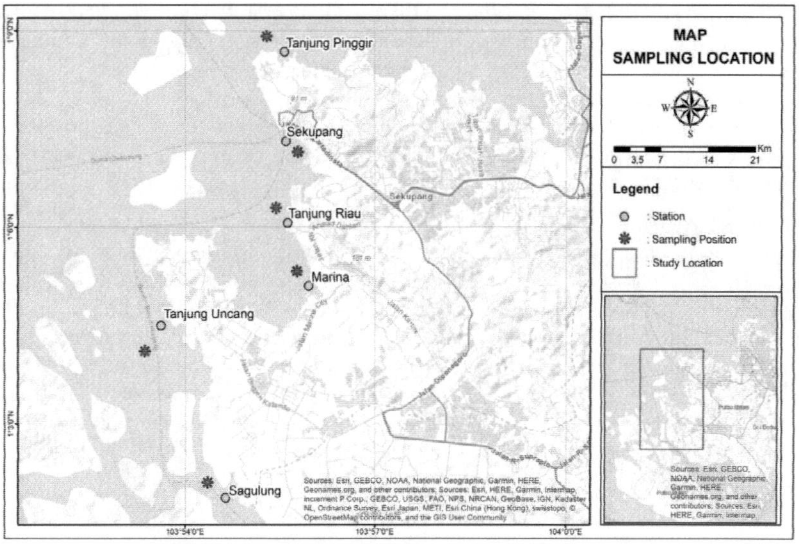

Figure 1. Sampling location

a. Sample treatment
Water samples were acidified with nitric acid until pH 2 then filtered with Whatman filter paper. The filtrate collected into 50 ml volume and ready to be analyzed using atomic absorption spectrophotometer.

The fish meat samples were separated using Dolphin surgical equipment. Seagrass and fish samples were dried using an oven Kirrin model KBO-250 RA at 105°C until constant weight. Five grams of the smooth samples was added 5 ml of nitric acid and heated for 30 minutes or until a clear liquid was obtained. The mixture was filtered using Whatman filter paper. The filtrate obtained then added aquadest up to 50 ml volume and ready to be analyzed.

2.2 Heavy metals analysis
The determination of copper and lead were performed as describe by [7].

2.3 Data analysis
Bioconcentration factors were analyzed based on metal concentration in water and on tissues of seagrass and fish. BCF analysis performed to determine the accumulated level of Cu and Pb in seagrass and fish tissues. BCF of heavy metals in the samples calculated as [9]:

$$BCF = \frac{C_{biota}}{C_{ambient\ medium}}$$

Where C_{biota} is the heavy metals concentration in the biota and $C_{ambient\ medium}$ for fish is the heavy metals concentration in water while for seagrass is both of heavy metals concentration in water and sediment.

3.Result and Discussion

3.1 Distribution of heavy metals
The result of distribution of heavy metals and the permission standard guidelines are shown in Table 2. In this study, the level of heavy metals varied in all different components. Level of Cu and Pb in sediment obtained greater than their level in seawater. The level of metals in sediments in the coastal area can be elevated due to high inputs from natural as well as anthropogenic sources [13]. Heavy metals have a high tendency to bind to suspended matter and through sedimentation. According to [4], the metal content of sediment was controlled by organic matter and Fe contents. It was supported by a factor of activity degree (pH) that was alkaline and normal temperature range resulted in insoluble Pb and deposited on sediments [14]. Sediments can act as source and sink for nutrients and heavy metals depending upon the physicochemical condition. Hence, sediments are potential secondary sources of contaminants in aquatic ecosystems [15]. The present study found that the levels of Cu and Pb in sediments still meet the level of heavy metals in USEPA standard for a category not polluted. However, levels of Cu and Pb in seawater have exceeded the permissible limit set by Indonesia Ministry of Environment.

Based on this study, the level of Cu and Pb in seagrass obtained greater than fish samples. The levels of Pb in seagrass was exceeded the permissible limit for food and seafood specified in SNI 7378, 2009 and FAO standard's 2003. However, the content of Cu still meets the FAO's standard that is 30 mg/kg. In addition, the level of Cu in all species of fishes ranged from 0.225 to 1.4033 mg/kg. The highest level of Cu in fish samples was found in *Sphyraena sp*. However, this value still meets the standard of FAO. Lead is the second element (after arsen) of top 20 list of the most poisoning heavy metals [1]. The levels of Pb in all species of fish have exceeded the permissible limit in SNI 7378, 2009 and FAO, 2003. It was ranged from 1.6325 to 3.1075 mg/kg. The highest level of Pb obtained for *Siganus sp*. Thus, it is revealed that there is a potential high risk to health if the peoples consume the fishes caught in this area.

Table 2. Concentration of heavy metals in seawater, sediment, seagrass and fish collected from coast Batam, Indonesia

Samples		Heavy metals (mg/kg)		Reference
		Cu	Pb	
Seawater	Mean	0.0423	0.0403	
	Sd	0.0032	0.0040	
Sediment	Mean	11.5472	12.9592	
	Sd	10.6384	5.1429	
E.accoroides	Mean	24.8983	10.7333	
	Sd	12.8629	4.3628	
Siganus sp	Mean	0.3950	3.1075	
	Sd	0.2040	1.6099	
Sphyraena sp	Mean	1.4033	1.6950	
	Sd	1.7579	0.7991	
Chaesionidae sp	Mean	0.2250	1.6325	
	Sd	0.2899	0.7275	
Chaetodontidae sp	Mean	0.5800	1.6700	
	Sd	0.3869	0.8242	
Permission limit in food and seafood in Indonesia (SNI 7378, 2009)		-	0.3	[8]
Permission standard limit in food (FAO, 2003)		30	0.5	[16]
Permission limit in seawater (Indonesia Ministry of Environment No. 5, 2004)		0.008	0.008	[7]
Standard Quality Sediment (USEPA)		25	30.2	[4]

replication (n) = 6

3.2 *Bioconcentration factor (BCF) of heavy metals in seagrass and fish*

In the present study, $BCF_{b/w}$ refers to the ratio of the heavy metals concentration in biota and seawater, while the $BCF_{b/s}$ refers to the ratio of the heavy metals concentration in biota and sediment. Bioconcentration is the accumulation of the contaminant by aquatic organisms through non-dietary uptake routes, e.q., from the soluble phase [17]. The results of BCF from the present investigation are shown in Table 3.

Table 3. The BCF values of the heavy metals of fishes and seagrass from coast Batam, Indonesia

Samples	$BCF_{b/w}$		$BCF_{b/s}$		BCF range	Category BCF[a]
	Cu	Pb	Cu	Pb		
Siganus sp	9.33	77.10			>1000	Very high
Spyreaenus sp	33.17	42.06			100-1000	High
Caesionidae sp	5.32	40.51			30-100	Moderate
Chaetodontidae	13.71	41.44			<30	Low
E. accoroides	588.61	266.34	2.16	0.82		

[a]Potipat et al., 2015

Level of heavy metals in fish closely related to the disposal waste near the fish live. Intake heavy metals by fish depending on ecological requirements, metabolisms, and other factors such as salinity, water pollution level, food and sediment [18]. According to [19], accumulation rate of heavy metals in fishes depends on the characteristic of chemical compounds, concentration and species of fish. The *Siganus sp, Sphyraena sp, Caesionidae sp* and *Chaetodontidae sp* were chosen because of their abundance in the sampling area and represented for calculation of the BCFs. If the values of BCF greater than 1.00 it is indicated that the metals were highly bio-accumulated and bio-magnified [12]. From this investigation, all species of fishes obtained in the moderate categories for Pb, and low categories for Cu, except *Sphyraena sp* in moderate category.

The result of $BCF_{b/w}$ and $BCF_{b/s}$ showed the ability of *E.accoroides* to accumulate heavy metals. This study found that *E.accoroides* is at high BCF category for Cu and Pb in water, but low BCF category for Cu and Pb in sediment. It is revealed that *E.accoroides* is a good accumulator for Cu and Pb from seawater. The BCF value for Cu obtained higher than Pb. Copper is the essential element of some enzymes for organisms [20] and serve as micronutrients for cellular metabolism [8]. According to [20], the great level of Cu intake via seafood consumption can cause adverse health problems such as liver and kidney damage, but not carcinogenic to humans and animals.

4. Conclusion

The result of this study found that the distribution of the essential element (Cu) in seawater, sediment and fishes from coast Batam, Riau Islands, Indonesia lower than the non-essential element (Pb), except seagrass (*E.accoroides*). The BCF obtained for Cu and Pb were higher than 1.00 revealed that the Cu and Pb were highly bio-accumulated and bio-magnified in the tissue of fishes and seagrass. Cu has higher BCF than Pb in *E.accoroides*. In another side, the BCFs of Cu in all of the fish species were lower than Pb. The accumulation of Cu in fish was decreasing as follows: *Siganus sp>Sphyraena sp>Chaetodontidae sp>Chaesionidae sp* as demonstrated in the BCF study. The accumulation of Pb in fish was decreasing respectively as follows: *Sphyreana sp>Chaetodontidae sp>Siganus sp>Chaesionidae sp.*

Acknowledgement

This research was funded by a research grant under the scheme PEKERTI. The authors greatly acknowledged the Directorate of Research and Community Services of Higher Education, Republic of Indonesia. The second author thanks to Indonesia Endowment Fund for Education (LPDP) for the scholarship to pursue the doctoral study.

References

[1] El Nemr A, Khaled A, Moneer AA and El Sikaily A. 2012 *J Aquat Res* **38** 67.
[2] Ali MM, Ali ML, Islam MS and Rahman MZ. 2016 *Environ Nanotechnology, Monit Manag* **5** 27.
[3] Nwamaka K and State A. 2013 *J Anim Res Int.* **10** 1792–8.
[4] Barakat A, Baghdadi M El, Rais J and Nadem S. 2012 *Res J Environ Earth Sci.* **4** 797.
[5] Hossain PKSMD.2011 *Environ Sci Technol.* **6** 384–8
[6] Annibaldi A, Illuminati S, Truzzi C, Libani G and Scarponi G. 2015 *Mar Pollut Bull.* **96** 441
[7] Ismarti I, Ramses R, Suheryanto S, and Amelia F. 2017 *Omniakuatika.* **13** 78.
[8] Ismarti I, Ramses R, Fitrah A and Suheryanto S. 2017 *DEPIK* **6** 23.
[9] Potipat J, Tangkrock-olan N and Helander H. 2015 *EnvironmentAsia.* **8** 118.
[10] Arifin Z. 2011 *J Ilmu dan Teknol Kelaut Trop.* **3** 104.
[11] Zeitoun MM, Sayed E-, and Mehana E. 2014 *Glob Vet.* **12** 219.
[12] Falusi B a, and Olanipekun EO.2007 *J Appl Sci Environ Manag.* **11** 51.
[13] Liu Ming; Zhang ALYCBFD 2015 *Mar Pollut Bull.* 1–10
[14] Tupan CI and Uneputty PA. 2017 *AACL Bioflux* **10** 1610.
[15] Iqbal J and Shah MH. 2014 *J Anal Sci Technol* **5** 28
[16] Baharom ZS and Ishak MY. 2015 *Procedia Environ Sci* **30**:320.
[17] Chaudhuri A, Mitra M, Havrilla C, Waguespack Y and Schwarz J.2007 *Bot Mar.* **50** 151.
[18] Ashraf MA, Maah MJ, and Yusoff I. 2011 Vol. 77, *Proceedings of the Indian National Science Academy*. 57.
[19] Sarong MA, Jihan C, Muchlisin ZA, Fadli N and Sugianto S. 2015 *AACL Bioflux.* **8** 1.
[20] Yap CK, Cheng WH, Karami A and Ismail A. 2016 *Sci Total Environ.* **553**:285.

Isolation of Ethyl P-Methoxy Cinnamate from *Kaemferia galanga* L

Aliefman Hakim[1], Yayuk Andayani[1], Baiq Deana Rahayuan[1]

[1]Study Program of Chemistry Education, University of Mataram
Jalan Majapahit 62, Mataram-Lombok 83125, Indonesia

E-mail: aliefmanhakim27@gmail.com

Abstract. *Kaempferia galanga* L is a tropical plant that grows in many areas in Indonesia known as medicinal plants and herbal ingredients by the community. There are many secondary metabolite compounds in *K. galanga* such as ethyl cinnamic, paraffin, borneol, p-methoxystyrene and ethyl p-methoxycinnamate. The ethyl p-methoxycinnamate compound is known to be the major compound in *K. galanga* and is widely used in the cosmetic and anti-fungal. Many benefits of the ethyl p-methoxycinnamate compound make the need for this compound constantly increasing and the need for an isolation method that can produce this compound in large quantities. The isolation method used in this research is the soxhletation method using ethanol solvent and recrystallization using n-hexane solvent. Samples of *K. galanga* were used as much as 50 gram and obtained crystalline ethyl p-methoxycinnamate as much as 0.49 gram with percentage of rendement as much 0.98%. The resulting crystalline ethyl p-methoxycinnamate has been identified using UV-vis spectroscopy, IR and H-NMR.

Keywords : Isolation, *Kaempferia galanga*, ethyl p-methoxycinnamate

1. Introduction

Kaempferia galanga L is a tropical plant that grows in many areas in Indonesia known as medicinal plants and herbal ingredients by the community [1]. There are many secondary metabolite compounds in *K. galanga* such as ethyl cinnamic, paraffin, borneol, p-methoxystyrene and ethyl p-methoxycinnamate [2]. The ethyl p-methoxycinnamate compound is known to be the major compound in *K. galanga* and is widely used in the cosmetic and anti-fungal [2]. Ethyl p-methoxycinnamate is classified as phenyl propanoid compound because based on structure [3]. In this article will describe the isolation procedure of ethyl p-methoxycinnamate from kencur (*kaempferia galanga L*).

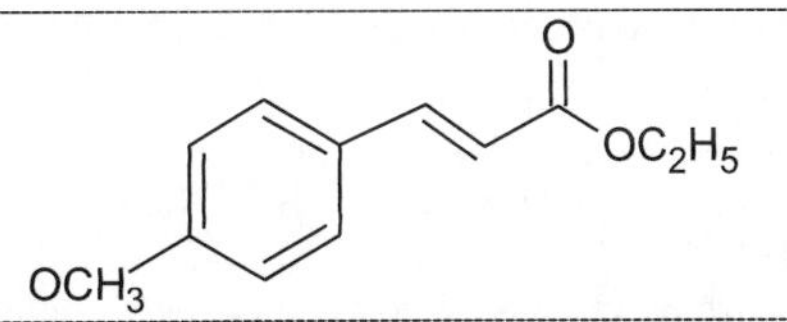

Figure 1. Structure of ethyl p-methoxycinnamate

IOP Conf. Series: Journal of Physics: Conf. Series **1095** (2018) 012039 doi:10.1088/1742-6596/1095/1/012039

2. Discussion and Conclusion

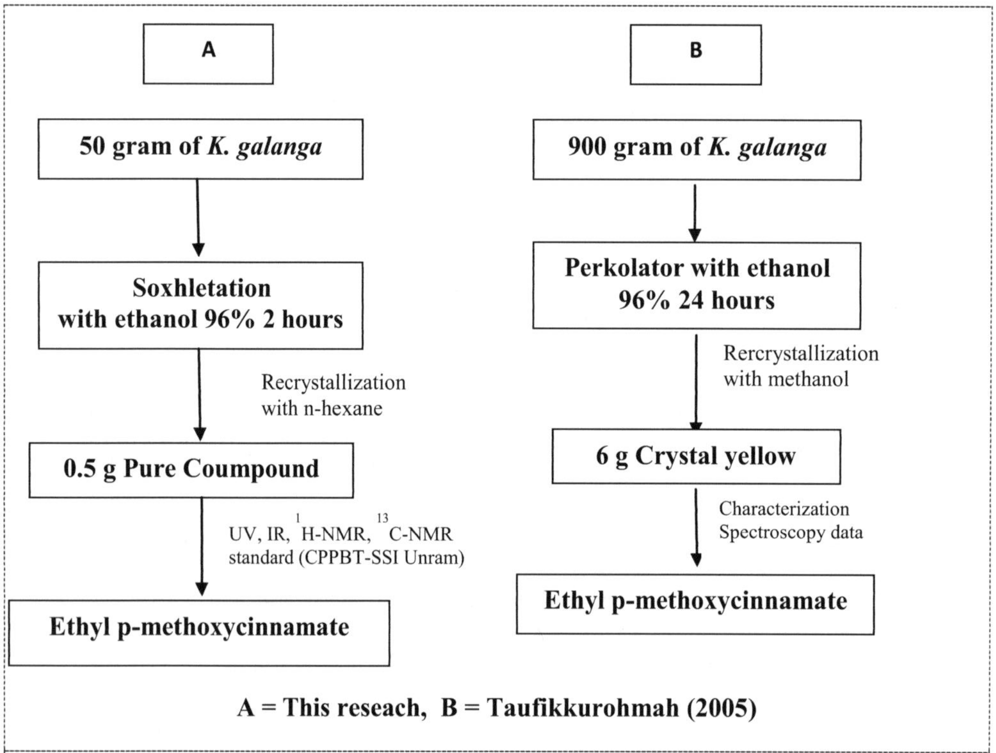

Figure 2. Comparison of isolation procedure of ethyl p-methoxycinnamate

The same secondary metabolite compounds can be isolated in different ways [4, 5, 6]. In this study, 50 gram of rhizome of *K. galanga* obtained 0.49 gram of white crystal ethyl p-methoxycinnamate. The obtained crystals were pure as indicated by the TLC giving a spot with the same distance as the standard solution spot when using n-hexane: ethyl acetate: acetone (9: 3: 1) as eluen. Ethyl p-methoxycinnamate has been produced in large quantities and commercialized through the Calon Perusahaan Pemula Senyawa standar Indonesia (CPPBT-SSI) cooperation of Ristekdikti with Mataram University.

Isolation procedures of ethyl p-methoxycinnamate from kencur previously have been published by Taufikkurohmah [2]. Comparison of ethyl p-methoxycinnamate isolation procedures performed in this study and reference [2] is shown in Figure 2. It is seen that Taufikkurohmah [2] procedures was yield 6 gram of total crystal of ethyl p-methoxycinnamate from 900 gram of *K. galanga* and in this study result 0,5 gram from 50 gram of *K. galanga*. On the other hand this study directly was recrystallization of total extract of *K. galanga* using n-hexane to produce ethyl p-methoxycinnamate. Based on the above explanation ethyl p-methoxycinnamate isolation procedures performed in this study is simpler than the ethyl p-methoxycinnamate isolation procedure have been published previously by Taufikkurohmah [2]. Ethyl p-methoxycinnamate has been produced in large quantities and commercialized through the Calon Perusahaan Pemula Senyawa standar Indonesia (CPPBT-SSI) cooperation of Ristekdikti with Mataram University.he first paragraph after a heading is not indented (Bodytext style).

IOP Conf. Series: Journal of Physics: Conf. Series **1095** (2018) 012039 doi:10.1088/1742-6596/1095/1/012039

References

[1] Umar Ihtisham M., Asmawi, Zaini M., Sadikun, Amirin and Majid. 2014 *CLINICS* **69(2**) *134*
[2] Taufikkurohmah T *2005 Indo. J. Chem.* **5 (3)** 193.
[3] Hakim A, Liliasari, Kadarohman A.and Syah Y M (2016). *IJC.* **16 (3)** 315.
[4] Hakim A and Jufri A W 2017 "*Applications of Isolation and Structure Elucidation of Secondary Metabolites in Natural Product Chemistry Laboratory*" (New York -NOVA Publisher).
[5] Hakim A., Liliasari, Kadarohman A and Syah Y M. 2016. *Journal of Turkish Science Education.* **13 (2)** 27.
[6] Hakim A., Liliasari, Kadarohman A and Syah Y M. 2016. *J. Chem. Educ.* 93 **(1)** 193.

IOP Conf. Series: Journal of Physics: Conf. Series **1095** (2018) 012040 doi:10.1088/1742-6596/1095/1/012040

High Conversion and Yield of Biodiesel using Electrolysis Method

S A Rachman, L N Komariah, A I Andwikaputra, NB Umbara

Department of Chemical Engineering, Faculty of Technology, Universitas Sriwijaya
Jl. Raya Palembang-Prabumulih Indralaya South Sumatera, 30862 Indonesia

e-mail: susila_arita@yahoo.com

Abstract. Electrolysis method is one of the newest method recently used for biodiesel production. It is hoped that the use of electrode may minimise energy usage in the process for producing biodiesel. In this research, electrolysis process has been tested using raw materials of crude palm oil (CPO) and waste cooking oil (WCO). Methanol is used as solvent with volumetric ratio of 12:1 to raw material, with catalyst of NaOH and KOH. Stirring is employed to quicken the contact between raw materials and the electrodes. The distance between the two electrodes is 15 mm. Electrolysis is started by stirring the mixture at 500 rpm for 2 hours with electrical voltage varied at 14 V, 16 V, 18 V, 20 V, and 24 V. By the use of this method, biodiesel with best quality has been found from raw material of WCO with maximal conversion of 98.6% and highest yield of 90.26%. This result is achieved at condition of 18 V with catalyst of KOH. According to Faraday laws, current density is proportional to the amount product from anode or cathode. Yet, this also depends on the space between the two electrodes. As conclusion, electrolysis method may make transesterification reaction between triglyceride and methanol to occur which gives yield to biodiesel product with high conversion.

1. INTRODUCTION

Biodiesel is an alternative fuel for diesel machine which is produced from transesterification reaction of vegetable oil or animal oil with short chain alcohol such as methanol. The reaction needs catalyst which normally is of strong bases, that will produce a new chemical called biodiesel. Biodiesel synthesis from waste frying oil using electrolysis method firstly was researched by Guan and Kusabe, et al., 2009 [1]. According to them, the merit of this method is the possibility of using raw materials containing quite high total mixture ($\leq$ 2% wt) which can be utilised directly at the electrolysis reaction.

At this research, besides WCO, CPO (Crude Palm Oil) is also used as raw material which has not yet been used by other reasearchers in the catalytic electrolysis reaction, even if this process is considered to be quick, easy, and giving high conversion. The homogeneous catalyst of NaOH and KOH has already generally been used in the synthesis of biodiesel using other method, yet it has never been used in the electrolysis method. By comparing biodiesel product from feedstock of Waste Cooking Oil (WCO) and Crude Palm Oil (CPO) using electrolysis method and catalytic reaction, it is hoped that there will be specific condition giving best conversion and yield.

1.1. Waste Cooking Oil (WCO)

Waste cooking oil (WCO) is waste oil originated from various cooking oil such as corn oil, frying oil, samin oil, etc. This oil comes from domestic use which can be reused for cullinaire, yet from its chemical composition, waste oil consists of substance with the characteristic of being carcinogenic, if it repeatedly heated up at high temperature, that it may cause cancer disease in the long time periode.

1.2. Crude Palm Oil (CPO)

Oil from palm fruit consists of crude palm kernel oil and crude palm oil (CPO) which is found from the kernel of palm fruit and the part of mesocarp of fruit meat *Elaeis guineensis*. *Elaeis guineensis* is an eternal plant which produces high amount of oil and has been become one of the most important vegetable oil for biodiesel feedstock due to its characteristic such as higher cetane number, viscosity and density, productivity, low price, high oxidation stability, good fatty acid content and having plasticity characteristic at room temperature [2]. Palm oil is oil which has saturated fatty acid and unsaturated fatty acid contents of high level, such as palmitic fatty acid, oleic, linoleic, etc. Other raw material used in this research is methanol. At the process of biodiesel production, methanol functions as solvent. Methyl alcohol, or commonly known as methanol, is a substance with molecular formula of CH_3OH. Methanol is the simplest alcohol. The catalyst used is natrium hydroxide (NaOH), which may accelerate reaction rate to quicken the reaction. Kalium hydroxide (KOH) can also be used as catalyst for producing biodiesel. Kalium hydroxide is an inorganic substance with the formula of KOH, and commonly called as potash fire.

1.3. Graphite Electrode

Graphite is a form of carbon allotroph which is grey in colour and has the most stable form under certain condition, where its atomic structure affect its phisycal and chemical characteristics. There are three types of graphite used in the industry, they are in the form of crystalline, lumps and amorphous. Graphite may function as electrical conductor.

1.4. Electrolysis Cell

Electrolysis cell is the main equipment used in the electrolysis process, consisting of dc current source and positive and negative electrodes. The substance to be electrolysed is an electrolyte, in the form of solution or liquid (melt) of pure substance. If a liquid or solution electrolyte is charged with electrical dc current via the electrode rod, the ions in the said liquid or solution will move to the electrode with different/opposite charge. Chemical reaction will occur when electric current is flown via electrolyte solution, when electrical energy is converted into chemical energy (redox reaction). There are 3 characteristics in the electrolysis process, which are:

a. There is electrolyte solution consisting of free ions.
b. These ions may give or take/accept electron so that the electron may flow via the solution.
c. There is a source of electricity from outside, such as battery, which supply direct current electricity (dc).

2. RESEARCH METHODOLOGY

This research is carried out in the Laboratorium of New Energy and Renewable Energy in Sriwijaya University, Palembang.

2.1. Equipment Used

A dc Stavolt potentiostat equipped with anode and cathode connectors, two graphite electrodes (P=15 cm), a beaker glass (capacity of 250 mL), and a magnetic stirrer are employed in this research.

Equipment Arrangement

Figure 1. Apparatus for Electrolysis with its arrangement

2.2. Experimental Procedure

Biodiesel Synthesis

- Electrolytic apparatus is set up precisely. The distance between electrodes was set at 15 mm, and it is confirmed that the Stavolt cable connectors to cathode and anode have been placed properly.
- The electrolyte cell is filled with mixtures of oil and methanol with volumetric ratio of 1:12.
- Stavolt is switched on, so does the magnetic stirrer which is adjusted to medium rate. Reaction time is set for 2 hours. The voltage is varied and controlled with dc Stavolt potentiostat at 14 V to 24 V.
- Reaction product is examined visually and the data are recorded and reported for every hour.
- After reaction time of 2 hours, the product mixture is filled into separation column to separate biodiesel from glycerol.
- Product of biodiesel is washed with hot water to eliminate the rest of glycerol and triglyceride.
- To completely free the product from water and methanol, evaporation was carried out at temperature of 120 C for 2 hours.
- The purified biodiesel was analysed using gas chromatography to get its composition and its water content. Picnometer is used to measure water content, whilst a viscometer is employed to measure the density of the product.

3) RESULTS AND DISCUSSION

The result and discussion of this research is presented as follows. Results from the analysis of the two raw materials of waste oil and crude palm oil using gas chromatography show different composition between the two materials. The composition of fatty acid in crude palm oil is depicted in Table 1, whilst the composition of fatty acid in waste cooking oil can be seen in Table 2.

The characteristics of the two raw materials are: the percentage of FFA is 1,33% wt in WCO, and 1,24% wt in CPO, whilst the water content is 0,1318% wt in WCO and 0,1209% wt in CPO. The reaction is carried out for 2 hours at STP condition. In this method, NaCl has also been added, which makes the solution became an electrolyte, so as the transfer of ion between anode and cathode may occur well.

Table 1 The composition of fatty acid in crude palm oil

Fatty acid	Result (%)
Lauric Acid (C12)	0,14
Myristic Acid (C14)	1,64
Palmitic Acid (C16)	32,45
Stearic Acid (C18)	3,97
Oleic Acid (C18:1)	40,45
Linoleic Acid (C18:2)	21,18
Arachidic Acid (C20)	0,17

Table 2 The composition of fatty acid in waste cooking oil

Fatty acid	Result (%)
Lauric Acid (C12)	0,07
Myristic Acid (C14)	1,35
Palmitic Acid (C16)	38,48
Stearic Acid (C18)	16,29
Oleic Acid (C18:1)	27,98
Linoleic Acid (C18:2)	15,72
Arachidic Acid (C20)	0,11

3.1. The Effect of Catalyst Variation, Electrical Voltage, and Raw Material to Conversion and Yield of Biodiesel Product

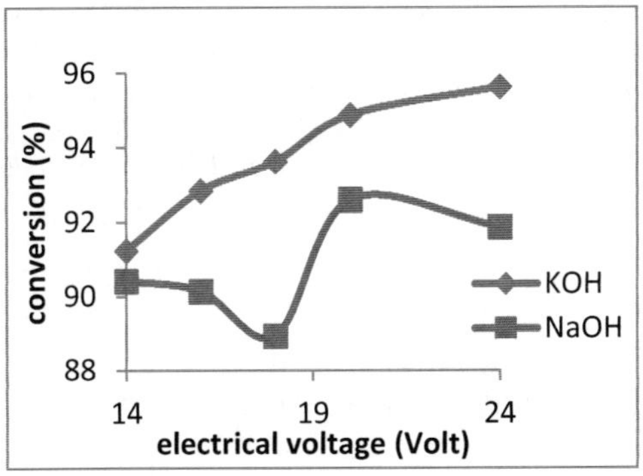

Figure 2a. The Value of Conversion from Raw Material of CPO

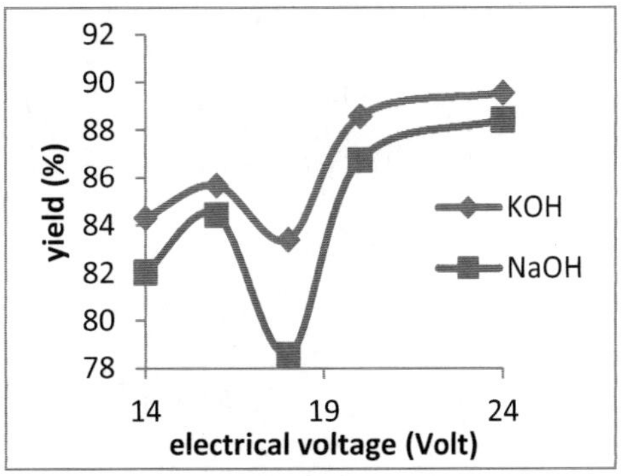

Figure 2b. The value of yield from raw material of CPO

From Figure 2a and 2b, the value of conversion is fluctuative and the yield of biodiesel from raw material of CPO increases with the increase of electrical voltage. The best conversion is 95.6% at voltage of 24 Volt, using catalyst of KOH, whilst the best yield is 89.55% at voltage of 24 Volt, with catalyst of KOH. According to Guan and Kusakabe, 2009 [1] in the electrolysis process, the higher the voltage causes the production of hydroxide ion become more and more. This ion is produced continuously at cathode and can react immediately with methanol and oil at transesterification reaction. But, there is decrease of conversion and yield at several points in Figure 2a and 2b. These were caused by saphonification reaction accompanying transesterification process. The saphonification reaction may occur due to the availability of hydroxyl group at the structure of the catalyst used, that can decrease the yield of biodiesel product. Even if NaOH and KOH are both have the characteristic of being catalyst, yet NaOH has higher effect in its catalyst performance caused by its higher base characteristic than KOH, so as it is easier to form soap as the side product of the reaction [3].

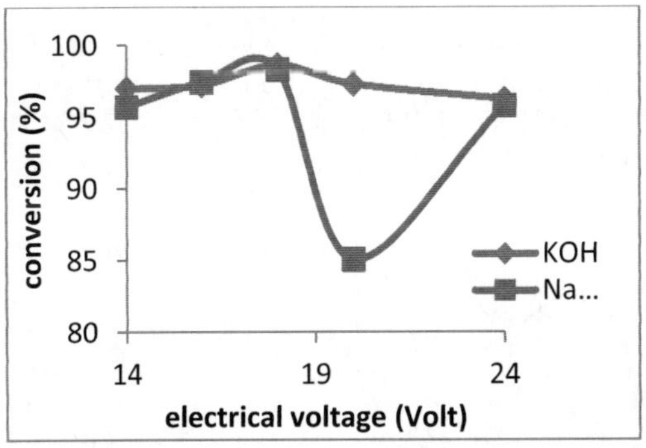

Figure 3a The value of conversion from raw material of WCO

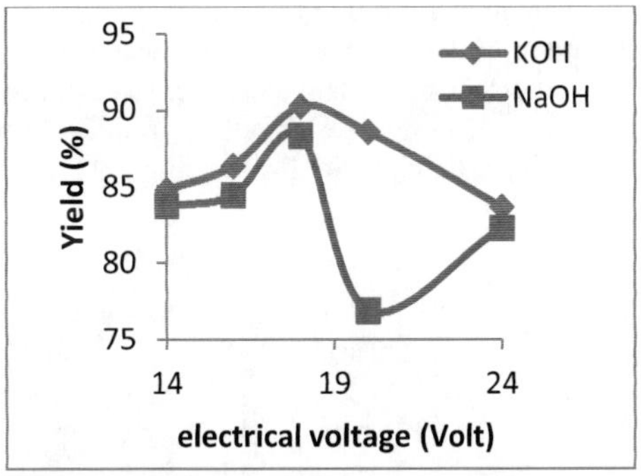

Figure 3b The value of yield from raw material of WCO

From Figure 3a and 3b, the value of conversion is quite fluctuative. The yield from raw material of WCO increases at potential up to 18 V, and decreases at voltage up to 24 V for reaction with KOH as catalyst. Product yield increases at potential up to 18 V, decreases at voltage up to 20 V and increases again at potential up to 24 Volt for catalyst of NaOH. The value of best conversion from raw material of waste cooking oil is 98.6%, and the value of best yield is 90.2% at voltage of 18 V, with catalyst of KOH. The decreases of conversion and yield may occur due to the formation of saphonification reaction.

According to research by Guan and Kusakabe 2009 [1] with WCO as raw material, using electrolysis method without catalyst, the process is able to give yield of 97%. Even if this WCO has high water content of 0.5% wt and acid number of 7.7 mg KOH/gr, which was afraitent may cause saphonification. Other raw material used in their research was corn oil which gives yield of 98.5%. Equipments and other chemicals used in that research had have high quality, such as platinum electrode (Pt), which has very much helped the reaction to achieve highest yield. Compare with this recent work which uses graphite elecrtrode and gives highest yield of 90.33%.

3.2. The Effect of Variation of Catalyst, Voltage, and Raw Material to the Viscosity of Biodiesel Product

Kinematic viscosity is one of the characteristic should be paid in attention for biodiesel fuel and petrodiesel. Viscosity must be lowered from its initial value in the raw material by conducting transesterification process to the raw material substance[4] (Knothe G and Steidley K.R., 2005). Tesfa B., et al., 2010 [5] said that high viscosity may increase the risk of machinery problem such as carbon deposit at the filter, bad atomisation of fuel at spraying, increase the need of energy for fuel pumping, and wearing of fuel pump and injector.

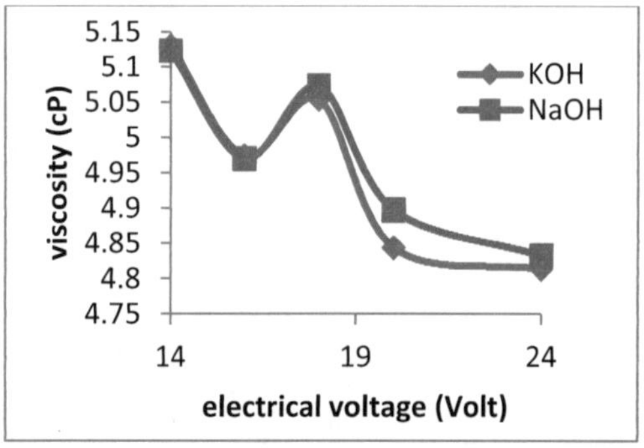

Figure 4 The value of viscosity of biodisel from CPO

Figure 4 shows that initially the viscosity has decreased with the increase of electrical voltage up to potential of 16 V, but the viscosity increases at voltage up to 18 V before it decreases again at voltage up to 24 V. This is may be due to the increase of tension which makes evaporation process in the solution mixture for reaction becomes more and thus increases the viscosity of the product. The presence of impurities in the product is due to the imperfect separation process of biodiesel from glycerol, or it may also be due to the unclean process during washing stage of the biodiesel. The lowest viscosity is 4.77 cSt for biodiesel product using voltage of 18 V and catalyst of NaOH. For all biodiesel products with raw material of CPO, the viscosity has meet the requirement in SNI standard 7182:2015 no 2 with kinematic viscosity range of 2.3-6.0 cSt.

In Figure 5, the value of viscosity decreases with the increase of electrical voltage up to potential of 18 V. But, when the potential is increased to 24 V, the viscosity of biodiesel tends to increase. The viscosity of the samples of biodiesel product from raw material of WCO are all in the range of the standard kinematic viscosity for biodiesel which is 3 – 6.0 cSt.

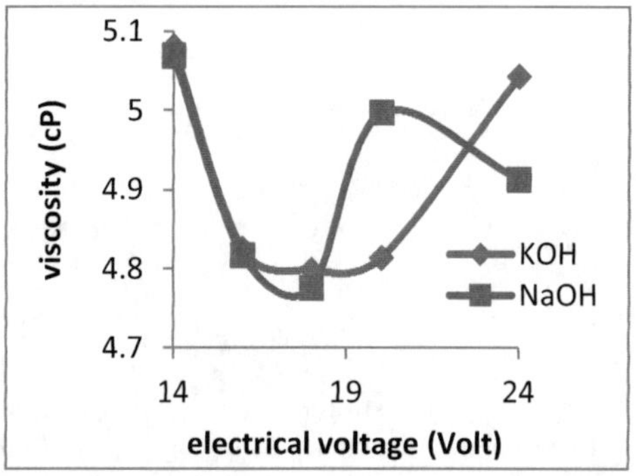

Figure 5 The value of viscosity of biodiesel from WCO

3.3. The Effect of Variation of Catalyst, Voltage and Raw Material to the Density of Biodiesel Product

Density is one of the most important parameters in the physical properties of biodiesel because it can affect the performance of machinery and the characteristics of its emission. According to Figure 6, the value of biodiesel density from CPO in this work tends to be stable with electrical voltage, except at voltage of 18 V with catalyst of KOH, which has lowest value of 0.873 gr/mL.

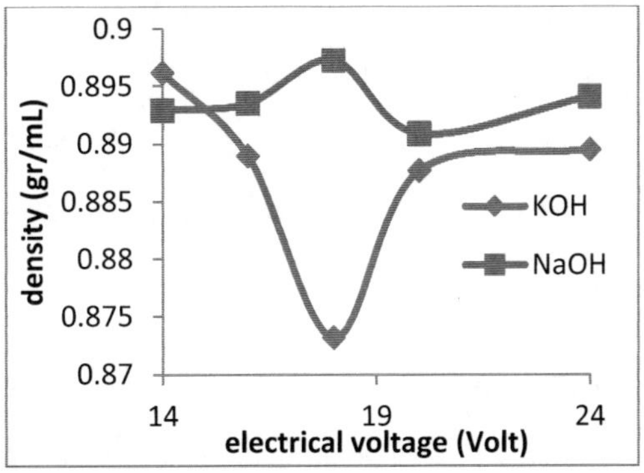

Figure 6 The Value of Biodiesel Density from CPO

According to the work by Nezihe A, et al., 2007 [6], biodiesel produced using catalyst of 1.5% KOH has density of 0.879 gr/cm^3. With catalyst concentration which are not far different between the two works (at 1% wt KOH for the present work), the density resulted from the two works are almost of the same, even if the previous work does not use electrolysis method. The standard quality for biodiesel density based on SNI (Indonesian National standart) 7182:2015, [7] is 0.85-0.89 gr/cm^3.

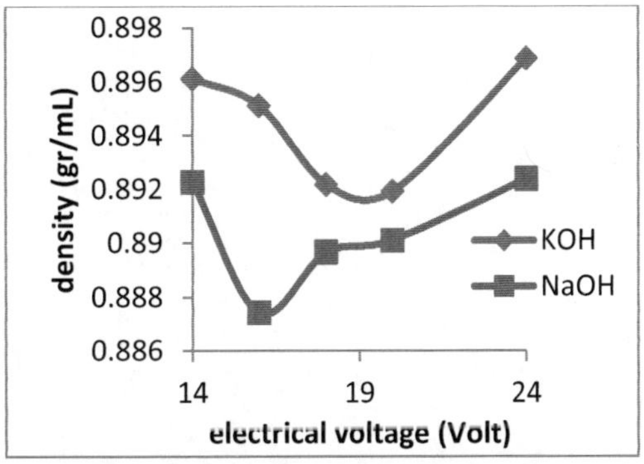

Figure 7 The Value of Biodiesel Density from WCO

Figure 7 shows that the density of biodiesel from raw material of WCO using catalyst of NaOH and potential between 16 V and 20 V, has been the best value, which fulfill the SNI requirement. When the voltage is increased to 24 V, the value of density is proportional to the value of kinematic viscosity, which is over 0.89 gr/cm^3, and does not meet the requirement of SNI 7182:2015, [7] .

3.4. The Effect of Variation of Catalyst, Voltage, and Raw Material to the Acid Number of Biodiesel Product

Figure 8 shows the acid number of biodiesel product from raw material of CPO. All samples of biodiesel product from CPO have low acid number. The value of acid number produced at electrical potential of 16 V to 24 V is about 0.024684 mg KOH/gr. The standard of acid number in SNI 7182:2015 is maximum of 0.5 mg KOH/gr. The lower the acid number, the better the quality of the product. Based on this, all biodiesel products from CPO have fulfill the requirement of acid number in the SNI 7182:2015, [7].

Acid number is connected with the pH of biodiesel produced. The degree of acidity will affect corrosion characteristic of the biodiesel. The small value of this acid number indicates that the biodiesel produced has good quality, with degree of corrosivity which is low. The acid number of biodiesel produced in Figure 9 has the value which is not far different with the acid number of biodiesel in Figure 8. The acid numbers in those two figures are far lower than the acid number of the raw material, WCO, which is at 0.029 mg KOH/gr. That makes the biodiesel product has fulfill SNI 7182:2015, [7].

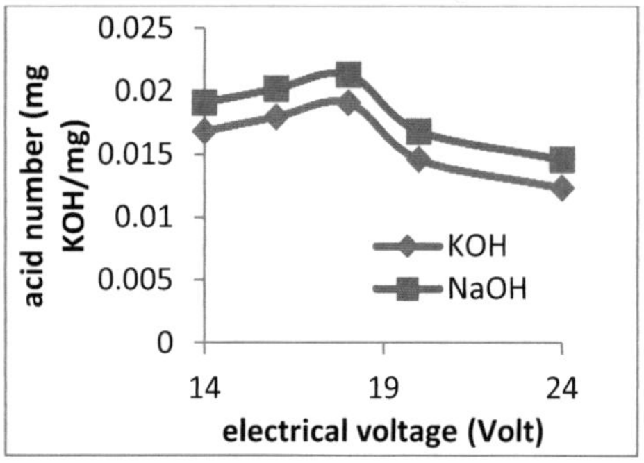

Figure 8 The Value of Acid Number of Biodiesel from CPO

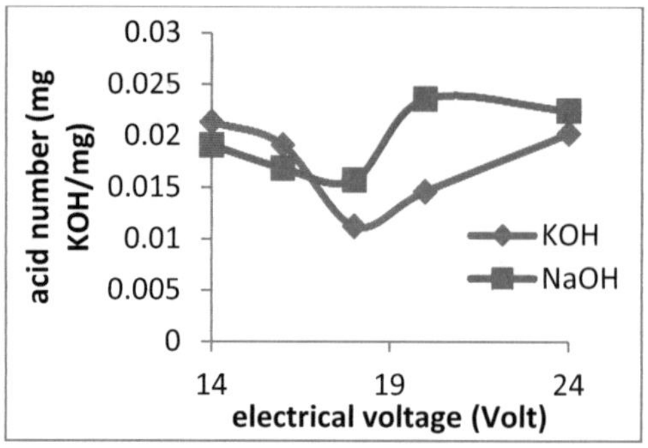

Figure 9 The Value of Acid Number of Biodiesel from WCO

High value of acid number, may cause problem in the combustion system of diesel machine, in the form of precipitate (carbon residue). Other than being precipitate, the value of acid number which is high also may affect the quality of biodiesel product of FFA Biodiesel. The content of FFA (free fatty acid) in the raw material will affect the type of reaction (chosen) which will be carried out. The CPO used has FFA content of 1.24 %.

3.5. The Effect of Variation of Catalyst, Voltage, and Raw Material to the Free Fatty Acid of Biodiesel Product

The performance of catalyst of NaOH and KOH to catalyse transesterification reaction may decrease the percentage of FFA in the biodiesel product. When connected with yield (in Figure 2 and 3) it can be seen that the product from raw material of CPO with catalyst of KOH in Figure 10 has % FFA lower producing yield biodiesel and higher conversion.

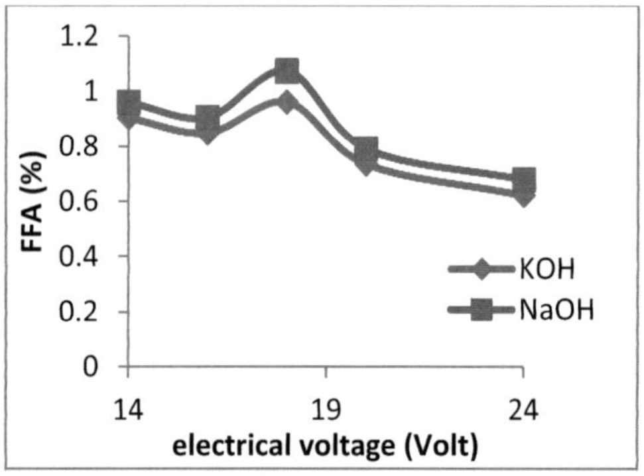

Figure 10 The Percentage of FFA in Biodiesel from CPO

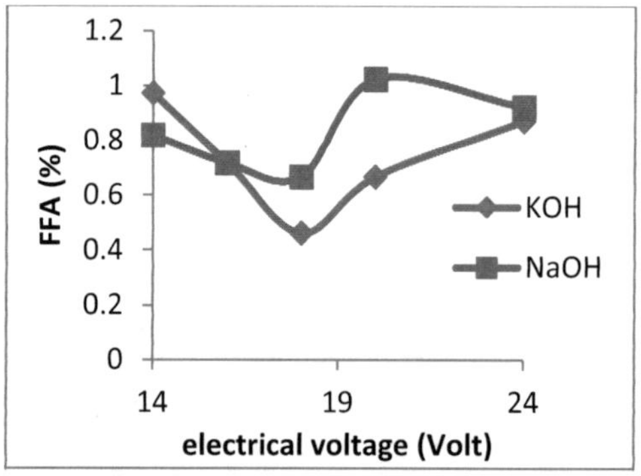

Figure 11 The Percentage of FFA in Biodiesel from WCO

From Figure 11, the percentage of FFA is not far different from the results from Figure 9 where the biodiesel product from KOH has lower percent of FFA. As a whole, the percentage of FFA content in the biodiesel product is lower compared with biodiesel from WCO which has FFA content of 1.3312 %. Based on SNI 7182:2015, part of the biodiesel from WCO has fulfill the requirement of standard quality.

3.6. The Effect of Catalyst Variation, Voltage, and Raw Material to Water Content of Biodiesel Product

Water content in the two raw materials for producing biodiesel in this research are quite high, they are 1.33 % and 0.98 % . After going through electrolysis reaction, some samples of the biodiesel product as in Figure 12, have water content which are quite fluctuatives. With catalyst of KOH and potential of 16 V, the water content in the biodiesel product has still been quite high. This is because the heating process of biodiesel has not yet been completed, or less long time. Heating has been done to decrease water content in the biodiesel product. Other source of water content in the product is from the water used for electrolysis process (at the amount of 2 % wt to the oil), which contributes in the residual water content. According to Guan and Kusakabe 2009 [1], the most active area where transesterification takes place in the electrolysis is in the cathode where the water is electrolysed. The presence of water content of 2 % affect the movement electrical charge between cathode and anode.

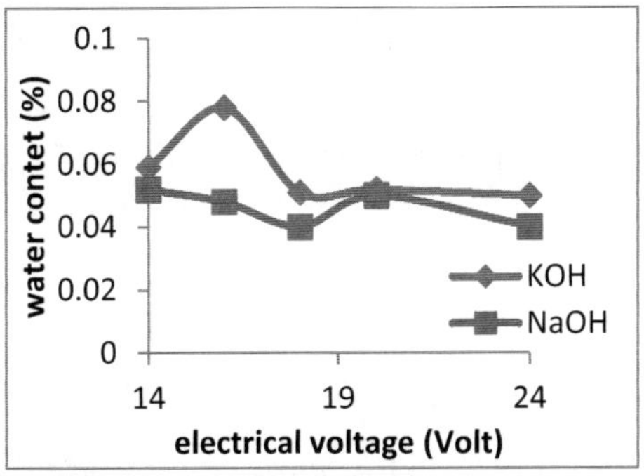

Figure 12 Water Content in Biodiesel from CPO

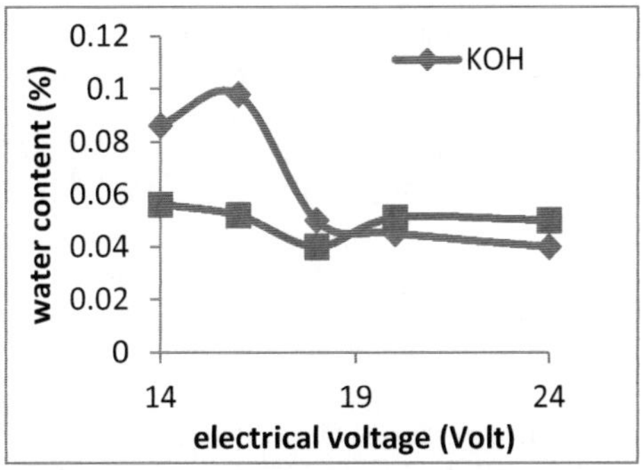

Figure 13 Water Content in Biodiesel from WCO

In Figure 13 the product of biodiesel from raw material of WCO is also not far different from raw material of CPO, only for product at 20 V and 24 V, KOH has lower water content than catalyst NaOH. According to Fregolente et al., 2012 [8] the existence of water in the biodiesel or in the diesel may cause water accumulation and the development of microbes in the water canal of fuel. Moreover, according to I.M Atadshi, et al., 2012 [9], the problem caused by water content in the biodiesel makes difficulty in the reaction for synthesis of biodiesel during alkali catalysation process in the transesterification, the decrease of biodiesel quality, the decrease of combustion heat, corrosion in the combustion system, and acceleration of hydrolysis reaction. This is why, for standard of biodiesel fuel, its water content is controlled by SNI 7182:2015 no 8 with maximum volume of 0.05 % vol. So, some biodiesel products from CPO and WCO in this work have fulfill the said standard.

3.7. The Effect of the Variation of Catalyst, Voltage, and Raw Material to the Total Glycerol in the Biodiesel Product

According to Bondioli P and Bella LD 2005 [10] some amount of glycerol contained in the biodiesel which is used as diesel blending is one of the cause failure in the machinery and combustion system in the automotives. That is why, in SNI 7182:2015 is fixed maximum glycerol total for biodiesel of 0.24 %.
For raw material samples, total glycerol measured, that is 10.4% raw material of CPO and 10.8% for raw material of WCO. Based on Figure 14 almost all product has decrease in glycerol which is very drastic, which is below the number ogf one. From all biodiesel bsed on Figure 14 there is none fulfill SNI from its total glycerol.

In Figure 15, even if has total glycerol which is decerase far from initial raw material, only one sample which has fulfill SNI 7182:2015, that is the biodiesel using catalyst KOH with potential of 18 V. To the other product which has not fulfill SNI, the cause of this happening is has not been maximum of the transesterification process, so that needs time longer so that the reaction may occur maximal.

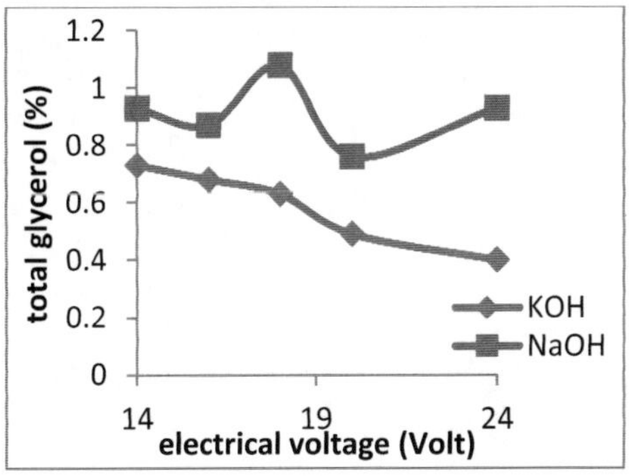

Figure 14 Total Glycerol in Biodiesel from CPO

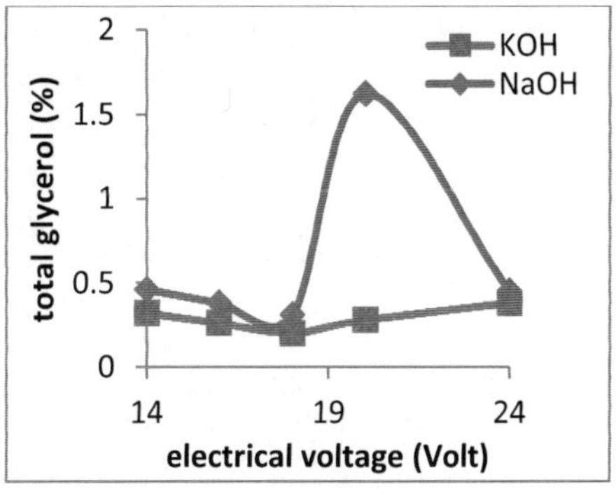

Figure 15 Total Glycerol in Biodiesel from WCO

CONCLUSION

The production of biodiesel with electrolysis method if it uses raw material of waste cooking oil gives highest conversion of 98.6%, with best catalyst of KOH at 18 V, whilst if it uses raw material of crude palm oil gives the conversion of maximal 98.3% with best catalyst of NaOH.
The product of biodiesel made from raw material of waste cooking oil shows better quality than that made from raw material of crude palm oil. This can be seen from the total biodiesel glycerol from raw material of waste cooking oil which is at the amount of 0.2, which has meet the requirement of biodiesel quality standard of SNI 7182:2015 which is at 0.24, whilst biodiesel from raw material of crude plam oil has total glycerol of 0.4.

REFERENCES

[1] Guan, G., & Kusakabe, K. (2009). *Synthesis of Biodiesel Fuel Using an Electrolysis Method.* Chemical Engineering Journal, 159-163.
[2] Susila A R, Leily N K, 2014 Optimum Excess Air for the Utilization of Palm Biodiesesl Blends in Fire Tube Boiler, *Applied Mechanics and Materials Journal*, **660**, 402-405.
[3] Laksono S 2013 Skripsi Makassar Fakultas Peternakan Universitas Hasanuddin.
[4] Gerhard K, Kevin R S 2005 Kinematic viscosity of biodiesel fuel components and related compounds-Influence of compound structure and comparison to petrodiesel fuel components *Fuel Journal* **84** 1059–1065
[5] Tesfa **B**, Mishra R, Gu F, Powles N 2010 Prediction models for density and viscosity of biodiesel and their effects on fuel supply system in CI engines, *Renewable Energy Journal*, **35** 2752-2760
[6] Nezihe A, Aysegul D 2007 Alkali catalyzed transesterification of cottonseed oil by microwave irradiation *Fuel* **86** 2639–2644
[7] Standar Nasional Indonesia. Biodiesel. SNI 7182:2015
[8] Fregolente P B L, Wolf M R M, Oliveira L S 2015 Removal Of Water Content From Biodiesel And Diesel Fuel Using Hydrogel Adsorbents *Brazilian Journal Of Chemical Engineering* **32** 895 - 901
[9] Atadashi I M, Aroua M K, Aziz A R 2012 The Effect of Water on Biodiesel Production and Refining Technologies: A Review. *Renewable and Sustainable Energy Reviews*, **16** 3456-3470.
[10] Bondioli P, Bella L D 2005 An Alternative Spectrophotometric Method for the Determination of Free Glycerol in Biodiesel *European Journal of Lipid Science and Technology* **107** 153-157.
[12] Fregolente, P. B., Fregolente, L. V., & Maciel, M. R. (2012). *Water Content in Biodiesel, Diesel, and Biodiesel-Diesel Blends.* Journal Chemical Engineering Data, 1817-1821.

Antibacterial Activity of Endophytic Fungi Isolated from the Leaves of Jambu Biji (*Psidium guajava L.*)

Susilawati[1], Ella Amalia[1], Desi Oktariana[1], Maulia Sari Khairunnisa[1]

[1]General Medicine Department of Faculty of Medicine of Sriwijaya University

Jl. DR. Moh. Ali Komp. RSMH Palembang 30126

Email:susilwt78@gmail.com

Abstract. Antibacterial resistance has been a major health issue and remains a threat to the global health care system. Endophytic fungi, particularly those isolated from medicinal plants are potential as a source of new antibiotics.The endophytic fungi were isolated from the surface of sterilized leaves of jambu biji (*Psidium guajava* L.) by potato dextrose agar (PDA) and the isolates were identified through their morphological characteristics. The antibacterial activities from ethyl acetate extracts of endophytic fungi were evaluated against *Escherichia coli* and *Salmonella tiphi*by Kirby Bauer method. Four endophytic fungi were isolated from the leaves of jambu biji and were identified as*Aspergillus niger Van Tieghem*, *Aspergillus flavus*, *Aspergillus fumigatus* Fres, and *Aspergillus* sp.(J4).The antibacterial activity of the extracts was classified as moderate and strong category as its inhibition zone was in the range of ≥50% compared to positive control.Therefore, this extract has the potential to be developed into an antibiotic drug.
The first section in your paper.

Keywords : Antibacterial activity, Endophytic Fungi, *Psidium guajava L*

1. Introduction

People from developing countries such as Indonesia are prone to suffer from diseases caused by poor environment and food intake like diarrhea. Due to the poor environment and lack of awareness on healthy lifestyle, the spread of the disease like diarrhea is rife in slum areas.Unhygienic lifestyle with no concern of sanitation could make colon vulnerable to the attack of pathogenic *Escherichia coli*and *Salmonella tiphi* bacteria that cause diarrhea (1,2).

Antibiotic resistance is a globally popular health problem (3,4). The usage of antibiotics that breaches prescribed rules could make pathogenic microbes resistant to the existing antibiotic (5,6,7. The time duration since the usage of an antibiotic being introduced to the event of resistance varies starting from 0 to 36 years (8). Furthermore, the average duration of an antibiotic to become resistant is about 6.6 years since the introduction. This fact shows the existing antibiotics could not forever be used to cure infection diseases and calls for sustainable development of existing antibiotics as well as newly-sourced antibiotics. The utilization of medicinal plants data as the result of ethnobotany research is one of the effective ways in discovering new chemical materials that could be useful for medication such as to gain new antibiotic compound (9,10).

Another way to gain new antibiotic compound is by utilizing endophytic fungi from each plant. Since there are many kinds of plants, there needs to be an approach to narrow down the search of endophytic fungi that give off antibiotic activity such as plants with history of ethnobotany like Jambu Biji (*Psidium guajava L.*), in which use is related to diarrhea medication so the chance to get new antibiotic compound is higher (11).

IOP Conf. Series: Journal of Physics: Conf. Series **1095** (2018) 012041 doi:10.1088/1742-6596/1095/1/012041

The leaves of Jambu Biji are believed to have the effects of astringent, wound healing, anti - allergy, damaged cell repair, and anti-microbe activities against bacteria that generally infect wounds of surgery and other soft tissues such as *Staphylococcus aureus, Streptococcus spp, Escherichia coli, Salmonella typhi, Proteus mirabilis,*and *Shigella dysenteria* (12).

2. Materials and Methods

2.1. Study Time and Location

The study was conducted in June to September 2017. The antibacterial activity in vitro testing was done at the Microbiology Laboratory in the Faculty of Mathematics and Natural Sciences of Sriwijaya University (FMIPA UNSRI), while as for the extraction and evaporation were done at the Organic Chemistry Lab of Chemistry Department of FMIPA UNSRI.

2.2. Materials

Materials required for the study consisted of the host plant jambu biji leaves while as for sample sterilization it took 70% of ethanol, 0.5% of NaOCl 0.5%, and aquadest sterile. The study used Potato Dextrose Agar (PDA) and Potato Dextrose Broth (PDB) as fungal growth media as well as Nutrient Agar (NA) and Mueller Hinton Agar (MHA) as the media in antibacterial activity testing. As for the testing bacteria, the study used *E.Coli* and *S. Tiphi* with the antibiotic standard of Ciprofloxacin, and ethyl acetate technical as the organic solvent in sample extraction.

2.3. Methods

2.3.1. Sample Sterilization and Isolation of Endophytic

Samples of jambu biji leaves were washed with flowing water for approx 5 minutes. Then the surface of the leaves were to be sterilized with 70% of alcohol solution poured into a cotton ball for approx. 3 minutes before being washed with aquadest sterile for approx. 1 minute. Next, the leaves were slathered with 5% (b/v) of $HgCl_2$ (mercury chloride) for approx. 1 minute to later be washed with aquadest sterile for approx 1 minute. The sterilized samples were cut vertically with aseptic technique then were planted in petri dishes containing PDA media and were incubated at 37^0C for 2 x 24 hours. Next, the observation was done every day for 3 to 7 days until growing fungi appeared (13).

2.3.2. Purification of Endophytic Fungi

The fungal colony grown in PDA medium that showed different morphological characteristics in shapes, colors, and sizes were later purified. The purification was done by relocating the colony into a new PDA medium with streak plate method, then being incubated at 37^0C for 2x24 hours. Next, the purified fungal colony was made into work and stock cultures by growing it in a slant PDA (14).

2.3.3. Identification of Endophytic Fungi

The endophytic fungal strain was identified based on the colony morphology and microscopic observation of mycelia and sexual spores according to the method described in the literature (15).

2.3.4. Cultivation of Endophytic Fungi

Medium (PDB) of 300 mL was used for the cultivation of endophytic fungi and was placed into 3 flasks. Fungal suspension containing 10^6 spores/mL was inoculated under sterile condition to each of 300 mL PDB medium (ratio 1:10). The cultures were incubated for six weeks in a static condition at room temperature. At the end of incubation period, each culture was filtered to separate the mycelium to later be extracted with ethyl acetate solvent as well as to be evaporated to get concentrated extract.

2.3.5. Extraction of Jambu Biji Leaves
The leaves of jambu biji used as the comparison in the study amounted to approx. 100 grams of fresh leaves, chopped, and extracted by maceration in methanol for 2 x 24 hours. Next, the extract was filtered and evaporated so it would produce concentrated extract towards which later be tested for its antibacterial activity using *E. Coli* and *S. Tiphi*, with the same method.

2.3.6. Antibacterial Activity Testing
The making of suspension with 0.5 McFarland standard and the making of testing bacteria suspension (16).

2.3.7. Preparation of Antibacterial Compound Concentration
The antibacterial sample concentration was made with1,000 μg/disk or 1,000 μg/10 μL and the concentration of ciprofloxacin antibiotic as the standard of 500 μg/disk atau 500 μg/10 μL.

2.3.8. Antibacterial Activity Test by Kirby Bauer Method
Each of the samples dissolved in methanol. The MHA (15 mL) was poured into petri dishes and inoculated with 100 μL of the suspension containing 1×10^5 CFU mL^{-1} of bacteria. Ciprofloxacin was used as the positive control and discs treated with methanol were used as the negative control. Sterile paper discs (6 mm) loaded with 10 μL of the samples were placed onto the surface of the agar. The plates were then placed in an incubator at 37°C for 24 h, after which the diameter of the zone of inhibition around each of the discs was measured and recorded. Each experiment was performed in triplicate (17).

2.4. Data Analysis and Management
The determination of antibacterial activity in the test sample (concentrated extracts from endophytic fungi) was done by comparing the inhibition zone in each concentration to the inhibition zone in the positive controls that are standard antibiotic and methanol extracts from the host plant, therefore the relative activity of antibacterial sample tested with positive controls could be determined. The criteria of each antibacterial compound concentration tested against positive controls were determined with the following formula (18).

Weak $:\frac{A}{B}$ x 100% < 50%; Moderate: 50 $<\frac{A}{D}$ x 100% < 70% ;

Strong: $\frac{A}{B}$ x 100% > 70%

A: zone of inhibition (mm) of tested antibacterial compound

B: zone of inhibition (mm) of positive controls (standard antibiotic and host plant extracts).

3. Results and Discussions
Out of jambu biji leaves, four endophytic fungi were isolated and were characterized in macroscopic and microscopic ways. The four fungi were identified as J1 to J4. The result of such characterization showed the four fungi were *Aspergillus niger Van Tieghem*, *Aspergillus flavus*, *Aspergillus fumigatus Fres*, and *Aspergillus sp.*(J4) (as shown on Figure 1).

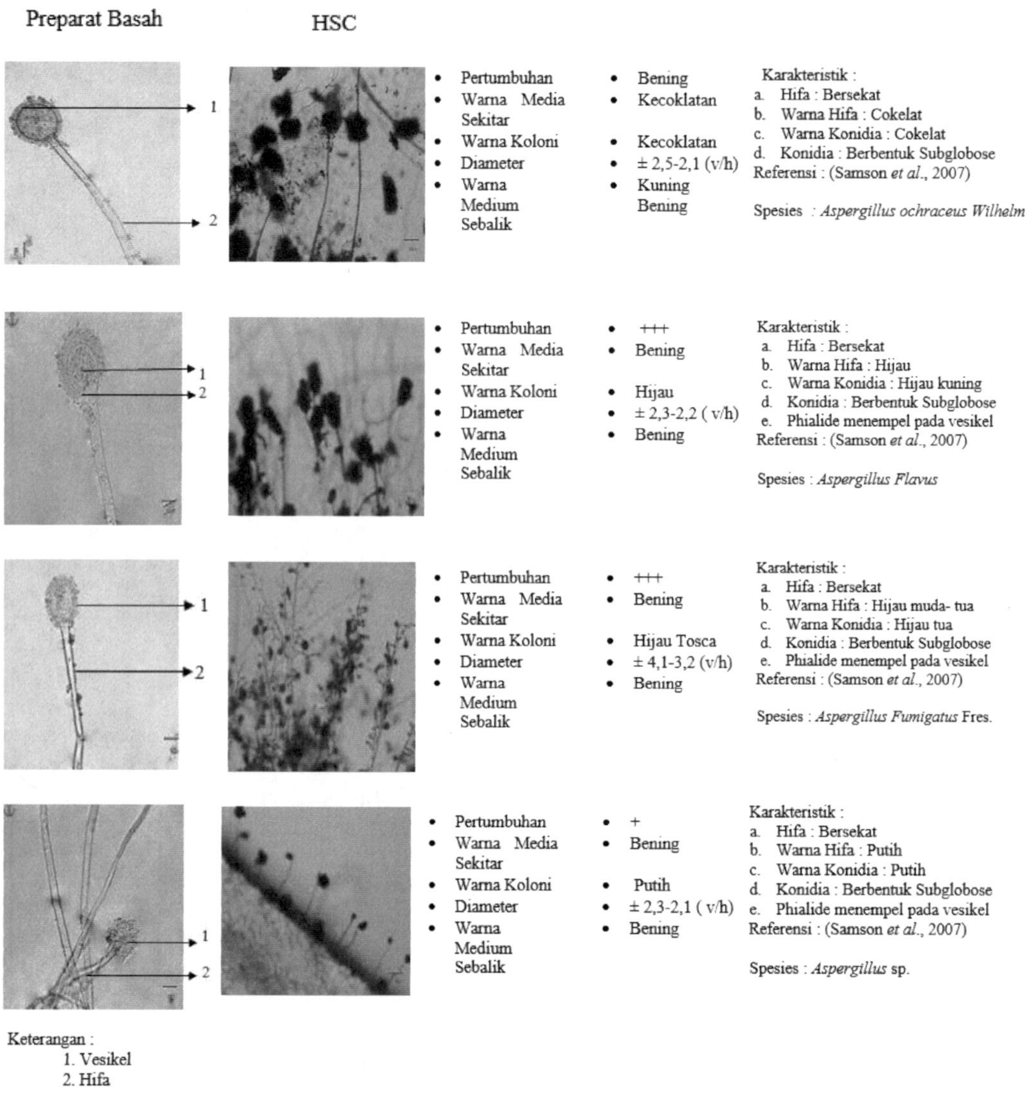

Figure 1. Macroscopic and microscopic observations of endophytic fungi of jambu biji plant

Each fungal isolates obtained was cultivated in 3x300 mL of PDB liquid medium for six weeks to later be filtered and extracted in organic solvent ethyl acetate. The ethyl acetate fraction was then separated and evaporated. Meanwhile, the antibacterial activity testing was conducted on condensed extracts using test bacteria *E. Coli* and *S. Tiphi* with Kirby-Bauer method in single concentration that was 1,000 µg/disc. The results of antibacterial testing on the ethyl acetate extracts of endophytic fungi against two test bacteria that have been compared with antibacterial activity of positive controls comprising ciprofloxacin antibiotic (500 µg/disk) and the methanol extract sof host plant jambu biji leaves (1,000 µg/disk) are shown on Table 1 and 2.

Table 1. Antibacterial activity percentages of ethyl acetate extracts of endophytic fungi from jambu biji leaves against test bacteria *E.Coli*, compared to the antibacterial activity of positive controls (ciprofloxacin antibiotic and methanol extract of jambu biji leaves)

No	Extracts of endophytic fungi (1,000 μg/disk)	Diameter of zone of inhibition on testing bacteria *E. Coli*	Percentages of antibacterial activity of methanol extracts on positive controls (%)	
			ciprofloxacin antibiotic(500 μg/disk) 3.53 cm	Extracts of jambubiji leaves(1,000 μg/disk) 1.31 cm
1	Fungal extract 1 (J1)	1.24	35.1 +	94.6 +++
2	Fungal extract2 (J2)	1.91	54.1 ++	146.0 +++
3	Fungal extract3 (J3)	1.38	39.1 +	105.3 +++
4	Fungal extract4 (J4)	0.89	25.2+	68.0++

Information: + = weak inhibition <50%
++ = moderate inhibition 50-70%
+++ = strong inhibition >70%

Table 2. Antibacterial activity percentages of ethyl acetate extracts of endophytic fungi from jambu biji leaves against test bacteria *S. Tiphi*, compared to the antibacterial activity of positive controls (ciprofloxacin antibiotic and methanol extract of jambu biji leaves)

No	Extracts of endophytic fungi (1,000 μg/disk)	Diameter of zone of inhibition on testing bacteria *S. tiphi*	Percentages of antibacterial activity of methanol extracts on positive controls(%)	
			ciprofloxacin antibiotic(500 μg/disk) 3.82 cm	Extracts of jambubiji leaves(1,000 μg/disk) 1.72 cm
1	Fungal extract1 (J1)	1.41	36.9 +	82.0 +++
2	Fungal extract2 (J2)	2.15	56.3 ++	125 +++
3	Fungal extract3 (J3)	0.94	24.6 +	54.6 ++
4	Fungal extract4 (J4)	0.83	21.7 +	48.2 +

Information: + = weak inhibition< 50%
++ = moderate inhibition50-70%
+++ = strong inhibition>70%

The study used ciprofloxacin (500 μg/disk) as a comparator antibiotic standard in the antibacterial test. The zone of inhibition of an antibiotic standard is classified as strong, moderate, and resistant categories with certain levels of antibiotic per disk (19). An antibiotic could be set as a standard if it produces inhibition zones in moderate and weak categories when tested at an inoculum equals to McFarlan 0.5. However, if an antibiotic's zone of inhibition belongs to the resistant category, it could not be used as a standard.

The zones of inhibition listed on Table 1 and 2 have shown that ciprofloxacin is classified as strong category for all testing bacteria so it can be used as a standard in the reference of antibacterial activity of test extracts. The zone of inhibition of jambu biji leaves (1,000 μg/disk) worked as a comparator of antibacterial activity power of endophytic fungi extracts against the host plant.

The results on Table 1 and 2 showed that ethyl acetate extracts of four endophytic fungi in which antibacterial activities were tested against 2 test bacteria in single concentration of 1,000 μg/disk produced varied zones of inhibition depending on the strength of the activities.

After the zone of inhibition of each fungal extract against both test bacteria being compared to the zone of inhibition of ciprofloxacin antibiotic standard, the study acquired fungal extract J2, the extract with the strongest level of activity that was at moderate category with the inhibition zone ranging from 50% to 70% (Table 1 and 2). Meanwhile, the other fungal extracts comprising J1, J3, and J4, were grouped into weak category against the two test bacteria. This showed that J2 fungal extracts contained active compound mixed with several other inactive compounds which limited the active compound ability so it would not work optimally to inhibit the test bacteria, as compared to such ability of antibiotic as a synthetic compound.

Next, the zones of inhibition of the four endophytic fungi extracts were compared to those of methanol extracts of the host jambu biji leaves. Jambu biji extract has been known for its healing power in diarrhea medication and has been used in industrial scale. The results on Table 1 and 2 show that fungal extract J2 possesses the strongest activity against test bacteria *E. Coli*, with it being in the strong category with the zone of inhibition range of>70% against the comparator jambu biji leaves extracts. Furthermore, fungal extracts J1 and J3 also showed strong activities. Mean while, the same cannot be said for fungal extract J4 as it being the only extract with activities in moderate category that is 68%. The similar result is shown by fungal extracts J1 to J4 against test bacteria *S. Tiphi* as fungal extracts J1 and J2 gave off strong activities, while J3 and J4 each showed moderate and weak activities. The work stages and potentials of endophytic fungi of jambu biji leaves as new antibacterial compound producer are shown on Figure 2.

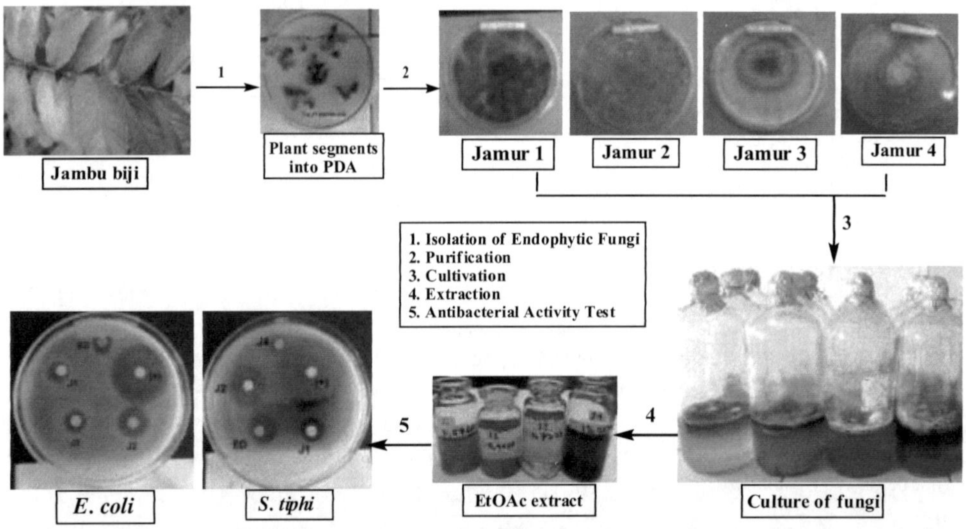

Figure 2. Brief isolation procedures of endophitic fungi from the leaves of jambu biji and antibacterial activity test from their ethyl acetate extracts

The study result shows that fungal extract J2 possesses the strongest antibacterial activity compared to the other three fungal extracts. This indicates the presence of antibacterial active compound in ethyl acetate extracts of J2 which was mixed with several other inactive compounds. If the active compound in J2 is purified, then there will be more potential for antibacterial activity.

The antibacterial activity of J2 extract is stronger than the extracts of the host plant that is jambu biji. This is a breakthrough in the field of medication and is worth to be researched further and in a more structured way to later be developed as a main ingredient of antibiotics in the future. Possible for endophytic fungi to obtain active compound of medicine as the number could be multiplied when necessary in a short time in line with the growth curve (approx. 4 to 8 weeks) and it does not take a wide space and expensive treatment to make it happen (20).

4. Conclusion

From the leaves of a jambu biji plant, there had been the isolation and characterization of four endophytic fungi namely *Aspergillus niger Van Tieghem* (J1), *Aspergillus flavus* (J2), *Aspergillus fumigatus Fres* (J3), and *Aspergillus sp.*(J4).Of all the four fungi, the ethyl acetate extract of *Aspergillus flavus* (J2) possessed the strongest activity, making it belong to the moderate category against ciprofloxacin antibiotic standard and stronger than the activity of methanol extracts of jambu biji leaves. The result shows that with further studies, the active compound in the fungi extract of *Aspergillus flavus* (J2) is potential to be developed as an active ingredient of antibiotics.

Acknowledgements

The author thanked the agency of research and community service of Sriwijaya University that has funded the study through the scheme of science, technology, and art research.

References

[1] Unicef . 2012. Air Bersih, Sanitasi & Kebersihan. *Ringkasan Kajian* 1
[2] Amin, LZ. 2014 *Medicinus*, 27(3) 40.
[3] Habib MR, MM Rahman, A Mannan, AH Zulfiker, Uddin ME and Sayeed MA 2011 *IJABPT* **2(1)** 420.
[4] Biswas SK, A Chowdhury, J Das, UK Karmakar, MC Shill and SZ Raihan 2011 *JAPS* **1 (6)** 112.
[5] Priya P, FG Shoba, M Parimala, and J Sathya,2014 *International Journal of Pharmaceutical and Clinical Research* **6(2)** 174
[6] Lau KY, Zainin NS, Abas F and Rukayadi Y 2014 *Int.J.Curr.Microbiol.App.Sci.*, 3(12) 499
[7] Malik A and Ahmad AR, 2013 Int. Res. J. Pharm., **4(4)** 106.
[8] Guifoile PG 2007 *Antibiotic-Resistant Bacteria. Deadly Diseases and Epidemic*, New York.
[9] Kaneria M and Chanda S, 2012 *Asian Pacific Journal of Tropical Biomedicine*, S1526.
[10] Bhargavi S, Kanakaiah B, Sowmya DK, Ravi B and Nama S, 2013. *International Journal of Phytopharmacology* **4(3)** 171.
[11] Gangadevi V. and Muthumary J. 2008 *Mycologia Balcanica*, **5** 1.
[12] Desiyana, L S., Husni, M A, Zhafira, S, 2016 *Jurnal Natur*, **16(2)** 23.
[13] Elfita, Munawar, Muharni, and Suprayetno. 2013 *Indonesian Journal of Chemistry*, **13(3)** 209
[14] Elfita, Muharni, Munawar, and Aryani S. 2012-b *Indonesian Journal of Chemistry.* **12(2):**195
[15] Liu K, Ding X, Deng B, and Chen W 2009 *Journal of Industry Microbiology and Biotechnology*, **36** 1171.
[16] *Clinical and Laboratory Standards Institute. 2009. Susceptibility Tests for Bacteria That Growically;* Approved Standard—8th Ed. Pennsylvania.
[17] Elfita, Munawar, Muharni, and Ivantri I. 2015 *Microbiology Indonesia*, **9 (2)** 82.
[18] Wong SK, Lim YY., and Chan EWC. 2010 *Ethnobotanical Leaflets*, **14** 781.
[19] Harley and Prescott. 2002. *Laboratory Exercises in Microbiology*, 5th Ed. (Newyork -The McGraw–Hill Companies)
[20] Strobel G., Daisy B., Castillo U., and Harper J. 2004 *J. Nat. Prod.* **67** 257.

Melimoluccanin, A new isoprenylated quinolone alkaloid from the leaves of *Melicope moluccana* T.G. Hartley

Tjitjik Srie Tjahjandarie[1*], Ratih Dewi Saputri[1], Ryan Ayub Wahjoedi[1] and Mulyadi Tanjung[1]

[1]Natural Products Chemistry Research Group, Organic Chemistry Division, Department of Chemistry, Faculty of Science and Technology, Universitas Airlangga, Surabaya, Indonesia.

tjitjiktjahjandarie@fst.unair.ac.id

Abstract. A new quinolone alkaloid, melimoluccanin (**1**) along with two known alkaloids, leptanoine C (**2**) and 7-*O*-isoprenyl-γ-fagarin (**3**) were isolated from the leaves of *Melicope moluccana* T.G. Hartley. Structures all of compounds were elucidated using spectroscopic methods such as UV, IR, HRESIMS, 1D and 2D NMR. Compounds **1-3** were evaluated for their cytotoxicity against P-388 cells and for antimalaria activity using *giemsa* methods. The cytotoxic data of compounds **1-3** showed IC_{50} of 0.63; 5.85 and 2.09 μg/mL, respectively. The antiplasmodial activity of compound **1-3** showed IC_{50} value of 4.28; 2,28 and 0.18 μg/mL, respectively. Based on the data, it can be concluded that compound **1** has high activity against P-388 cells and compound **3** is potential as antimalaria.

Keywords : Melimoluccanin, *Melicope moluccana*, antimalaria

1. Introduction

Malaria is a major cause of death in the word caused by a protozoan of the genus *Plasmodium* and transmitted by *Anopheles* mosquito vectors, especially in tropical countries. Recently, chloroquine and artemisinin have been used as antimalarial drug and showed resistance against *Plasmodium* parasites in Indonesia.

Melicope is one of the genus of family Rutaceae, consisting of about 280 species. The plants have been known produce a variety of alkaloids [1,2], flavonoids [3], coumarins [4], acetophenones [5] and lignans [4], which exhibit various biological activities including antioxidant [4], anticancer [1,2], and antiinflammatory [6].

This study is part of our research on the chemical constituents of *Melicope* species found in Indonesia. In continuation of our research for alkaloid compound in this medicinal plant, we report the isolation of 4-methoxy-3-(3-methylbut-2-en-1-yl)-7-((3-methylbut-2-en-1-yl)oxy)quinolin-2(1H)-one (**1**) from the methanol extract of the leaves of *Melicope moluccana* T.G. Hartley. The chemical structure of compound **1** were established by UV, IR, HRESIMS, 1D and 2D NMR, as well as by comparison with those related compounds previously reported. The cytotoxic and antioxidant activities of isolated compound from this species are also briefly described.

In this paper we report the isolation of melimoluccanin (**1**) melineurine (**2**) and 7-*O*-isoprenyl-γ-fagarin (**3**) from the leaves of *Melicope moluccana* T.G. Hartley. In addition, preliminary cytotoxic and antiplasmodial activities of the isolated compounds cells will also be described.

2. Materials and Methods

2.1 *General experimental*

^{1}H and ^{13}C NMR spectra were measured on a JEOL JNM-ECA spectrophotometer operating at 400 (^{1}H) and 100 (^{13}C) MHz using residual and deuterated solvent peaks (δ_H 7.26 and δ_C 77.1 for $CDCl_3$) as reference standards. High-resolution mass spectra were obtained with an ESI-TOF Waters LCT Premier XE mass spectrometer. UV spectra were recorded in MeOH on a Shimadzu series 1800 UV-VIS spectrophotometer. IR spectra were recorded in KBr on a One Perkin Elmer instrument. Column chromatography and planar radial chromatography were carried out using silica gel 60 and silica gel 60 PF_{254}.

2.2 *Plant Material*

The leaves of *M. moluccana* T.G. Hartley were collected in July 2015 from the conserved forest of Weda, Central Halmahera, North Maluku, Indonesia. The plant was identified at the Herbarium Bogoriense, Bogor Botanical Garden, Bogor, Indonesia.

2.3 *Extraction and isolation*

The dried and powdered leaves of *M. moluccana* T.G. Hartley (2.0 kg) was macerated with methanol at room temperature for 24 h two times to give a dark methanol extract (108 g) after solvent evaporation. The methanol extract was partitioned into *n*-hexane (15 g) soluble fractions. The methanol extract was added 3% sulfuric acid and adjusted to pH 3-4. Furthermore, the acid extracts partitioned with ethyl acetate and then evaporated by rotavapor to yielded ethyl acetate extract (8 g). The acid extracts were basified with ammonia solution and adjusted to pH 8-9 and then partitioned with ethyl acetate to give crude alkaloid. The crude alkaloid (5 g) of *M. moluccana* was separated by coloumn chromatography on silica gel. Elution with *n*-hexane-ethyl acetate mixture containing increasing amount of ethyl acetate (9:1, to 1:1) to give four fractions A-D. On TLC analysis, fraction A (250 mg) showed two major spots. On separation of this fraction using column chromatography eluted with *n* hexane-EtOAc (from 4:1 to 1:1) gave three subfractions, A_1-A_3. Purification of subfraction A_2 by planar radial chromatography with eluent *n* hexane-$CHCl_3$ (from 7:3 to 1:4) yielded compound **1** (12 mg). Fraction C (470 mg), purification using planar radial chromatography with eluent *n* hexane-$CHCl_3$ (from 9:1 to 7:3) yielded compound **2** (18 mg) and **3** (19 mg).

2.4 *Cytotoxic assay*

Cytotoxic properties of the isolated compounds **1-3** against murine leukemia P-388 cells was evaluated according to the MTT method as previously described [7-9]. Artonin E was used as the positive control.

2.5 *Plasmodial activity*

Antiplasmodial activity of the isolated compounds **1-3** against *Plasmodium falciparum* strain 3D7 was evaluated according to Trager and Jensen method as previously described [10,11]. Chloroquine was used as the positive control.

3. Results and discussion

Phytochemical study on the MeOH extract from the leaves of *M. moluccana* yielded melimoluccanin (**1**), leptanoine C (**2**) and 7-*O*-isoprenyl-γ-fagarin (**3**).

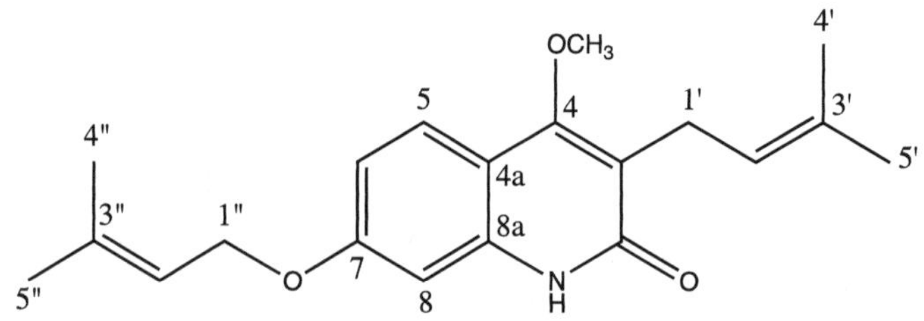

(1)

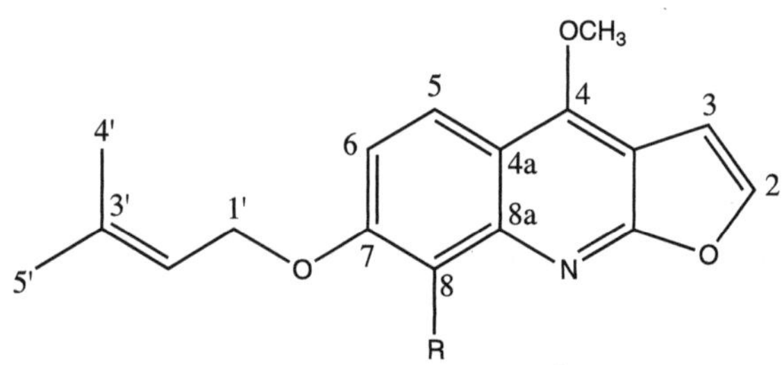

(2) R=H
(3) R=OCH_3

Figure 1. Alkaloid compounds **1-3** isolated from the leaves of *M. moluccana.*

Melimoluccanin (**1**), isolated as yellow solid has a molecular formula of $C_{20}H_{25}NO_3$ based on HRESIMS spectrum ($[M+H]^+$ 328.1928, calcd *m/z* 328.1918). The UV spectrum exhibited absorption maxima λ_{max} (nm) (log ε): 227 (4.57), 269 (3.86), 282 (3.85), 322 (4.09) and 336 (3.98) typical for a 2-quinolone structure [1]. The IR absorptions showed vibrations for conjugated carbonyl (1645 cm^{-1}), aromatic (1562, 1512, 1461 cm^{-1}) and ether (1093 cm^{-1}) groups, respectively. The ^{1}H NMR spectrum of **1** (Table 1) showed the presence of ABX systems at δ_H 7.65 (d, *J* = 8.8 Hz; H-5), 6.83 (dd, *J* = 8.8; 2.4 Hz; H-6) and 6.79 (d, *J* = 2.4 Hz; H-8) characteristic for 2-quinolone with three substituents. Furthermore, the ^{1}H NMR spectrum also showed an isoprenyl (3-methyl-2-butenyl) group [δ_H 1.83 (3H, s, H-4'), 1.79 (3H, s, H-5'), 3.39 (2H, d, *J* = 6.8 Hz, H-1'), 5.29 (1H, t, *J* = 6.9 Hz, H-2')], an oxyisoprenyl (3-methylbut-2-en-1-yl)oxy) group [δ_H 1.83 (3H, s, H-4"), 1.70 (3H, s, H-5"), 4.59 (2H, d, *J* = 6.8 Hz, H-1"), 5.52 (1H, t, *J* = 6.9 Hz, H-2")] and a singlet proton of methoxy group at δ_H 3.93. The ^{13}C NMR spectrum

(Table 1) of **1** showed 20 carbon signals. The assignment of ^{13}C NMR spectrum was confirmed by HMQC and HMBC spectra. The placement of isoprenyl, oxyisoprenyl and methoxy groups in 2-quinolone skeleton were established by HMQC and HMBC spectra. Based on HMBC spectrum of **1**, a methoxyl signal at δ_H 3.93 correlated to δ_C 162.5 (C-4). The methylene proton signal of isoprenyl group at 3.39 (H-1') correlated to a carbonyl of amide carbon (δ_C 166.0; C-2), an oxyaryl carbon (δ_C 162.5; C-4), two quarternary carbons [(δ_C 132.2, C-3'), (δ_C 119.3, C-3)] and a methine carbon signal (δ_C 121.8, C-2'), indicating that isoprenyl group attached at C-3 and methoxyl at C-4. The proton signal of aromatic region at δ_H 7.65 (H-5) showed long-range correlations with an oxyaril carbon signals (δ_C 160.6) and a quarternary carbon (δ_C 139.1). The result revealed δ_C 160.6 has position at C-7 and δ_C 139.1 at C-8a. Furthermore, the methylene proton signal of an oxyisoprenyl group at δ_H 4.59 (H-1") correlated to a oxyaryl carbon (δ_C 160.6, C-7), a quarternary carbons (δ_C 139.0, C-3") and a methine carbon signal (δ_C 118.9, C-2"), showing that oxyisoprenyl group attached at C-7. On the basis of the above spectral evidence, the structure of **1** was elucidated as melimoluccanin (**1**). Other HMBC correlations consistent with the structure **1** are shown in Table 1 and Figure 2. The structure of **1** is a novel compound.

Table 1. NMR spectroscopic data of melimoluccanin (**1**) in $CDCl_3$.

No.C	δ_H (mult, *J* Hz)	δ_C	HMBC
2	-	166.0	-
3	-	119.3	-
4	-	162.5	-
4a	-	111.0	-
5	7.65 (d, 8.8)	124.2	C-7, C-8a
6	6.83 (dd, 8.8; 2.4)	112.4	C-4a, C-8
7	-	160.6	-
8	6.79 (d, 2.4)	98.8	C-4a, C-7, C-8a
8a	-	139.1	-
1'	3.39 (d, 6.8)	23.2	C-2, C-3, C-4, C-2', C-3'
2'	5.29 (t, 6.9)	121.8	C-4', C-5'
3'	-	132.2	-
4'	1.83 (s)	18.0	C-2', C-3', C-5'
5'	1.79 (s)	25.7	C-2', C-3', C-4'
1"	4.59 (d, 6.8)	65.1	C-7, C-2", C-3"
2"	5.52 (t, 6.9)	118.9	C-4", C-5"
3"	-	139.0	-
4"	1.83 (s)	18.3	C-2", C-3", C-5"
5"	1.70 (s)	25.9	C-2", C-3", C-4"
4-OCH_3	3.93 (s)	61.8	C-4
NH	11.29 (s)	-	-

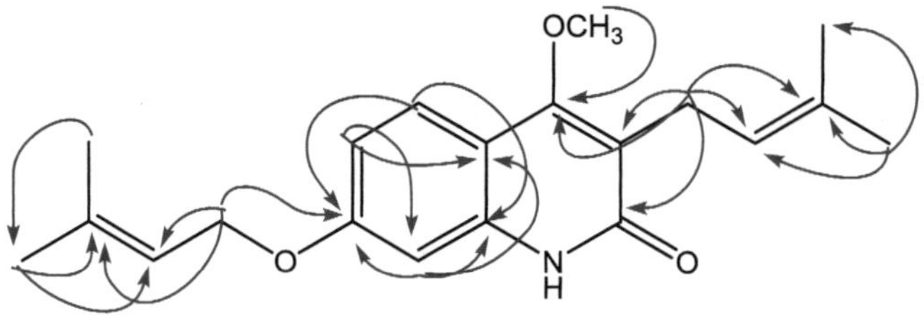

Figure 2. Selected HMBC correlations for **1**.

Leptanoine C (**2**) was isolated as a yellow solid, showed a quasimolecular ion $[M+H]^+$ at *m/z* 314.1407 corresponding to the molecular formula $C_{18}H_{20}NO_4$. The UV spectrum exhibited absorption maxima λ_{max} (nm) (log ε): 243 (4.50), 323 (3.77) and 334 (3.65) typical for a furoquinoline alkaloid [6]. The IR absorptions showed vibrations for aromatic (1622, 1587 cm^{-1}) and ether (1236 cm^{-1}) groups, respectively. The ^1H NMR spectrum (Table 2) of **1** showed the presence of a pair (*J* – 2.8 Hz) of proton signal of furo at δ_H 7.55 (H-2), δ_H 7.02 (H-3) and two singlets of isolated aromatic proton at δ_H 7.48 (H-5) and δ_H 7.33 (H-8) which is typical for a furoquinoline alkaloid with three substituents [6]. The ^1H NMR spectrum also showed an oxyisoprenyl group [δ_H 1.79 (3H, s, H-4'), 1.77 (3H, s, H-5'), 4.73 (2H, d, *J* = 6.6 Hz, H-1'), 5.59 (1H, t, *J* = 6.7 Hz, H-2')], and two singlets of methoxyl group at δ_H 4.42 (4-OCH_3), and 4.00 (6-OCH_3). The ^{13}C NMR spectrum (Table 1) of **2** showed 18 carbon signals were confirmed by HMQC and HMBC spectra. The placement of oxyisoprenyl at C-7, and two methoxyl groups at C-4 and C-6 in furoquinoline skeleton were established by HMQC and HMBC spectra. Based on HMBC spectrum of **2**, a methoxyl signal at δ_H 4.42 correlated to δ_C 155.5 (C-4) and a methoxy signal at δ_H 4.00 correlated to δ_C 148.0 (C-6). The methylene proton signal of an oxyisoprenyl group at δ_H 4.73 (H-1') correlated to a oxyaryl carbon (δ_C 151.8, C-7), a quarternary carbons (δ_C 138.3, C-3') and a methine carbon signal (δ_C 119.2, C-2'), showing that oxyisoprenyl group attached at C-7. Other HMBC correlations consistent with the structure **2** are shown in Table 2 and Figure 3.

Table 2. NMR spectroscopic data of leptanoine C (**2**) and 7-*O*-isoprenyl-γ-fagarin (**3**).

	2			**3**		
No.C	δ_H (mult, *J* in Hz)	δ_C	HMBC	δ_H (mult, *J* in Hz)	δ_C	HMBC
2	7.55 (*d,* 2.8)	142.3	C-3, C-3a, C-9a	7.52 (*d,* 2.8)	142.8	C-3, C-3a, C-9a
3	7.02 (*d,* 2.8)	104.6	C-2, C-3a, C-9a	6.98 (*d,* 2.8)	104.5	C-2, C-3a, C-9a
3a	-	102.0		-	101.8	
4	-	155.5		-	157.0	
4a	-	112.8		-	114.8	
5	7.48 (*s*)	101.1	C-4, C-4a, C-6	7.92 (*d,* 9.3)	117.8	C-4, C-7, C-8a
6	-	148.0		7.17 (*d,* 9.3)	114.0	C-4a, C-7, C-8
7	-	151.8		-	151.2	

8	7.33 (*s*)	107.6	C-7, C-8a	-	142.6	
8a	-	142.5		-	141.4	
9a	-	163.0		-	164.1	
1'	4.73 (*d*, 6.6)	65.7	C-7, C-2', C-3'	4.72 (*d*, 6.7)	66.6	C-7, C-2', C-3'
2'	5.59 (*t*, 6.7)	119.2	C-4', C-5'	5.53 (*t*, 6.7)	120.0	C-4', C-5'
3'	-	138.3		-	137.8	
4'	1.79 (*s*)	25.9	C-2', C-3', C-5'	1.75 (*s*)	25.7	C-2', C-3', C-5'
5'	1.77 (*s*)	18.3	C-2', C-3', C-4'	1.74 (*s*)	18.1	C-2', C-3', C-4'
4-OCH_3	4.42 (*s*)	58.8	C-4	4.36 (*s*)	58.8	C-4
6-OCH_3	4.00 (*s*)	56.0	C-6	-	-	
8-OCH_3	-	-		4.00 (*s*)	61.4	C-8

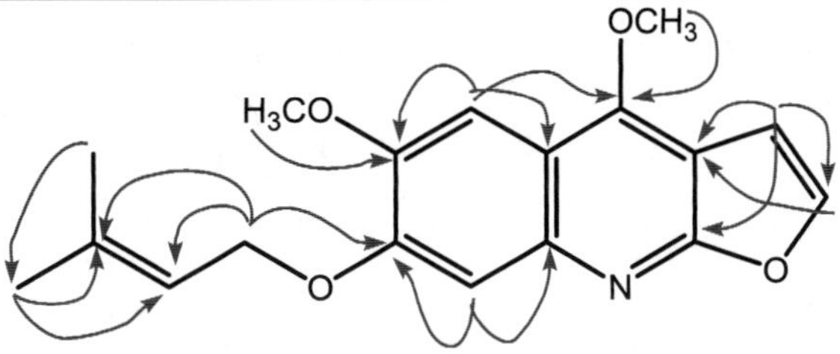

Figure 3. Selected HMBC correlations for **2**.

7-*O*-Isoprenyl-γ-fagarin (**3**) was isolated as a yellow solid, showed a quasi-molecular ion $[M+H]^+$ at *m/z* 314.1403 corresponding to the molecular formula $C_{18}H_{20}NO_4$. The UV spectrum (λ_{max} 250, 269, 320 and 333 nm) and IR spectrum (1616, 1502 and 1263 cm^{-1}) absorptions were very similar to those of **2**.

The ^{13}C NMR spectrum of **3** showed 18 carbon signals and the structure of **3** were confirmed by HMQC and HMBC spectra (Table 2 and Figure 4). The 1D and 2D NMR spectra data are consistent for 7-*O*-isoprenyl-γ-fagarin (**3**).

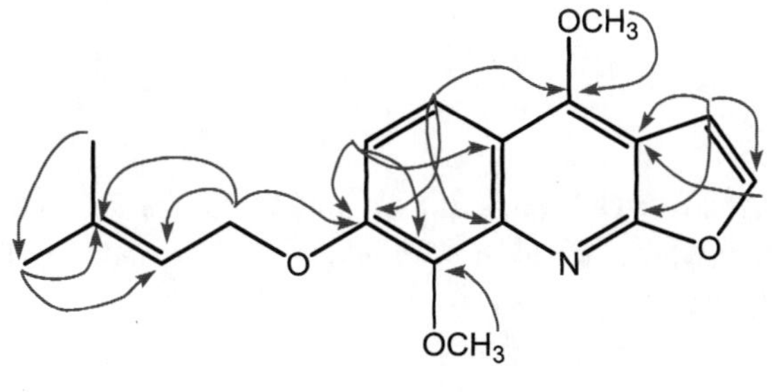

Figure 4. Selected HMBC correlations for **3**.

The cytotoxic activity (Table 3) of compounds **1-3** were evaluated for their cytotoxicity by MTT assay against murine leukemia P-388. Artonin E was used as the positive control. The cytotoxic activity of **1** showed high activity with IC_{50} 0.63 μg/mL. Those cytotoxic data for alkaloids **2-3** suggested to have moderate activity. Methoxyl group at C-8 of compound **3** showed higher activity than **2** (methoxyl group at C-6). The antiplasmodial activity (Table 3) of compounds **1-3** were evaluated according to Trager and Jensen method against against *Plasmodium falciparum*. Chloroquine was used as the positive control. The antiplasmodial activity of **2** showed high activity with IC_{50} 0.18 μg/mL. Those antiplasmodial for alkaloids **1** and **3** suggested to have moderate activity.

Table 3. Cytotoxic and antiplasmodial activities of compounds **1-3**

No.	**Compound**	**P-388 cells IC_{50} (μg/mL)**	**Antiplasmodial IC_{50} (μg/mL)**
1	Melimoluccanin (1)	0.63	4.28
2	Leptanoine C (2)	5.85	2.28
3	7-*O*-Isoprenyl-γ-fagarin (3)	2.09	0.18
4	Artonin E	1.33	-
5	Chloroquine	-	1.03

4. Conclusions

The phytochemical investigation of the leaves of *M. moluccana* to yielded compounds, melimoluccanin (**1**), leptanoine C (**2**) and 7-*O*-isoprenyl-γ-fagarin (**3**). Melimoluccanin (**1**) showed high activity against murine leukemia P-388 cells. 7-*O*-Isoprenyl-γ-fagarin (**3**) showed high activity against *Plasmodium falciparum* strain 3D7.

Acknowledgments

This research was supported by Directorate of Higher Education, Ministry of National Education, Republic of Indonesia (Penelitian Hibah Mandat, Universitas Airlangga, 2017).

References

[1] Tanjung M, Saputri R D, Wahjoedi R A, and Tjahjandarie T S 2017 *Molbank.* **2** M939.
[2] Nakashima K, Oyama M, Ito T, Witono J R, Darnaedi D, Tanaka T, Murata T, and Iinuma M 2011 *Tetrahedron. Lett.* **52** 4694.
[3] Sultana N, Hartley T G, and Waterman P G 1999 *Phytochem.* **50** 1249.
[4] Kassim N K, Rahmani M, Ismail A, Sukari M A, Ee G C L, Nasir N M, and Awang K 2013 *Food. Chem.* **139** 87.
[5] Kamperdick C, Van N H, Sung T V, and Adam G 1997 *Phytochem.* **45***(5)* 1049.
[6] Chen J, Cho J, Hwang T, and Chen I 2008 *J. Nat. Prod.* **71** 71.
[7] Tanjung M, Hakim E H, and Syah Y M 2017 *Chem Nat. Compd.* **53** 215.
[8] Tanjung M, Rachmadiarti F, Prameswari A, Agyani V U W, Saputri R D, Tjahjandarie T S, Syah Y M 2017 *Nat.Prod. Res.* **32** 1493.

[9] Tanjung M, Saputri R D, and Tjahjandarie T S 2016 5*Molbank.* **3** M906.
[10] Tjahjandarie T S, Pudjiastuti P, Saputri R D, and Tanjung, M 2014. *J. Chem. Pharm. Res.* **6** 786.
[11] Tanjung M, Fitriati F F, Saputri R D, and Tjahjandarie, T S 2016 *J Biol Active Prod from Nature.* **6** 95.

IOP Conf. Series: Journal of Physics: Conf. Series **1095** (2018) 012043 doi:10.1088/1742-6596/1095/1/012043

Synthesis and Characterization of Zeolite NaY from kaolin Bangka Belitung with variation of synthesis composition and crystallization time

Y K Krisnandi, I.Y Parmanti[1], R.T. Yunarti[1], R. Sihombing[1], I.R Saragi[1]

Department of Chemistry, Faculty of Mathematics and Natural Sciences (FMIPA), Universitas Indonesia, Kampus UI Depok, Depok 16424, Indonesia

Corresponding author's e-mail: yuni.krisnandi@sci.ui.ac.id

Abstract. Synthesis NaY zeolite using natural kaolin (Si/Al ratio~1) as an alternative source for silica and alumina is still challlenging. The aims of this study is to find out an effective and enviromentally friendly route to synthesis NaY zeolite from different source of silica and alumina. Pre-treatment kaolin like purification and fragmentation were conducted in order to reducing major impurity in kaolin such as quartz. Activation kaolin by Sub-Molten Salt System (SMS) using NaOH was employed because structure of Si-O and Al-O are inert and inactive. The kaolin was produced then used to synthesis NaY zeolite through hydrothermal method with a seeding technique using the mole ratio of (seed gel (5% of Al): 10,67Na_2O: Al_2O_3: 10SiO_2: 180H_2O) [1]. In order to find out optimum condition, the variation of crystallization time was performed. By the various characterization technique, the experimental result reported that NaY zeolite has the structure with mesoporous pore 3.6 nm and Si/Al ratio 3.8.

Keywords: ***fragmentation, kaolin, sub-molten salt system, NaY zeolite***

1. Introduction

Kaolin is a natural mineral that mainly composed of kaolinite (Si / Al = 1) with the chemical composition $Al_2Si_2O_5(OH)_4$ but also contains some impurities such as quartz, illites, smectites, and feldspars [2]. It belongs to phyllosilicate group with layer tetrahedral adn octahedral structure [3]. In this study, kaolin was utilized as an alternative source of silica and alumina for zeolite synthesis, with expectation of providing a more affordable source of silica and alumina for zeolite synthesis.

Zeolite is a microporous alumina silicate material (pore size < 2nm) with different types of pore shapes and structure. One of the as-synthesized zeolites is NaY zeolite belonging to the faujasite class, with a Si/Al ratio of 2-4 and main frame of two 6 rings that bound through a sodalite cage with an average pore diameter of 7.4 Å [3]. Sunardi, et al reported the treatment of kaolin powder with alkaline salt at 500-950°C to convert kaoline to metakaolin, the more reactive form, before being used as Al-Si source [4]. However this process was shown ineffective, as some part of unreacted kaolin became impurities during synthesis. In 2014, Yue et.al reported the fragmentation of natural aluminosilicate minerals structure through sub-molten salt system and use it as precursor for synthesis of ZSM-5[5].

In this work, the attempt to synthesized NaY zeolite using fragmented kaolin as Si and Al source was reported. In order to reach the desired Si amount, suspension of colloidal silica (LUDOX 40) was also added. Several characterization techniques were applied to the as-synthesized materials, and show that the synthesis was successful to produce mesoporous NaY.

IOP Conf. Series: Journal of Physics: Conf. Series **1095** (2018) 012043 doi:10.1088/1742-6596/1095/1/012043

2. Materials and methods

2.1. General

Bangka Belitung natural kaolin, colloidal silica -LUDOX 40 (Sigma Aldrich), NaOH (pellet, 99%,Merck), CH_3COOH (glacial, Merck), $CH_3COOHNa.3H_2O$ (Merck), H_2O_2 (30%, Merck), $Na_3C_6H_5O_7.2H_2O$ (Merck), $NaHCO_3$ (Merck), deionized water.

2.2. Modification of Natural Kaolin

Modification of Bangka Belitung natural kaolin consisted of physical activation, purification and fragmentation. Natural kaolin was polished using a grinder and filtered with a sieve. Activation of natural kaolin was performed by washing the kaolin with deionized water (1:3 w/v) under stirring for 3 h. The solid phase then was dried at 300ºC for 1 h. The purification process was conducted following Ming and Dixon, 1987. The activated natural kaolin was treated with 1 M NaOAc buffered to pH 5, 30% H_2O_2, and dithionite-citrate-bicarbonate. After purification, fragmentation was done by sub-molten salt system (SMS) [6] technique using strong base (NaOH) with 2: 1 ratio in 250ºC. Characterization of natural zeolite was done using XRD, FTIR, and SEM-EDX [7].

2.3. Synthesis of NaY Zeolite

The synthesis of NaY zeolite refers to the method of Ginter et al [1] with the ratio of the mole ratio (Seed gel: Na_2O: Al_2O_3: SiO_2: H_2O = (5% Al): 10,67: 1: 10: 180). Fragmentated kaolin was used as Al and Si source and certain amount of colloidal silica (LUDOX 40) was an additional Si source. The crystalization was carried out in polypropylene bottle at 100 ºC for 16 and 24 hours. Extensive characterization techniques such as XRD, FTIR, SEM-EDX, surface area analyzer were carried out on kaolin, fragmented kaolin and the as-synthesized zeolites. The cation exchange capacity of raw kaolin and as-synthesized NaY were also conducted

3. Results and discussion

Characterization of Pre-treated Kaolin

The synthesis process of zeolite NaY begins with activation and purification to remove impurities on the raw material of the synthesis of zeolite.

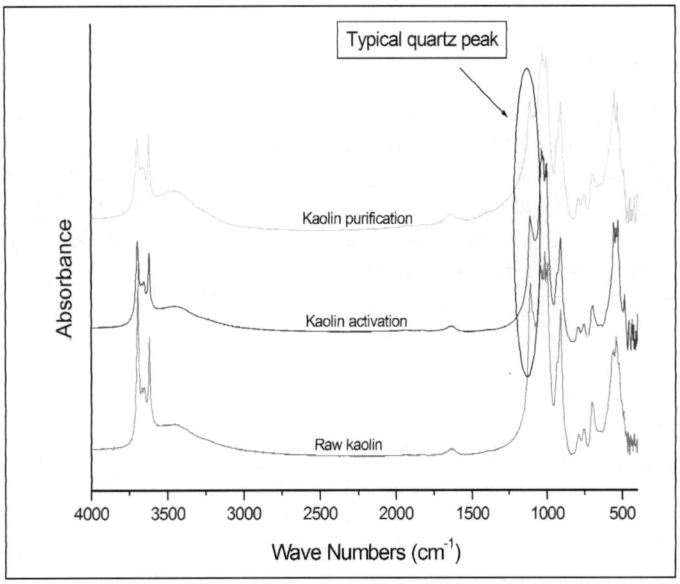

Figure 1. Spectra FTIR kaolin pre-treated Bangka Belitung

Figure 1 shows the comparison between FTIR spectra of raw, activation and purification kaolin. The peaks around 1105-1117 cm^{-1} indicate the purity difference since it belongs to quartz that shifts from 1115 cm^{-1} to 1108 cm^{-1} [4]. It confirmed that purified kaolin has smaller absorption in that area than the others, estimated quartz in kaolin reduced.
From Figure 2a, it is found that there are significant differences between raw kaolin and kaolin after fragmentation such as fragmented kaolin has peak at wavenumber 1646 cm^{-1}, while in the raw kaolin it does not appear.

The XRD pattern from Figure 2b shows comparison of raw diffraction patterns of kaolin [7] and kaolin after fragmentation, in which there is a peak shift at 13.90 ^{o} And 24.86 ^{o}. As well as the disappearance of some kaolin peaks in the two theta of 12.34 ^{o}; 19,87 ^{o}; 35.6 ^{o}; and 38.4 ^{o}. So that it can be concluded that the process of fragmentation or submolten salt system (SMS) successfully done because it has destroyed kaolin framework.

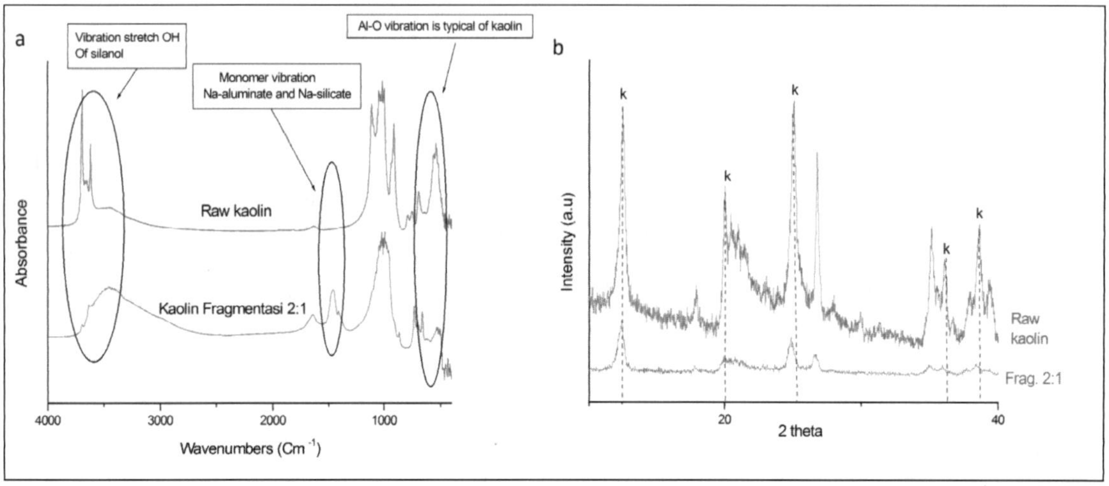

Figure 2. a. FTIR spectra of kaolin fragmentation b. XRD pattern raw kaolin and kaolin fragmentation

Zeolites NaY Characterization

Figure 3 illustrates the FTIR spectra of two synthesis variations (NaY 2116 and NaY 2124) indicate a double six ring absorption or a secondary builder 4 and 6 external linkage rings with a faujasite structure (FAU). Peaks at 3750-3500 cm^{-1} shows the presence of hydroxyl groups (-OH) in the silanol group (Si-OH). The absorption band at 1500-1700 cm^{-1} belongs to H-O-H bending vibration of water molecules and also there is a peak of Si-O / Al-O asymmetry vibration at 1250-950 cm^{-1} wavenumber which is the typical peak of the silica alumina compound. The results from FTIR spectra confirm that synthesis of zeolites NaY has been successful.

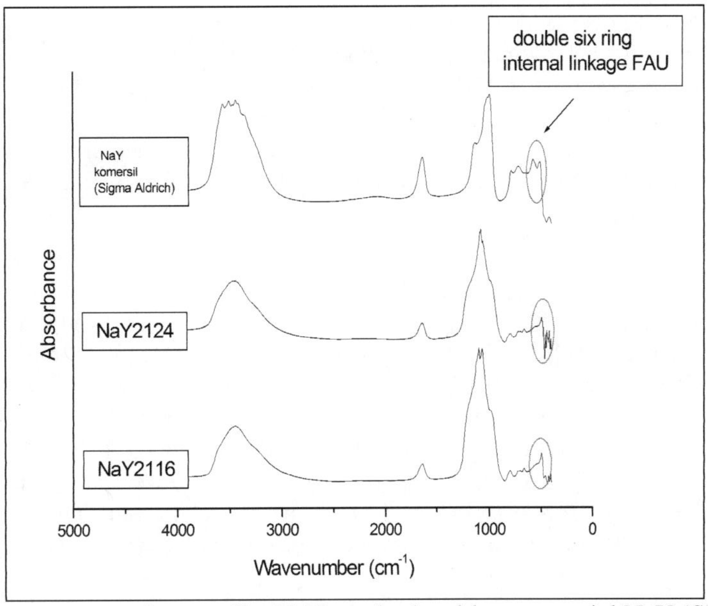

Figure 3. FTIR Spectra comparison zeolite NaY synthesis with commercial NaY (*Sigma Aldrich*)

Figure 4 shows that characterization of as-synthesized NaY zeolite with XRD instrument, which was carried out to find out crystallinity and prove that the zeolite formed is NaY zeolite. For that reason, the comparison of NaY diffraction, resulted from synthesis, and standard NaY zeolite from *IZA-online* was carried out. Differences in diffraction pattern are possible because there are still some impurities such as Fe^{2+}, Ca^{2+}, K^{+} and Mg^{+} ions in the fragmented kaolin, the source used in this synthesis. However, it can be seen that on the addition of NaOH 2:1 during fragmentation in NaY2124 resulted in the formation of pattern of resemblance with more NaY zeolite compared to the 2:1 fragmentation of NaY2116. This indicates that 24 hour crystallization (of NaY2124) has given more NaY crystals compared to 16 hour crystallization.

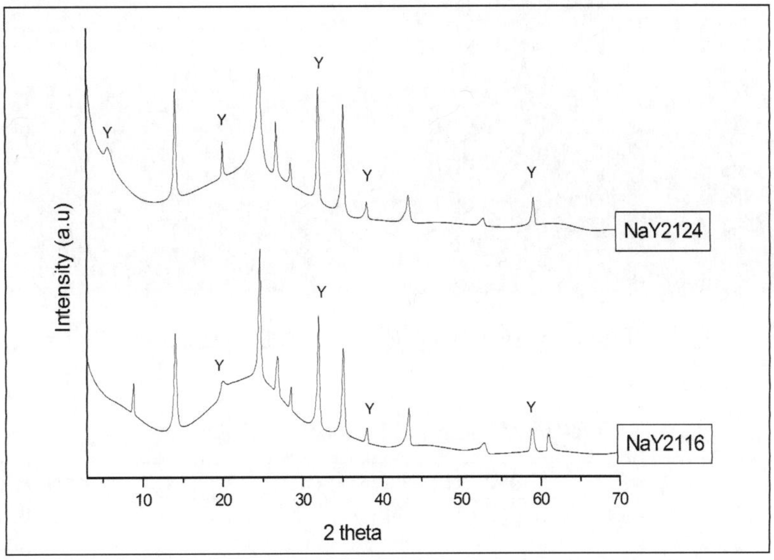

Figure 4. Summary of the comparison XRD patterns of as-synthesized NaY zeolites

IOP Conf. Series: Journal of Physics: Conf. Series **1095** (2018) 012043 doi:10.1088/1742-6596/1095/1/012043

After matching the patterns of the zeolite peaks with those from IZA-online, as shown in Figure 4, it is found that the Sodalite octahydrate type is closest to as-synthesized zeolites. The impurities of zeolite P (Sodalite) are often found in NaY synthesis [8] . The peaks that have similarities with those in NaY zeolites are with 2θ of14.42 °; 24.74 °; 35.26 °; and 43.52°. This resemblance indicates that the as-synthesized zeolite NaY is formed over the dominant sodalite framework. Characterization using EDX was also carried out to find out the Si/Al ratio contained in the zeolites. Supporting the XRD data, the most closely related to the Si/Al ratio of zeolite NaY is the as-synthesised zeolite with code NaY2124, with Si/Al ratio of 3.8, while the Si/A; ratio of NaY2116 is 4.68.

In Figure 5, some images of the as-synthesized zeolites NaY2116 and NaY2124 are displayed. Rugged morphology on the crystalline shapes may indicate the effect of mesoporosity on the zeolite. Further characterized using BET surface area analysis (summarized in Table 1) shows that the NaY2124 has lower surface area than the NaY2116, and its pore size is smaller, although the average size is within mesopore range. The time of crystallization affects the formation of crystals. It was observed at higher magnification (10.000x) of the two as-synthesized zeolites showed that at 24 hours of crystallization (Figure 5D) the NaY zeolite crystal habit is better developed, compared to that of the NaY zeolite synthesized for 16 hours (Figure 5C). It can be seen from their morphology that the crystals were intergrown and built through growing phases derived from the larger crystals.

Figure 5. SEM Imaging of NaY zeolite synthesized

TABEL 1. Textural Parameters of Zeolites

Samples	Surface area (S_{BET}) (m^2 g^{-1})	Average Pore Radius (nm)
NaY2116	990.4	5.012
NaY2124	162.3	3.601

4. Conclusions

The as-synthesized zeolite that closely similar to NaY zeolite is NaY2124 with 2: 1 fragmentation composition and a 24 hours crystallization time. This is evidenced by the XRD results addressing some typical zeolite NaY peaks and the Si/Al ratio of 3.8. The zeolite is mesoporous with an average pore of 3.601 nm.

Acknowledgments

The authors would like to thank Universitas Indonesia for funding this research through PITTA Grant from Universitas Indonesia with contract number No. 702/UN2.R3.1/HKP.05.00/2017.

References

[1] D M Ginter, A T Bell and C J Radke 1992 *Synthesis of Microporous Material Molecular Sieve* vol 1 ed M L Occelli and H E Robson (New York: Van Nostrand Reinhold)
[2] Murray H H et al 1993 *Kaolin Genesis and Utilization* Ed Boulder CO (The Clay Minerals Society)
[3] Cundy C S and Cox P A 2005 *The Hydrothermal Synthesis of Zeolites: Precursors, Intermediates and Reaction Mechanism.* (Microporous and Mesoporous Materials) vol 82 pp 1–78
[4] Sunardi, Arryanto Y and Sutarno 2009 *Adsorption Gibberelic Acid Onto Natural Kaolin From Tatakan, South Kalimantan* vol 9(3) (Indonesia : Indonesian Journal of Chemistry) pp 373-379
[5] Yue, Y Liu, Haiyan,Yuan, et al 2014 *From Natural Aluminosilicate minerals to hierarchical ZSM-5 Zeolite: A Nanoscaledepolimerization-reorganization approach.* (Journal of Catalyst: 319), pp 200-210
[6] Zhang Y, LI Zuo-Hu and et al 2005 *Sub-Molten Salt Environmentally Benign Technology* (China)
[7] Ekosse G E 2005 *Fourier Transform Infrared Spectrophotometry and X-ray Powder Diffractometry as Complementary Technique in Characterizing Clay Size Fraction of Kaolin* vol 9 (2) (Nigeria : University of Port Harcourt) pp 43-48
[8] Wang L, Zhang Z, Yin C, et al 2010 *Hierarchical Mesoporous Zeolits with Controllable Mesoporosity Templated from Cationic Polymers.* Microporous and Mesoporous Materials, vol 131(1–3) (Amsterdam: Elseiver) pp 58–67

Synthesis and Characterization of ZSM-5 Zeolite from Dealuminated and Fragmentated Bayat-Klaten Natural Zeolite

Y K Krisnandi, I Mahmuda, D U C Rahayu and R Sihombing

Department of Chemistry, Faculty of Mathematics and Natural Sciences (FMIPA), Universitas Indonesia, Kampus UI Depok, Depok 16424, Indonesia

Corresponding author's e-mail: yuni.krisnandi@sci.ui.ac.id

Abstract. Synthesis of ZSM-5 zeolite was conducted using natural zeolite of Bayat-Klaten by employing fragmentation method of sub-molten system through breaking zeolite framework into its monomer in alkaline condition at 250°C. Prior to fragmentation, one part of zeolite was treated with dealumination and the other one without dealumination, labeled as ZSM-5_a and ZSM-5_wa, respectively. Characterization with FTIR, XRD, and EDX showed that the fragmentation was successful, in which the structure of zeolite was damaged and the infrared bands due to AlO_4 and SiO_4 were appeared. The synthesis of ZSM-5 was performed hydrothermally using tetrapropylammonium hydroxide (TPAOH) as a directing agent of pore structure and addition of LUDOX (40% silica suspension in water). The XRD characterization showed the as synthesized materials had the structure of ZSM-5 zeolites with high crystallinity. Imaging with SEM showed the characteristic of hexagonal ZSM-5 crystals. The BET surface area analysis showed a typical surface area of microporous zeolite, 262 m^2/g, and pore diameter of 1.496 nm; indicating that the ZSM-5 zeolite obtained was micropore size. The Si/Al ratio of ZSM-5_a and ZSM-5_wa were 18 and 21, respectively. These results indicated that Bayat-Klaten natural zeolite can be used as silica and alumina source for high crystalline ZSM-5 with medium Si/Al ratio and having good thermal stability. Thus, these materials are potential to be tested as fluid hydrocracking catalyst.

Keywords: dealumination, fragmentation, ZSM-5, Bayat-Klaten natural zeolite

1. Introduction

Indonesia has large potential of natural zeolite production, such as that located in Lampung, West Java, Central Java, East Java, East Nusa Tenggara, and Sulawesi. Zeolites are aluminosilicate material with a three-dimensional framework consisting of SiO_4 tetrahedral and AlO_4 arranged regularly where every oxygen atom is shared between two tetrahedral. The structure of the zeolite framework and make the zeolite has many uses, unique properties, and easy to modify, so that the rapidly growing research on zeolites. However, the natural zeolite has some disadvantages, including having a low purity because it contains a lot of impurities and poor crystallinity. Therefore, its utilization becomes less optimal compared with synthetic zeolite [1].

The large of potential resource of natural zeolite in Indonesia affected this natural zeolite is cheaper than synthetic one. Based on the acidity properties of the natural zeolite and the ability to catalyze of synthetic zeolite, we need some process and modification of natural zeolite in order to compete with synthetic zeolites [2]. To improve the utilization of natural zeolite, silica and alumina contents presented in the natural zeolite can be used as a source of Si and Al source to form a zeolite with a high purity level. Acid treatment to zeolite should be conducted in order to eliminate Al and cause deformation on its structure. Furthermore, depolymerization of zeolite structure through sub-molten technique was required to obtain aluminate and silicate fractions. The defective zeolite mixture was then used as the Si and Al source for ZSM-5 zeolite synthesis.

2. Materials and methods

2.1. General

Natural zeolite was obtained from Bayat, Central Java, Indonesia. The chemicals used were hydrochloric acid (HCl) p.a., sodium hydroxide (NaOH) pellet, acetic acid glacial (CH_3COOH) p.a., hydrogen peroxide (H_2O_2) 30% p.a., tetrapropylammonium hydroxide (TPAOH) 1M, colloidal silica (LUDOX 40%), and tetraethylorthosilica (TEOS) from Sigma Aldrich, sodium ditionate ($Na_2S_2O_4$), sodium-citrate-bicarbonate, and sodium-acetate buffer (NaOAc) from Merck.

2.2. Pretreatment of Natural Zeolite

Modification of Bayat natural zeolite consisted of physical activation, purification, dealumination, and fragmentation. Natural zeolite was crushed using a grinder and filtered with a sieve. Activation of natural zeolite was performed by washing the zeolite with deionized water (1:3 w/v) under stirring for 3 h. The solid phase then was dried at 300°C for 1 h. The purification process was conducted following Ming and Dixon, 1987 [3]. The activated natural zeolite was treated with 1 M NaOAc buffered to pH 5, 30% H_2O_2, and dithionite-citrate-bicarbonate. After purification, dealumination process was conducted by stirring zeolite in 1 M HCl solution (10 g zeolite/100 mL solution) at 100°C for 2 h under reflux. The solid phase was washed with demineralized water and dried at 65°C. Pretreated zeolite, with or without acid treatment, was fragmented using sub-molten salt system (SMS) technique, by mixing it with strong base (NaOH pellet), and ratio of zeolite:NaOH was 2:1. Incipient amount of water was poured into the mixture and then heated at 250°C with stirring.

2.3. Synthesis of ZSM-5 Zeolite

The fragmented zeolite (with or without acid treatment) was added to deionized water contained 1 M TPAOH and stirred using a magnetic stirrer at 100° C. LUDOX 40% and TEOS were also added to the mixture until the composition of molar ratio in the mixture was 64.35 SiO_2: 1 Al_2O_3: 10.08 TPAOH: 6 Na_2O: 3571.67 H_2O [4]. CH_3COOH was added to regulate the pH of the mixture (pH 11 ± 0.2). Stirring was done for 3 to 24 h. The gel formed then was transferred into the autoclave to be further heated at 150°C for 144 h. The resulting product was collected by filtration and washed with aquades, then dried at 60°C for 12 h, and 105°C for 1 h. Furthermore, calcination was done at 550°C for 8 h to remove the templates. The ZSM-5 synthesized from zeolite that undergo acid treatment before fragmentation was labeled ZSM-5_a. While the counterpart that was synthesized using fragmented zeolite without acid treatment was named ZSM-5_wa.

2.4. Characterization of zeolites

Characterization of natural zeolite and ZSM-5 were done using XRD, FTIR, and SEM-EDX. BET was also conducted for characterization of ZSM-5.

3. Results and discussion

Characterization using XRD was conducted to find out the crystallinity and typical peak pattern of the Bayat natural zeolite. Bayat raw zeolite diffraction pattern (Figure 1A) indicated peaks at 2θ of 9.85°, 22.21°, and 25.61°. The resulting peak did not have a sharp intensity, since the crystallinity contained in Bayat zeolite was still low or amorphous. Furthermore, these peaks were similar to the typical peak of the mordenite type zeolite (JCPDS Card No.29-1257) main peak of 9.77°, 22.20°, and 25.63°.

IOP Conf. Series: Journal of Physics: Conf. Series **1095** (2018) 012044 doi:10.1088/1742-6596/1095/1/012044

From the diffraction pattern (Figure 1B(a-c)) of the Bayat natural raw zeolite showed the same pattern with the pattern of active zeolite diffraction and purification. This diffraction pattern indicated that the activation and purification process of Bayat natural zeolite did not alter its crystal structure. Figure 1B(d-e) showed that peaks appeared at 2θ of 20.90°, 26,68°, and 36.57° were the peak for quartz (Quartz). This is because the fragmentation process aims to destroy the zeolite frame structure, breaking the Si-O-Si and Si-O-Al bonds into Si-O and Al-O active, therefore the tetrahedral structures of Si-O and Al-O will be damaged and their structural shape is no longer intact [5].

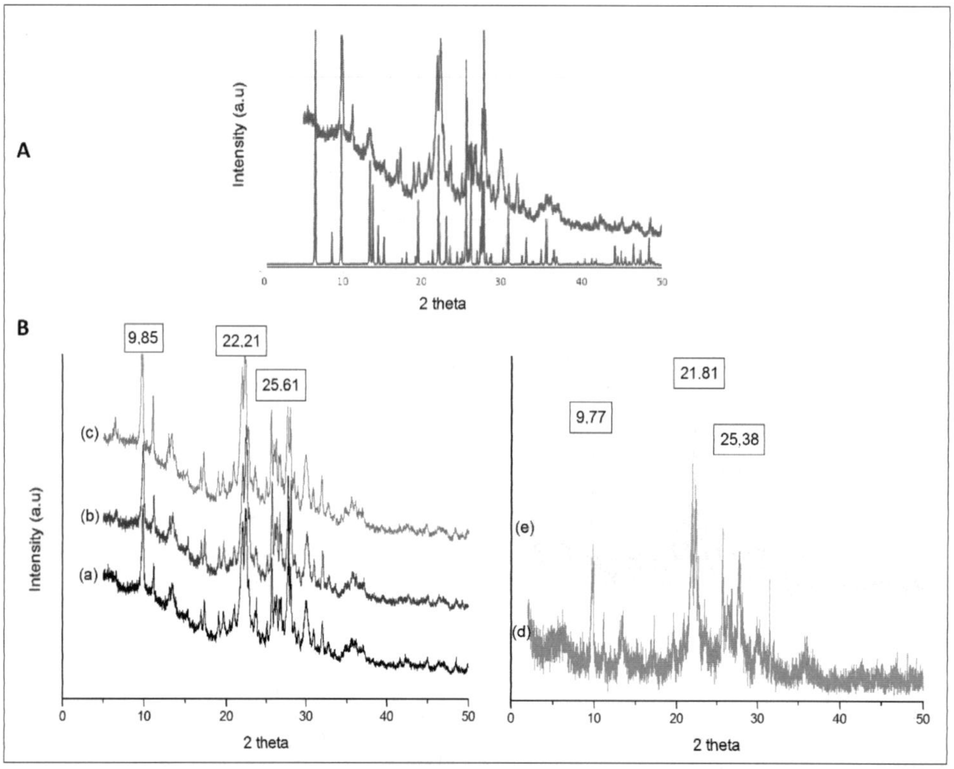

Figure 1. XRD patterns of Bayat natural zeolite A. raw and simulated mordenite in IZA-online, B. (a) raw, after pretreatment: (b) purification and (c) activation, and after fragmentation: (d) acidic and (e) without acidic treatment

Figure 2 showed the FTIR spectra of as-synthesized ZSM-5 before and after calcination. ZSM-5 FTIR spectra before calcination showed peaks at 2960-2850 cm^{-1} belongs to vibration band of C-H and 1470-1350 cm^{-1} belongs to C-H bend vibration. This absorption band came from a template used in the process of zeolite synthesis, TPAOH. The calcination process was performed to remove the organic template used in the synthesis. With the loss of this template, the pores and channels on the zeolite become empty. The loss of template in ZSM-5 after calcination was showed by the loss of the vibration band C-H vibration absorption and C-H bending vibration [6]. With the calcination process, the absorbing bands at 3600 cm^{-1} was increased which indicated the absorption bands of silanol (Si-OH) groups on the zeolite surface and located within the zeolite framework. The peaks at 700-1100 cm^{-1} showed the vibrations of Si-O and Al-O and 950-1250 cm^{-1} showed the vibration of the asymmetry Si-O and Al-O from the alumina-silica framework.

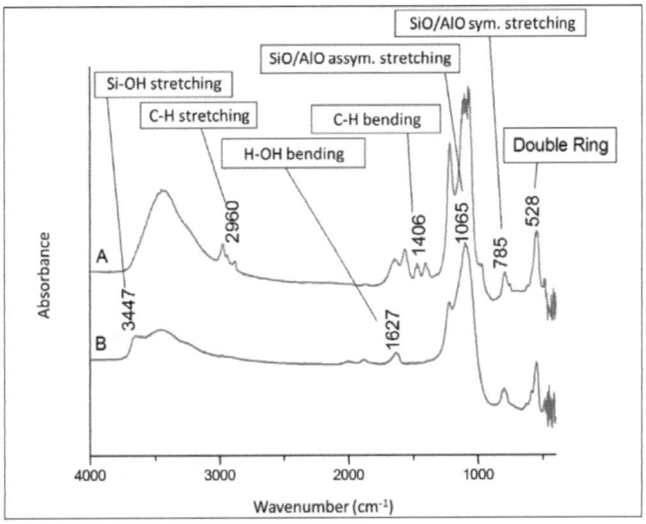

Figure 2. FTIR spectra of as-synthesized ZSM-5_a zeolite (a) before and (b) after calcination

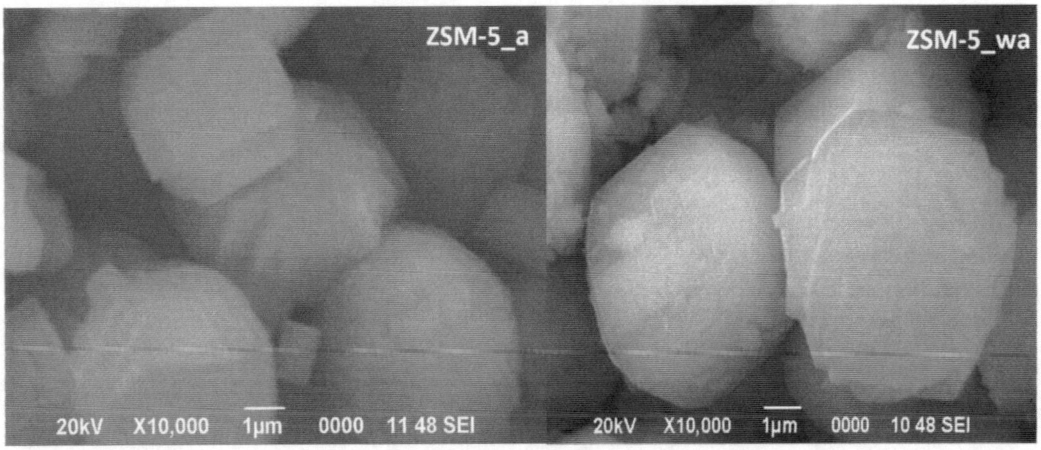

Figure 3. SEM Images of as-synthesized ZSM-5 zeolites using Bayat natural zeolite, with (ZSM-5_a) and without acidic (ZSM-5_wa) treatment

Surface morphology and crystal size of calcined ZSM-5 were characterized by SEM. ZSM-5 zeolite synthesized using TPAOH template had coffin-like morphology and hexagonal crystal geometry. Figure 3 showed SEM imaging of ZSM-5_a which clearly visible ZSM-5 zeolite-shaped hexagonal crystal resembling tablets rather than ZSM-5_wa.

The Si/Al ratio of ZSM-5_a and ZSM-5_wa were 18 and 21, respectively. In Figure 4, ZSM-5 zeolite XRD patterns at 2θ between 5-10° indicated a significant peak difference. ZSM-5_a compared with ZSM-5_wa showed low peak while the highest peak appeared in ZSM-5_wa. This result was similar with SEM and EDX results. It can be summarized that ZSM-5_wa has two distinctive ZSM-5 peaks at 2° position of 7.9-9.11°, three diffraction peaks at the region about 22-25°, and based on SEM-EDX data ZSM-5_wa formed a crystalline structure of ZSM-5. Furthermore, ZSM-5 zeolites were characterized on its pore and surface area using BET instrumentation. According to Table 1, both zeolites had high surface area and based on their pore diameter, they fall into microporous zeolites.

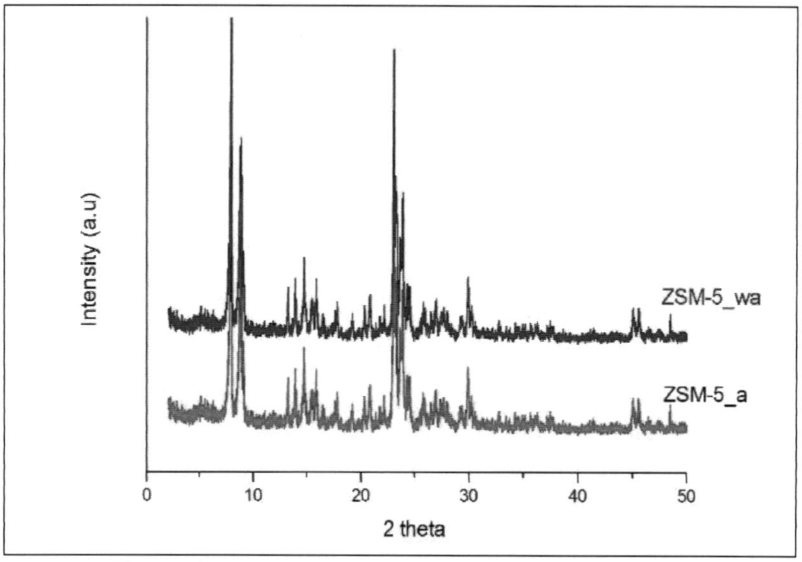

Figure 4. XRD pattern of synthesized ZSM-5 zeolites

Table 1. Textural Parameters of Zeolites

Samples	Surface area (S_{BET}) ($m^2\ g^{-1}$)	Average Pore Radius (nm)
ZSM-5_a	262	1.496
ZSM-5_wa	309	1.306

4. Conclusions

Microporous ZSM-5 was successfully synthesized using Bayat natural zeolite as partial source of Si and Al. Acidic treatment was not critical as it gave similar ZSM-5 structure and even lower surface area.

Acknowledgments

The authors would like to thank Universitas Indonesia for funding this research through PITTA Grant Universitas Indonesia with contract number No. 702/UN2.R3.1/HKP.05.00/2017.

References

[1] Yue Y, Liu H, Yuan P, Li T, Yu C, Bi H and Bao X 2014 From natural aluminosilicate minerals to hierarchical ZSM-5 zeolites: a nanoscale depolymerization–reorganization. *J. Catal.***319** pp200-210.

[2] Kusdarto 2008. Potensi Zeolit di Indonesia *J. Indones. Zeolites* **7(2)** 78-87.

[3] Ming D W and Dixon J B. 1987 Quantitative Determination of Clipnotilolite in Soils by A Cation Exchange Capacity Method *Clay. Clay Miner.* **35(6)** 463-468.

[4] Wang L, Zhang Z, Yin C, Shan Z and Xiao F-S 2010 Hierarchical mesoporous zeolites with controllable mesoporosity templated from cationic polymers *Micropor. Mesopor. Mat.* **131(1-3)** 58-67.

[5] Rohayati, Krisnandi Y K and Sihombing R 2017 Synthesis of ZSM-5 using Bayat Natural Zeolite as Silica and Alumina Source.(*AIP Conf. Proc.***1862** 030094-1-030094-2).

[6] Nurliati G, Krisnandi Y K, Sihombing R and Salimin Z 2015 Studies of Modification of Zeolite by Tandem Acid -Base Treatments and Its Adsorption Performance towards Thorium *Atom Indones.* **41(2)** 87-95.

Analysis of the Level of Conceptual Understanding

Y Andayani[1], S Hadisaputra[1], H Hasnawati[1]

[1]Postgraduate Program of Science Education, University of Mataram. Jalan Majapahit No 62. Mataram, 83125, Indonesia.

Email : yayukmtr@gmail.com

Abstract. The study aimed to analyze the level of chemistry student's understanding of chemical concepts and critical thinking disposition that learn using problem based learning model. This research is a descriptive research. The subjects of the study were 35 students in the sixth semester of chemistry education program, University of Mataram, Indonesia. The student level of conceptual understanding data was obtained through three levels multiple choice tests, whereas the level of student critical thinking disposition data was collected by CTD questionnaire. The student level of conceptual understanding results showed that 58.57 % of students understood the concept well, 4.99 % understood the concept, 23.46 % were unsure and had misconceptions and 12.98 % had not understood the concept correctly. The analysis of the level of critical thinking disposition indicates that the student critical thinking disposition was categorized as positive. Correlation test indicated a significant correlation between conceptual understandings with critical thinking disposition. In conclusion, the level of conceptual understanding and the critical thinking disposition of students after learning using problem-based learning model were categorized as positive and both have a significant correlation.

Keywords : Chemical concept, learning model, problem-based

1. Introduction

Education in Indonesia continues to improve its quality to meet the challenges of the 21st century. Teacher quality improvement is one of the main targets. Teachers must hold appropriate skills, one of them is the ability of high-level understanding [1,2]. Teachers must have a correct understanding of the concept. Understanding of the content of knowledge plays an important role for teachers in order to be able to teach effectively. In addition, the level of teachers understanding of the concept may determine the level of professionalism. Improving the professionalism of teachers is an important effort in improving the quality of education [3,4].

High quality education produces highly skill teachers that in line with its demands. There are nine learning criteria for producing 21st century skill teachers: curriculum relevant to the learner's daily life, learning involving multiple disciplines, developing high-level thinking skills, applying knowledge gained to other areas that intersect with daily life day learner, learn how to learn to learners, learning that prevents misconception, team learning, utilizing technology to support learning, encourage creativity of learners [5-8].

The constructivism learning approach is one of the learning approaches that meet the above criteria. Constructivism approaches include problem based learning model, project-based learning model, inquiry learning model and discovery learning model. Problem-based learning (PBL) model is developed to help learners develop thinking skills, problem solving, and intellectual skills, learn various roles of adults, through their involvement in real-life experiences or simulations, and become independent learners. PBL focuses on learning on selected issues so that learners not only learn the concepts related to the problem but also the scientific method to solve the problem [9,10].

In the chemistry context, the physical chemistry subject has many abstract concepts. It involves the macroscopic (observable), microscopic (particle constituent) and symbolic (substance) aspects so that the physical chemistry course is difficult for students to understand. It is reported that students enter the classroom with negative perceptions of the chemistry course and have low success expectations for this course. Students' difficulties in studying chemistry are triggered by difficulty understanding the term, difficulty understanding chemical concepts, and difficulty in numbers. These learning difficulties can be overcome through the management of good chemistry learning, especially in the learning planning stage and learning process [9-14]. On the one hand, the students' ability to understand concepts is related to the level of student thinking, such as the tendency to think critically. The purpose of this research is to analyze the level of understanding of chemistry concept and critical thinking deposition of prospective teachers whom teaches by problem-based learning model.

2. Methods

This research is descriptive research type. The dependent variable is conceptual understanding and critical thinking disposition. Attribute variable is a problem-based learning model. Research subjects are chemistry education students of Faculty of Teacher Training and Education, University of Mataram. The subjects consisted of 35 students in the third years that programmed Physics Chemistry 3 courses.

Data on concept comprehension level was obtained through three-level double choice test and analyzed using modified Certainty of Response Index technique. The criteria for concept comprehension level are: 1). understand the concept correctly, 2). understand concept but not sure, 3). misconceptions and do not know the concept [15].

Data on the level of critical thinking was obtained through questionnaires. The critical thinking tendency questionnaire consists of 75 statements representing 7 indicators of critical thinking disposition. These indicators are: truth-seeking, open mindedness, analyticity, systematicity, self-confidence, inquisitiveness, and maturity of judgment. The questionnaire comes with a selection of answers that use Likert scale 1 to 6, from strongly agree to strongly disagree. Each flat indicator has

10 statements and each indicator has a minimum score of 10 and a maximum score of 60. The overall score is at least 70 and a maximum of 420 with the score ≥ 349 enters in a strong category, 280-348 positive, 211-279 ambivalent, and ≤ 210 negative [16].

The data obtained were analyzed using descriptive statistical test in the percentage using formula:

$$P = \frac{\Sigma x}{\Sigma x1}\, x\ 100\ \% \qquad 1$$

Where P = Percentage of concept level understanding per category; Σx = Frequency in each category;

$\Sigma X1$ = Total frequency. The relationship between the levels of understanding of the concept with the level of critical thinking tendency is analyzed using correlation test.

3. Results and Discussion

The level of understanding of concepts was obtained through the three-tier select test at the end of the lecture. It is categorized into 4 levels: understanding the concept correctly (MB), understanding concepts but not convincing (M), misconceptions (MSK) and not knowing the concept (TM). The distribution of the concept comprehension level is expressed in percent (%) as in Table 1.

Table 1. Data on conceptual understanding

Sub topic	Level of understanding Concept (%)			
	MB	M	MSK	BM
Mechanism of liquid phase reaction	68.57	6.86	19.43	5.14
Photochemistry	47.62	2.86	26.66	22.86
Irreversible processes.	59.52	5.24	24.29	10.95
Total Average	58.57	4.98	23.46	12.98

The level of understanding of chemical concepts of prospective teacher students was identified using a three-tier select test with 17 items that included sub-materials on the mechanisms of liquid phase reaction, photochemistry and irreversible processes. The results indicated that there are 58.57 % of students who understand the concept correctly. Percentages of students who understand the concept correctly are over 50 %. This indicates that the applied PBL model facilitate the students to be able to understand the concept correctly.

This result is in line with previous conducted research which states that the mastery of learning concepts of learners can be improved using problem-based learning model. Implementation of problem-based learning model includes learning activities in the form of: giving questions, simple experiments, study of literature, test hypothesis and discussion. All these things can help students understand the correct concept and minimize misconceptions [17-19].

However, there are still 23.46 % students experiencing misconceptions. Learning with constructivism approach also allows students to experience misconception. Some students construct their knowledge according to the concepts that experts agree on, but there are also students who misconstruction their knowledge. This misconception constructed by students is what causes misconceptions.

Based on the learning theory of Piaget [20], students will examine every new concept with existing concepts, for example, students are faced with a natural phenomenon, then the students are asked to create a hypothesis, then the lecturer and student test the hypothesis with the lab. If the student's hypothesis does not match the result of the practicum (misconceptions are wrong), then the student will experience a cognitive conflict that may result in a change of cognitive structure, so that misconception can be corrected.

The results of the correlation test between the critical thinking disposition and conceptual understanding is depicted in Table 2. It shows that the significance of value (2-tiled) is $0.019 < 0.05$. It means that there is a positive correlation between understanding the concept with the critical thinking disposition with the value of r is 0.395. The results of this study are in line with the currently reported research that there is a positive and weak relationship between the critical thinking tendency and academic achievement of the prospective teacher students in Turkey [21]. Students with a high critical thinking tendency will have a better level of conceptual understanding than students with low critical thinking disposition.

\

Table 2. The relationship between conceptual understanding and critical thinking disposition

		Critical thinking disposition	Conceptual understanding
Critical thinking disposition	Pearson Correlation	1.00	0.39
	Sig. (2-tailed)	-	0.02
	N	35.00	35
Conceptual understanding	Pearson Correlation	0.39	1.00
	Sig. (2-tailed)	0.02	-
	N	35.00	35.00

Identification of the indicator of the tendency of critical thinking and its categorization is shown in Figure 1. In general, students have an average critical thinking tendencies with positive categories and the largest indicator (85.7%) contribute to positive category is the indicator of inquisitiveness. The indicator that categorized as the strongest (37.14%) is open-mindedness. Both of these indicators seem to have developed quite well on the application of problem-based learning models, thereby encouraging students' thinking to achieve a good level of conceptual understanding.

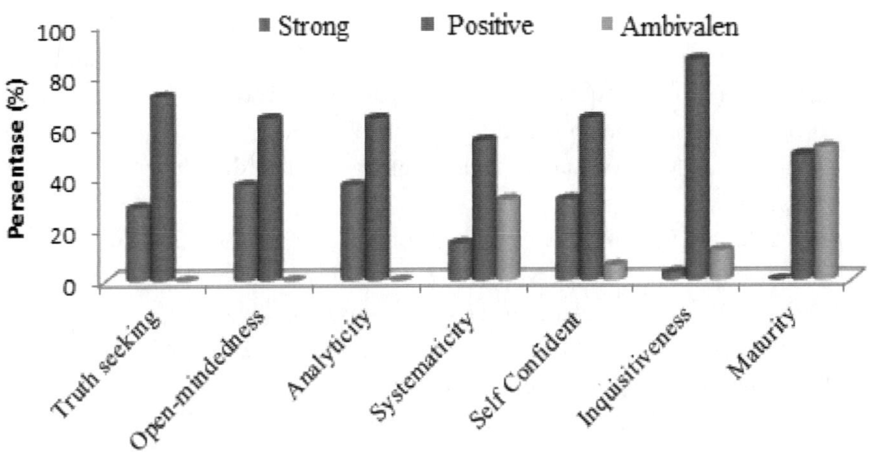

Figure 1. Tendering Profile of Student Critical Thinking.

The results showed that the level of understanding of the concept of prospective teachers is categorized as high. In addition, it is influenced by the PBL model and the level of critical thinking disposition. The result of critical thinking tendency analysis shows that all students have critical thinking tendency level in positive category. It means that students have a positive tendency to seriously undergo the learning process that gives positive outcome.

4. Conclusion

The level of understanding of the concept and the students' critical thinking disposition after applying problem based learning model within the learning process were categorized as positive and both have a significant positive correlation.

References

[1] Stronge J. H. 2018. *Qualities of effective teachers*. ASCD.
[2] Barnhart T., & van Es E. 2015. *Teaching and Teacher Education* **45** 83.
[3] Covay Mino, E., Desimon L, Caines Lee J and Hochberg E D 2016 *Education Policy Analysis Archives/Archivos Analíticos de Políticas Educativas, 24.*
[4] Gess Newsome, J. 2015. *Re-examining pedagogical content knowledge in science education*, 28.
[5] Williams L, Nixon S., Hennessy C, Mahon E and Adams G 2016 *Cogent Education*, **3(1)** 1154259.
[6]Ertmer P A and Ottenbreit-Leftwich A T 2010. *Journal of research on Technology in Education*, **42(3)** 255.
[7] Bell, S. 2010. *The Clearing House* 83(2) 39..
[8] National Research Council. 2013. *Education for life and work: Developing transferable knowledge and skills in the 21st century*. National Academies Press.

[9] Dasna, I Wayan. 2005. Penggunaan Model Pembelajaran Problem-based Learning dan Kooperatif learning untuk meningkatkan kualitas proses dan hasil belajar kuliah metodologi penelitian. Malang: Lembaga Penelitian UM.
[10] Jauhar, M. 2011. Implementasi Paikem dari Behavioristik sampai Konstruktivistik. Sebuah Pengembangan Pembelajaran Berbasis CTL. Jakarta: Prestasi Pustakarya.
[11] Gabel, D. 1999. *Journal of Chemical education*, *76*(4), 548.
[12] Taber, K. S. 2001. *Chemistry Education Research and Practice*, *2*(2), 123-158.
[13] Anwar, M. 2012. Pembelajaran Aktif Kooperatif dalam Perkuliahan Kinetika Kimia untuk meningkatkan Keterampilan Generik Sains Calon Guru. Tesis. Universitas Pendidikan Indonesia. Online repository.upi.edu. (diakses tanggal 29 juli 2016)
[14] Sözbilir, M. 2004. *Journal of chemical education*, *81*(4), 573.
[15] Hakim, A. Liliasari. dan Kadarohman, A. 2012. Internasional Online Journal of Educational Sciences. 3(3), 544-553.
[16] Tiwari, A., Alan, A., dan Patrick, L. 2003 Journal of Advanced Nursing. 44(3), 298–307
[17] Utami, N. R., Andayani, Y., Muntari, M. 2013 *Jurnal Pijar Mipa*, *8*(2).
[18] Nurwahidah, N., Andayani, Y., Loka, I. N. 2014. *Jurnal Pijar Mipa*, *9*(2).
[19] Andayani, Y., & Savalas, L. R. T. 2015 *Jurnal Penelitian Pendidikan IPA*, *1*(2).
[20] Piaget, J. 2001. The Psychology of Intelligence. New York: Routledge.
[21] Genç, G. 2017 Learners. *Educational Research Quarterly*, *41*(2), 43-73.